The Water's Edge:
Critical Problems
of the Coastal Zone

The MIT Press
Cambridge, Massachusetts,
and London, England

The Water's Edge: Critical Problems of the Coastal Zone

edited by
Bostwick H. Ketchum

The Coastal Zone Workshop was held 22 May–3 June 1972 in Woods Hole, Massachusetts, and was cosponsored by The Institute of Ecology and Woods Hole Oceanographic Institution.

This book was designed by The MIT Press Design Department.
It was set in Linotype Baskerville
by The Colonial Press Inc.,
printed on Decision Offset,
and bound in Interlaken AL1 539–7
by The Colonial Press Inc.
in the United States of America.

Library of Congress Cataloging in Publication Data

Coastal Zone Workshop, Woods Hole, Mass., 1972.
The water's edge.

The workshop was held 22 May–3 June 1972 and was cosponsored by the Institute of Ecology and Woods Hole Oceanographic Institution.
1. Coasts. 2. Marine resources. 3. Marine biology. I. Ketchum, Bostwick H., 1912– ed. II. Institute of Ecology. III. Woods Hole, Mass. Oceanographic Institution. IV. Title.
GB451.C67 333.9'17 72–7067
ISBN 0–262–11048–2
ISBN 0–262–61016–7 (pbk.)

Contents

Preface

The Workshop on Critical Problems of the Coastal Zone was held in Woods Hole, Massachusetts, 22 May–3 June, 1972. The Workshop was cosponsored by The Institute of Ecology and the Woods Hole Oceanographic Institution, and supported by funds supplied by the National Science Foundation and by the Rockefeller Foundation. This volume presents the conclusions and recommendations of the meeting, and summarizes the extensive background material that was available at the workshop and which formed the basis for the decisions reached.

In June 1971, The Founding Institutions of The Institute of Ecology convened a Coastal Ecosystems Study Conference in Houston, Texas. A major conclusion of this conference was that an intensive multidisciplinary workshop on the coastal zone was needed to define the critical problems, to examine and evaluate available information, and to recommend interdisciplinary research needed to achieve an understanding of these complex problems. It was concluded that necessary decisions concerning multiple use of coastal zone resources require a sound factual base, and it has been the goal of the Workshop to provide a compilation and evaluation of available information to define the present status of this factual base. It has been recognized from the start that existing information is inadequate to reach firm and final decisions on many of the problems, and that the Workshop would be one step on the way to develop, improve, and refine both the factual information and the understanding needed for the wise and effective use of the coastal zone.

A steering committee of five was appointed at the June meeting in Houston, Texas, and included Bostwick H. Ketchum, chairman; David Jameson, deputy chairman; Rezneat Darnell, Gerard Mangone, and Ruth Patrick. Planning meetings were held in Durham, N. H., in August and in Washington, D. C., in December with the goal of identifying potential workshop participants and establishing a workshop structure. The steering committee was enlarged in January by the inclusion of eight additional working group chairmen including: John Armstrong, Richard Bader, Donald Hood, J. Laurie McHugh, Dennis O'Connor, Robert Ragotzkie, John Teal, and Robert Warren. Dr. Richard Kolf of NSF/RANN actively participated in most of the planning meetings. Subsequent planning meetings for the purpose of devising a comprehensive set of interdisciplinary topics and for the selection of workshop participants were held in Philadelphia in January and Miami in April.

In preparation for the interdisciplinary workshop deliberations, the participants developed discipline-oriented position papers, which were used as a basis for discussions. At the mid-April planning meeting, the steering committee selected general topics to be addressed during the workshop by interdisciplinary panels. It was felt that deliberations of these topics during an intensive two-week session could provide scientists, government agencies and private groups with an authoritative assessment of coastal problems which could be used as a starting point in the process of making more rational management decisions.

Approximately sixty professionals representing many disciplines were involved in the workshop and in the preparation of the final report. In addition, about forty other contributors were involved in the pre-workshop preparations or attended the workshop for shorter periods of time. The specialists who cooperated in this interdisciplinary project were drawn from a broad range of disciplines including biology, chemistry, ecology, law, economics, engineering, oceanography, and sociology, and from research institutions, universities, government agencies (federal, regional, and state) and private enterprise. Participants at the workshop acted as individuals and not as representatives of their agencies or organizations.

This report represents the professional judgment of the participants on the problems created by man's use of coastal resources. Guidelines and recommendations are presented for actions which can be taken to ameliorate, correct or improve present or predicted deterioration of the coastal environment. The specific conclusions and recommendations of the workshop are summarized in Chapter 1 of this book. All participants at the workshop were involved in the discussions which led to these conclusions and a majority of workshop participants have endorsed these conclusions and recommendations.

The subsequent sections of the report contain the justification for the conclusions and recommendations presented in the summary. These sections were developed through intensive discussion and study by the participants who were divided among the interdisciplinary panels. They were then edited for form and smoothness in a post-workshop session by the steering committee, but great care was taken not to alter emphasis or meaning of important points. The participants who attended the workshop have not had an opportunity to examine the final wording of these expanded sections; therefore, it should not be assumed that every participant subscribes to every statement presented in Parts II and III of the report.

The interdisciplinary environmentally oriented workshop which prepared this book is a relatively new social invention and is an extension of such programs as the National Academy of Sciences Summer Study Groups. Much credit for its development and for its application to the environmental field must be given to Carroll Wilson and William Matthews of MIT, whose Study on Critical Environmental Problems (SCEP) has been a model for other organizations. The workshop concept is unique because it can mobilize and focus the energies of individuals from a variety of scientific and applied fields and thus clarify important questions which would otherwise remain obscure. This type of workshop bypasses the traditional departments of our universities which tend to unduly restrict the free-ranging intellect of its faculty. Also, it does not encourage the reward system that too often credits individuals rather than group efforts which can make individual insights possible. The interdisciplinary workshop encourages cooperation, and it attempts to give equal voice to all participants in the expression of varied opinions and ideas. Such workshops have been productive, so far, by specializing and designing the task that must be accomplished. The goal now is to develop and implement equally productive organizational relationships for conducting the work; work such as that spelled out by the several recommendations of this workshop. This is the task immediately confronting us if the fullest contributions of the natural and social sciences are to be realized by society.

We gratefully acknowledge financial support for the Coastal Zone Workshop which was provided by a grant from the National Science Foundation through the Research Applied to National Needs (RANN) Program and by a grant from The Rockefeller Foundation. Many workshop participants attended the full two-week session with the financial support of their own organizations. Without this broad organizational support, and without the interest and enthusiasm that was shown by other contributors, the participants could not have developed a report of this depth. We wish to thank all those who gave of their time and energy to this workshop.

Coastal Zone Workshop

22 May–3 June, 1972

Participants

George W. Allen
U. S. ARMY CORPS OF ENGINEERS
ATLANTA, GEORGIA

John M. Armstrong*
DIRECTOR OF SEA GRANTS PROGRAM
UNIVERSITY OF MICHIGAN
ANN ARBOR, MICHIGAN

William Aron
NOAA
DEPARTMENT OF COMMERCE
WASHINGTON, D. C.

Richard G. Bader*
ROSENSTIEL SCHOOL OF MARINE AND
ATMOSPHERIC SCIENCE
UNIVERSITY OF MIAMI
MIAMI, FLORIDA

Henyo T. Barretto
PETROBRAS—CENPES/DEPRO
RIO DE JANEIRO, GB, BRASIL

W. Brian Bedford
THE NATURE CONSERVANCY
ARLINGTON, VIRGINIA

Gerard A. Bertrand
OFFICE OF THE CHIEF OF ENGINEERS
FORRESTAL BUILDING
WASHINGTON, D. C.

Robert L. Bish
UNIVERSITY OF WASHINGTON
SEATTLE, WASHINGTON

Harold Bissell
JONES AND STOKES ASSOCIATES, INC.
SACRAMENTO, CALIFORNIA

Dayton E. Carritt
A-305 GRADUATE RESEARCH CENTER
UNIVERSITY OF MASSACHUSETTS
AMHERST, MASSACHUSETTS

John Clark
THE CONSERVATION FOUNDATION
WASHINGTON, D. C.

Lyle E. Craine
SCHOOL OF NATURAL RESOURCES
UNIVERSITY OF MICHIGAN
ANN ARBOR, MICHIGAN

L. Eugene Cronin, Director
NATURAL RESOURCES INSTITUTE
CHESAPEAKE BIOLOGICAL
LABORATORY
SOLOMONS, MARYLAND

Gabriel T. Csanady
DEPARTMENT OF MECHANICAL
ENGINEERING
UNIVERSITY OF WATERLOO
WATERLOO, ONTARIO, CANADA

Rezneat M. Darnell*
COLLEGE OF GEOSCIENCES
TEXAS A & M UNIVERSITY
COLLEGE STATION, TEXAS

William Gaither, Dean
COLLEGE OF MARINE STUDIES
UNIVERSITY OF DELAWARE
NEWARK, DELAWARE

Edward D. Goldberg
DEPARTMENT OF EARTH SCIENCES
SCRIPPS INSTITUTION OF
OCEANOGRAPHY
LA JOLLA, CALIFORNIA

Joel Goodman
COLLEGE OF MARINE STUDIES
UNIVERSITY OF DELAWARE
NEWARK, DELAWARE

Howard Gould
ESSO PRODUCTION RESEARCH
COMPANY
HOUSTON, TEXAS

Clare A. Gunn
COLLEGE OF AGRICULTURE
TEXAS A & M UNIVERSITY
COLLEGE STATION, TEXAS

C. W. Hart
THE ACADEMY OF NATURAL SCIENCES
PHILADELPHIA, PENNSYLVANIA

Joel W. Hedgpeth
MARINE SCIENCE CENTER
MARINE SCIENCE DRIVE
NEWPORT, OREGON

Frank Herrmann
U. S. ARMY CORPS OF ENGINEERS
WATERWAYS EXPERIMENT STATION
VICKSBURG, MISSISSIPPI

Marc J. Hershman
SEA GRANT LEGAL PROGRAM
LOUISIANA STATE UNIVERSITY
BATON ROUGE, LOUISIANA

Donald W. Hood*
INSTITUTE OF MARINE SCIENCE
UNIVERSITY OF ALASKA
FAIRBANKS, ALASKA

David L. Jameson*
UNIVERSITY OF HOUSTON
CULLEN BOULEVARD
HOUSTON, TEXAS

Bostwick H. Ketchum*
ASSOCIATE DIRECTOR
WOODS HOLE OCEANOGRAPHIC
INSTITUTION
WOODS HOLE, MASSACHUSETTS

Robert Lankford
UNIVERSIDAD NACIONAL DE MEXICO
INSTITUTO DE BIOLOGIA
MEXICO 20, D. F., MEXICO

Gerard J. Mangone*
WOODROW WILSON INTERNATIONAL
CENTER FOR SCHOLARS
SMITHSONIAN INSTITUTION BUILDING
WASHINGTON, D. C.

William D. Marks
MICHIGAN WATER RESOURCES
COMMISSION
LANSING, MICHIGAN

William V. McGuinness, Jr.
CENTER FOR THE ENVIRONMENT AND
MAN, INC.
HARTFORD, CONNECTICUT

J. L. McHugh*
MARINE SCIENCE RESEARCH CENTER
STATE UNIVERSITY OF NEW YORK
STONYBROOK, NEW YORK

H. Crane Miller
ALVORD AND ALVORD
WASHINGTON, D. C.

J. Robert Moore
MARINE STUDIES CENTER
UNIVERSITY OF WISCONSIN–MADISON
MADISON, WISCONSIN

William Niering
DEPARTMENT OF BOTANY
CONNECTICUT COLLEGE
NEW LONDON, CONNECTICUT

Dennis M. O'Connor*
LAW BUILDING
UNIVERSITY OF MIAMI
CORAL GABLES, FLORIDA

John Padan
MARINE MINERALS TECHNOLOGY
CENTER
TIBURON, CALIFORNIA

Patrick L. Parker
MARINE SCIENCE INSTITUTE
UNIVERSITY OF TEXAS
PORT ARANSAS, TEXAS

Ruth Patrick*
THE ACADEMY OF NATURAL SCIENCES
PHILADELPHIA, PENNSYLVANIA

William Perret
LOUISIANA WILDLIFE AND FISHERIES COMMISSION
NEW ORLEANS, LOUISIANA

Garrett Power
UNIVERSITY OF MARYLAND
SCHOOL OF LAW
BALTIMORE, MARYLAND

A. T. Pruter
NATIONAL MARINE FISHERIES SERVICE
SEATTLE, WASHINGTON

Robert A. Ragotzkie*
MARINE STUDIES CENTER
UNIVERSITY OF WISCONSIN–MADISON
MADISON, WISCONSIN

Robert W. Risebrough
BODEGA BAY MARINE LABORATORY
BODEGA BAY, CALIFORNIA

Claes Rooth
ROSENSTIEL SCHOOL OF MARINE AND ATMOSPHERIC SCIENCE
UNIVERSITY OF MIAMI
MIAMI, FLORIDA

Saul B. Saila
GRADUATE SCHOOL OF OCEANOGRAPHY
UNIVERSITY OF RHODE ISLAND
KINGSTON, RHODE ISLAND

Demetri Shimkin
NAVAL WAR COLLEGE
NEWPORT, RHODE ISLAND

Jens Sorensen
INSTITUTE OF URBAN AND REGIONAL DEVELOPMENT
UNIVERSITY OF CALIFORNIA
BERKELEY, CALIFORNIA

John S. Steinhart
UNIVERSITY OF WISCONSIN–MADISON
MADISON, WISCONSIN

George C. Steinman
MARITIME ADMINISTRATION
CODE M 720.7
DEPARTMENT OF COMMERCE
WASHINGTON, D. C.

Russell L. Stogsdill
DEPARTMENT OF ARCHITECTURE
TEXAS A & M UNIVERSITY
COLLEGE STATION, TEXAS

Robert Straus
HARVARD LAW SCHOOL
CAMBRIDGE, MASSACHUSETTS

H. Dixon Sturr
OCEANOGRAPHER
NAVAL WAR COLLEGE
NEWPORT, RHODE ISLAND

Thomas H. Suddath
COASTAL STATES ORGANIZATION
EXECUTIVE DIRECTOR
OCEAN SCIENCE CENTER OF THE ATLANTIC COMMISSION
SAVANNAH, GEORGIA

Durbin C. Tabb
ROSENSTIEL SCHOOL OF MARINE AND ATMOSPHERIC SCIENCE
UNIVERSITY OF MIAMI
MIAMI, FLORIDA

John M. Teal*
DEPARTMENT OF BIOLOGY
WOODS HOLE OCEANOGRAPHIC INSTITUTION
WOODS HOLE, MASSACHUSETTS

Gordon Thayer
NATIONAL MARINE FISHERIES SERVICE
ATLANTIC ESTUARINE FISHERIES CENTER
BEAUFORT, NORTH CAROLINA

Anitra Thorhaug
UNIVERSITY OF MIAMI
MIAMI, FLORIDA

Hermann Ugarte
TECHNICAL CONSUL OF MARITIME SOVEREIGNTY DIVISION
MINISTRY OF FOREIGN AFFAIRS
LIMA, PERU

Robert Warren*
UNIVERSITY OF SOUTHERN CALIFORNIA
CENTER FOR URBAN AFFAIRS
LOS ANGELES, CALIFORNIA

Contributors

Lewis M. Alexander
LAW OF THE SEA INSTITUTE
UNIVERSITY OF RHODE ISLAND
KINGSTON, RHODE ISLAND

Richard T. Barber
DUKE MARINE LABORATORY
BEAUFORT, NORTH CAROLINA

William Beller
COASTAL ZONE MANAGEMENT COMMITTEE
MARINE TECHNOLOGY SOCIETY
WASHINGTON, D. C.

F. Herbert Bormann
YALE SCHOOL OF FORESTRY
YALE UNIVERSITY
NEW HAVEN, CONNECTICUT

Norman H. Brooks
DEPARTMENT OF CIVIL ENGINEERING
CALIFORNIA INSTITUTE OF TECHNOLOGY
PASADENA, CALIFORNIA

Ruben Brown
NATIONAL ACADEMY OF SCIENCES
2101 CONSTITUTION AVENUE
WASHINGTON, D. C.

Dean Bumpus
DEPARTMENT OF PHYSICAL OCEANOGRAPHY
WOODS HOLE OCEANOGRAPHIC INSTITUTION
WOODS HOLE, MASSACHUSETTS

David Burack
NEW ENGLAND RIVER BASINS COMMISSION
270 ORANGE STREET
NEW HAVEN, CONNECTICUT

J. H. Connell
DEPARTMENT OF BIOLOGICAL SCIENCES
UNIVERSITY OF CALIFORNIA AT SANTA BARBARA
SANTA BARBARA, CALIFORNIA

J. D. Costlow
DUKE MARINE LABORATORY
BEAUFORT, NORTH CAROLINA

Dorian A. Cowan
UNIVERSITY OF MIAMI
MIAMI, FLORIDA

Russell Davenport
MASSACHUSETTS DEPARTMENT OF NATURAL RESOURCES

Paul K. Dayton
SCRIPPS INSTITUTION OF OCEANOGRAPHY
LA JOLLA, CALIFORNIA

Edward S. Deevey, Jr.
FLORIDA STATE MUSEUM
UNIVERSITY OF FLORIDA
GAINESVILLE, FLORIDA

Robert E. Eisenbud
NATIONAL PARKS AND CONSERVATION ASSOCIATION
WASHINGTON, D. C.

Edward W. Fager
SCRIPPS INSTITUTION OF OCEANOGRAPHY
LA JOLLA, CALIFORNIA

R. F. Ford
SCRIPPS INSTITUTION OF OCEANOGRAPHY
LA JOLLA, CALIFORNIA

Charles Foster
EXECUTIVE SECRETARY FOR ENVIRONMENTAL AFFAIRS
STATE OF MASSACHUSETTS

Charles Francisco
EG&G
ENVIRONMENTAL EQUIPMENT DIVISION
WALTHAM, MASSACHUSETTS

William J. Hargis, Jr.
COASTAL STATES ORGANIZATION
DIRECTOR, VIRGINIA INSTITUTE OF MARINE SCIENCE
GLOUCESTER POINT, VIRGINIA

Harold D. Hess
MARINE MINERALS TECHNOLOGY CENTER
TIBURON, CALIFORNIA

Douglas Inman
SCRIPPS INSTITUTION OF OCEANOGRAPHY
LA JOLLA, CALIFORNIA

Robert E. Johannes
DEPARTMENT OF ZOOLOGY
UNIVERSITY OF GEORGIA
ATHENS, GEORGIA

Bruce Johnson
FLORIDA COASTAL COORDINATING COUNCIL
TALLAHASSEE, FLORIDA

John A. Knauss, Dean
GRADUATE SCHOOL OF OCEANOGRAPHY
NARRAGANSETT MARINE LABORATORY
UNIVERSITY OF RHODE ISLAND
KINGSTON, RHODE ISLAND

Richard C. Kolf
DIVISION OF ENVIRONMENTAL SYSTEMS AND RESOURCES
NATIONAL SCIENCE FOUNDATION
WASHINGTON, D. C.

Robert B. Krueger
NOSSAMAN, WATERS, SCOTT, KRUEGER AND RIORDAN
LOS ANGELES, CALIFORNIA

G. Fred Lee
WATER CHEMISTRY LABORATORY
UNIVERSITY OF WISCONSIN
MADISON, WISCONSIN

William H. Matthews
DEPARTMENT OF CIVIL ENGINEERING
MASSACHUSETTS INSTITUTE OF TECHNOLOGY
CAMBRIDGE, MASSACHUSETTS

E. L. Mollo-Christensen
DEPARTMENT OF METEOROLOGY
MASSACHUSETTS INSTITUTE OF TECHNOLOGY
CAMBRIDGE, MASSACHUSETTS

A. Conrad Neumann
ROSENSTIEL SCHOOL OF MARINE AND ATMOSPHERIC SCIENCE
UNIVERSITY OF MIAMI
MIAMI, FLORIDA

W. J. North
CALIFORNIA INSTITUTE OF TECHNOLOGY
PASADENA, CALIFORNIA

Howard T. Odum
DEPARTMENT OF ENVIRONMENTAL ENGINEERING
UNIVERSITY OF FLORIDA
GAINESVILLE, FLORIDA

Frederick Offensend
WOODS HOLE OCEANOGRAPHIC INSTITUTION
WOODS HOLE, MASSACHUSETTS

J. Gordon Ogden, III
DEPARTMENT OF BIOLOGY
LIFE SCIENCES CENTER
DALHOUSIE UNIVERSITY
HALIFAX, NOVA SCOTIA, CANADA

R. T. Paine
UNIVERSITY OF WASHINGTON
SEATTLE, WASHINGTON

Edward Plumley, Vice President
NEW ENGLAND POWER COMPANY
WESTBORO, MASSACHUSETTS

John S. Rankin
THE COLLEGE OF LIBERAL ARTS AND SCIENCES
UNIVERSITY OF CONNECTICUT
STORRS, CONNECTICUT

D. J. Reish
UNIVERSITY OF CALIFORNIA AT LONG BEACH
LONG BEACH, CALIFORNIA

Theodore R. Rice
NATIONAL MARINE FISHERIES SERVICE
BEAUFORT, NORTH CAROLINA

Felix J. Rimberg
PEAT, MARWICK, MITCHELL & CO.
WASHINGTON, D. C.

Melvin Rosenfeld
WOODS HOLE OCEANOGRAPHIC INSTITUTION
WOODS HOLE, MASSACHUSETTS

Rogers Rutter, Director
STATE TECHNICAL SERVICES
STATE OF NEW HAMPSHIRE

John S. Schlee
U. S. GEOLOGICAL SURVEY
WOODS HOLE, MASSACHUSETTS

Paul H. Shore, Director
NEW ENGLAND ENERGY POLICY STAFF
NEW ENGLAND REGIONAL COMMISSION
BOSTON, MASSACHUSETTS

Paul Ferris Smith
STAFF OCEANOGRAPHER
GEODYNE
WOODS HOLE, MASSACHUSETTS

James Sullivan
DEPARTMENT OF ECONOMICS
UNIVERSITY OF CALIFORNIA
SANTA BARBARA, CALIFORNIA

Thomas L. Thompson
AMOCO PRODUCTION COMPANY
TULSA, OKLAHOMA 74102

Gary H. Toenniessen
ROCKEFELLER FOUNDATION
NEW YORK, NEW YORK

Robert Troutman, Jr.
ATLANTA, GEORGIA 30303

Karl Turekian
KLINE GEOLOGY LABORATORY
YALE UNIVERSITY
NEW HAVEN, CONNECTICUT

George Woodwell
BIOLOGY DEPARTMENT
BROOKHAVEN NATIONAL
LABORATORY
UPTON, LONG ISLAND, NEW YORK

Staff

James Bolan
BOSTON COLLEGE OF LAW
BOSTON, MASSACHUSETTS

Kathy Burns
DEPARTMENT OF BIOLOGY
WOODS HOLE OCEANOGRAPHIC
INSTITUTION
WOODS HOLE, MASSACHUSETTS

Richard H. Burroughs
W.H.O.I./M.I.T. JOINT PROGRAM
WOODS HOLE OCEANOGRAPHIC
INSTITUTION
WOODS HOLE, MASSACHUSETTS

Brad Butman
W.H.O.I./M.I.T. JOINT PROGRAM
22 MAGAZINE STREET
CAMBRIDGE, MASSACHUSETTS

Nancy Green
WOODS HOLE OCEANOGRAPHIC
INSTITUTION
WOODS HOLE, MASSACHUSETTS

Eleanor Kelley
INSTITUTE OF MARINE SCIENCE
UNIVERSITY OF ALASKA
FAIRBANKS, ALASKA

Walter McNichols
HEED UNIVERSITY
BOX 311
HOLLYWOOD, FLORIDA

Mitchell L. Moss
UNIVERSITY OF SOUTHERN
CALIFORNIA
CENTER FOR URBAN AFFAIRS
LOS ANGELES, CALIFORNIA

Patricia Pykosz
WOODS HOLE OCEANOGRAPHIC
INSTITUTION
WOODS HOLE, MASSACHUSETTS

William Robertson IV
ENVIRONMENTAL STUDIES BOARD
NATIONAL ACADEMY OF SCIENCES
WASHINGTON, D. C.

Peter Rosendahl
ROSENSTIEL SCHOOL OF MARINE AND
ATMOSPHERIC SCIENCE
UNIVERSITY OF MIAMI
MIAMI, FLORIDA

Nancy Serotkin
WOODS HOLE OCEANOGRAPHIC
INSTITUTION
WOODS HOLE, MASSACHUSETTS

Malcolm Shick
DEPARTMENT OF BIOLOGY
TEXAS A & M UNIVERSITY
COLLEGE STATION, TEXAS

Bruce W. Tripp*
WOODS HOLE OCEANOGRAPHIC
INSTITUTION
WOODS HOLE, MASSACHUSETTS

Linda Weimer
MARINE STUDIES CENTER
UNIVERSITY OF WISCONSIN
MADISON, WISCONSIN

* Steering Committee

Part I

A Summary of Results and Conclusions

Chapter 1
A Summary of Results and Conclusions

PREPARED BY
Bostwick H. Ketchum, Bruce W. Tripp

WITH CONTRIBUTIONS FROM
John Armstrong, Rezneat M. Darnell, David L. Jameson, Dennis M. O'Connor, Robert A. Ragotzkie, John Teal, Robert Warren

1 A Summary of Results and Conclusions

1.1 Introduction

One of the objectives of the coastal zone workshop was to identify the critical problems of the coastal zone, and this proved to be one of the easiest tasks. The workshop did not invent these problems, and the ones identified are probably no different from those that would have been defined by any other group with long and concerned interests about this part of our environment. The coastal zone was viewed as a unique national resource which is of great importance to a majority of our population. Conflicts result from a combination of population pressures combined with major and multiple demands upon the coastal zone area. Difficulties arise because of conflicting uses of the coastal zone. A good argument can be presented for a variety of uses, some of which are not compatible with others. We made no pretense about having all of the answers for these conflicting demands but hope that our conclusions and recommendations will help to clarify the areas of conflict and to identify those uses that are clearly a necessary part of coastal zone activities or that could be displaced to other parts of our environment and lessen the intensive pressures upon the coastal zone.

Other objectives of the workshop were to provide an interdisciplinary assessment of the effects of man's various activities on coastal zone processes, a definition of what is known and what needs to be learned, both about man's activities and about the natural processes which are affected, and the identification of scientific, legal, social, or economic constraints that prevent the rational management of coastal zone resources. The coastal environment constitutes a complex ecosystem that is an important and unique resource of our nation and that must be maintained for the benefit and use of mankind. Neither complete destruction of the natural environment nor complete prohibition of development and exploitation are acceptable goals for a national policy on coastal zone management. A balance of use, conservation, and preservation of the coastal zone should be maintained so as to optimize man's use of coastal resources through the long-term future which requires that the natural environmental processes on which most of the long-term continuing uses depend must also be maintained. Thus, maximum rational use of coastal resources consistent with the retention of life-support systems, beauties, and amenities of the coastal zone for the enjoyment of future generations must be the objectives of coastal zone management. Severe complexities have arisen, because maximum use, even though rational, and maximum preservation of the natural ecosystems and amenities can rarely, if ever, be achieved in the same units of space and time. It is hoped that our conclusions and recommendations will aid in setting guidelines for future development of coastal management plans which will be able to achieve these contradictory but mutually desirable objectives for its utilization.

One of the complexities facing a coastal zone workshop is an acceptable definition of what constitutes the coastal zone. There is no generally agreed upon and acceptable definition, in a legal sense; and we have heard public presentations of definitions that range from a landward width of 300 feet beyond the mean high tide for certain legal actions in California, to a statement that public requirements and demands would define a 5-mile-wide strip of the eastern coast of Florida as the coastal zone belt. The seaward limits of the coastal zone are equally ambiguous or arbitrary. It is generally conceded that, although we all understand each other when we speak of the coastal zone, it is not possible to place precise boundaries, either landward or seaward of the high tide mark, because of the marked differences not only within our own national boundaries but also on an international scale. A working definition, which combines demographic, functional, and geographical considerations, was adopted for the purposes of the workshop as follows:

The coastal zone is the band of dry land and adjacent ocean space (water and submerged land) in which land ecology and use directly affect ocean space ecology, and vice versa. The coastal zone is a band of variable width which borders the continents, the inland seas, and the Great Lakes. Functionally, it is the broad interface between land and water where production, consumption, and exchange processes occur at high rates of intensity. Ecologically, it is an area of dynamic biogeochemical activity but with limited capacity for supporting various forms of human use. Geographically, the landward boundary of the coastal zone is necessarily vague. The oceans may affect climate far inland from the sea. Ocean salt penetrates estuaries to various extents, depending largely upon geometry of the estuary and river flow, and the ocean tides may extend even farther upstream than the salt penetration. Pollutants added even to the freshwater part of a river ultimately reach the sea after passing through the estuary.

The seaward boundary is easier to define scientifically, but it has been the cause of extensive political argument and disagreement. Coastal waters differ chemically from those of the open sea, even in areas where man's impact is minimal. Generally, the coastal water can be identified at least to the edge of the Continental Shelf (depth of about 200 meters), but the influence of major rivers may extend many miles beyond this boundary. For the purposes of the Coastal Zone Workshop, the seaward boundary has been defined as the extent to which man's landbased activities have a measurable influence on the chemistry of the water or on the ecology of marine life. (See Figure 1.1.)

We do not present any prophecy of impending doom, even though such prophecies have become extremely popular in this age of intense concern about environmental quality. We do recognize the present degraded state of the coastal zone ecosystem over large areas, but we also recognize that even larger areas are still in a natural condition. We have recommended preservation of some of these unmodified areas, and this will require changes in our present activities. Unregulated continuation of present activities will lead to insidious spread of degradation, and controls and modifications of man's actions will be necessary to preserve the essential characteristics of the coastal zone. It is necessary to recon-

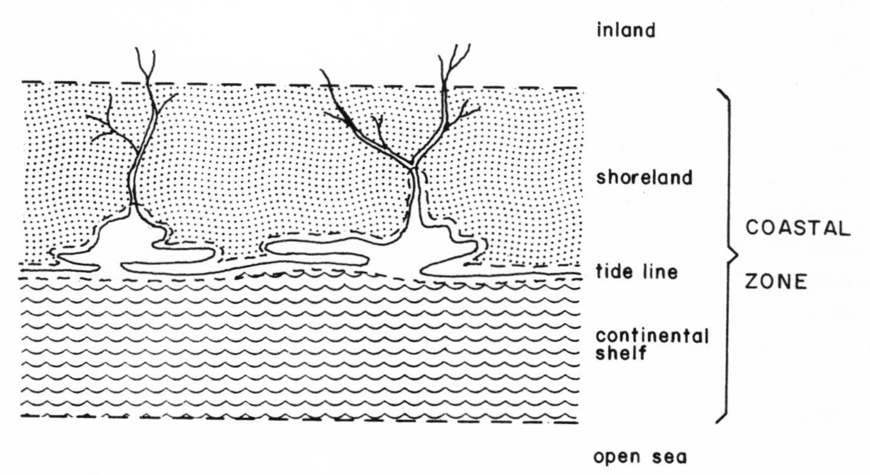

Figure 1.1 Schematic diagram of the coastal zone.

cile conflicting demands on coastal resources in such a way that full use of these resources is realized without destruction of natural processes required to renew them.

1.2 Characteristics of the Coastal Zone

In the coastal zone the land meets the sea, and this is an area in which processes depending on the interaction between land and sea are most intense. It is both the occurrence of these processes, and for human beings—which are land animals—the simple presence of the expanse of water, that make the coastal zone unique.

Since the coast is a junction of two environments, it is linear in nature, and the length is an essential characteristic. The area is small in comparison to length, since a large amount of land lies behind every bit of coastline, and it is faced by an even greater amount of sea. The effect of all of the events occurring at sea or on land are concentrated where they meet; thus the first important characteristic of the coastal zone is that it is an edge, limited in area, at which events are concentrated.

Where the land and the sea meet, the sea is shallowest and the land lowest. Flooding is common, both by daily tidal flooding of sand and mud flats and by storm tides that are individually unpredictable but certain to occur from time to time. Sea and tidal flooding is often augmented by river runoff and river flooding. In the coastal zone, new land is frequently claimed from the sea by rivers and currents and by strong winds piling sand above sea level or by human activities of diking or filling the shallows.

The shallowness of the sea, beginning some distance from the actual tide line, has an effect on the concentration of the sea's energy at the coast. Much of the energy of wind put into the sea surface is carried with little diminution until it meets the land where it is quickly dissipated, for example, in the breaking of waves against the shore. This concentration of energy collected by the water can have very large effects. A characteristic of coasts formed of mud and sand is their rapid change in shape as a result of the dissipation of this collected energy. The position of bars may be changed; channels may fill in; beaches move up and down the shore and may even disappear completely.

The shallower the shore and the more gradual its slope, the greater will be the distance over which the sea's energy is spent. On very gradually sloping coasts, large storms (hurricanes or typhoons) may raise the sea level by several feet and effectively move the coastline inland. The effects of breaking waves are then moved in over low-lying land behind the normal tideline. On naturally steep rocky coasts, the concentrated energy will act only very slowly. Where the steepness is man made, the effects depend on the strength and resistance of the materials used.

In contrast to water in the sea, runoff from land does not concentrate its energy at the coast. It is mostly lost upstream. Rivers most typically are flowing rather slowly when they enter the sea. At sea level the water flows downward no further and loses what little momentum it had left. Any particles held in suspension will tend to settle at the freshwater-saltwater interface. Thus, there is a concentration of the products of land erosion at the coast where they encounter the concentration of the sea's energy, and these particles may eventually become distributed to wider areas where they settle out.

Fine silts and clays settle only in quiet waters where they may be held firmly in place by plant roots in marshes. Sand may settle more rapidly in protected areas but is heavy enough to form beaches where waves break. In many places there is a net sand transport to the coast from the coastal seabed as well as from the land.

Some of the changes of land form along the coast, produced by these concentrations of energy and materials, are cyclical, such as the seasonal movements of beaches and bars produced by changes in the patterns of waves and currents. Other changes are unidirectional, such as the filling of protected shallows. Another important unidirectional change characteristic of the coast is the rise in sea level which is significant even in human lifetimes; it is nearly as much as an inch in five years along the southern New England coast. Sea level rise plus river erosion produces both large and small estuaries. These partially sheltered bodies of mixtures of fresh and sea water serve as efficient sediment traps.

The currents that move and collect sediments along the coast will often sort them according to size and weight. As a result, in many places along the coast, deposits of uniformly sized sands and gravels occur. Concentrations of useful mineral deposits may be increased by the same winnowing processes. Some of these are produced by the shore processes themselves, while others are ordinary stream placer deposits which are subsequently placed in the coastal zone by rising sea level. Lime deposits of high purity occur in the coastal zones of the Bahamas and Florida. These oolites, as they are called, are deposited in place, principally by chemical processes in warm shallow seas rather than as a result of the winnowing; but they are another mineral resource characteristic of the coastal zone. Shell deposits, whose formation depends on the high productivity of the zone, are characteristic lime resources along some parts of the coast.

Many chemical characteristics of the coastal zone result from the meeting of land, sea, and freshwater runoff. As fresh water is diluted by seawater, current patterns are produced so that seawater tends to flow into estuaries near the bottom and fresh water to flow out along the surface. This layered circulation helps to concentrate particulate substances heavier than water at the coast. Dissolved substances are also concentrated because of absorption by floating organisms and particulate matter. Once the organism dies, it becomes a particle that sinks and is carried into the estuary in the upstream flowing seawater.

In places of low rainfall and runoff, seawater may become more, rather than less, concentrated. Water evaporated by the sun's heat may concentrate salt which accumulates as a solid in natural or man-made shallows that are not flushed out with each tide. Since most of the water present in the coastal zone is seawater, which contains salt and other chemicals in high enough concentrations to constitute a resource, they can be utilized when further concentrated in this way.

Since natural sediments and some dissolved substances useful to organisms are concentrated at the coastal zone, we can expect, and unfortunately find, that many contaminants are also concentrated there by the same processes. Heavy metals tend to become trapped by the sulfides formed by the anoxic decomposition of organic matter. Substances dispersed in fresh water are flocculated by sea salts and settle to the bottom. Many industrial contaminants are adsorbed on particles and flocculants and so become trapped along the coast. But the water is also a dilutant for some contaminants and may disperse them so that they are made less harmful. Which process predominates depends on the behavior of the particular compound. The more it acts like water, the better it is diluted; the more it acts like a nutrient or particle, the more it will be retained and concentrated.

The most important biological characteristic of the coastal zone is high productivity, resulting from the concentration of nutrients for plant growth. This leads to a large concentration of organisms, including people. Tides and currents bring nutrients to plants, and the shallowness of the water permits the penetration of light to the bottom, so that fixed plants, growing in an area of flowing, nutrient ladened water, grow exceedingly well. Plant communities such as salt marshes and mangrove swamps, which grow intertidally, and fixed algae or seaweeds like rockweeds and kelp, turtle and eel grasses, that grow in shallow water, are highly productive. Floating algae, phytoplankton, which can grow throughout the aquatic part of the coastal zone, contribute the rest of the plant growth and are the principal producers in the deeper waters and the open sea.

The productivity is so large that it is not all consumed *in situ* but settles on the bottom, where it quickly accumulates, and the water and sediments become anoxic. In geological time, such organic deposits have produced and preserved oil and gas resources which are characteristic of the coastal zone.

Animals of the coastal zone feed on the fixed and floating plants so that animal productivity is also characteristically high. Many animals eat plants directly, including many fishes, clams, and crustacea. Because of this food resource, there are also many carnivores, including fishes, shellfish, birds, fur-bearing mammals, and humans concentrated in the coastal zone.

Coral reefs are characteristic of some stable tropical shores. These are very complex marine communities, which maintain their productivity by reusing their nutrients rather than by depending on a rapid flow of nutrients through the reef system.

Highly productive upland soils may be a feature of the coastal zone. Much of the silt load of rivers is dropped on floodplains near river mouths; and these floodplains, along with river deltas, contain some of the finest agricultural land and productive forest swamps in the country. The latter are the home of various mammals and birds, including game and fur-bearing species.

Various animals benefit from other characteristics of the coast as well as from its high production. The shallow edges of the waters, especially sheltered bays, protect juvenile stages from aquatic predators. Some animals take advantage of seasonal changes in salinity in estuaries to avoid predators which will not, or cannot, enter brackish water.

The coastal zone is the last area where migrating birds can rest on solid ground before launching out over water. Large concentrations of shorebirds and waterfowl are found on the beaches and marshes partly because of the high concentration of animal food. Small birds rest in forests and fields near the coastline before starting their migration over the sea.

People also cluster on the coast for many reasons. Transport by ship must give way there to land transport; the best fishing is there; the presence of the sea provides a large water resource for the removal of wastes and the cooling of industrial plants. Many of the world's largest cities are located in the coastal zone, and where there are cities and industries, there are jobs. The aesthetic characteristics also attract people. Great numbers of people retire near the sea, and the numbers that visit the shore for recreation are tremendous, as can be seen by the pictures of Coney Island beach on a hot summer day. It is a cliché that people go either to the mountains or to the beach during their summer holidays. Both areas have features in common—a cool or cooler climate, a view, and the feeling of open space.

The climate of the coast is tempered by water. It is both cooler in summer and warmer in winter than inland areas on the same latitude. The difference in heating and cooling between the land and sea produces sea breezes, which aid in cooling and provide dependable winds for sailing. There is water to play in and sunny beaches to lie on. Historical sites abound along the coast, for here are all the original sites of discovery and settlement. There is a certain excitement associated with the mysterious sea, as ships appear on the horizon. A feeling of spaciousness exists so that, even with a city at your back, you can feel uncrowded while looking out to sea.

The coastal zone is thus a limited border between the two major parts of the earth's surface, land and water. It is rich in resources, from minerals and petroleum, to ducks, fish, oysters, and recreation. It is the site of much human activity and business and the recipient of much of man's pollution. It is fragile in many ways, resilient in others, and worth planning and managing so that it will continue to have attraction and usefulness in the future.

1.3 Man's Uses of the Coastal Zone

Man has only recently come to realize the finite limitations of the coastal zone as a place to live, work, and play and as a source of valuable resources. This realization has come along with overcrowding, overdevelopment in some areas, and destruction of valuable resources by his misuse of this unique environment.

More than 50 percent of the population of the United States lives in the counties bordering the Great Lakes and the ocean, and the percentage is increasing. By the year 2000, it is estimated that 200 million people may live in the coastal zone of the United States. This high population along our coasts must dispose of its waste products; and, whether these are put into rivers or into ocean outfalls, they ultimately reach the sea.

These people are concentrated in large urban centers. In many cities or clusters of cities, the coastal waters have been ruined for other uses by the mismanagement of industrial and domestic waste disposal in the sea. Fisheries, recreation areas, and prime ecological habitats have all been destroyed by mismanagement or, more accurately, nonmanagement. Man has not consciously decided to wreak this destruction. It has come about by an exponential growth of human activities in the coastal zone in the absence of any planning at all. These processes are continuing, and although heroic attempts are being made in a few places to stem the flow, reversal of the process seems almost hopeless. The rising human population, with its attendant industrial and commercial activity and the resulting wastes, has become one of our most critical coastal zone problems.

Man also harvests food from the coastal zone. Last year, United States fishermen harvested nearly 5 billion pounds of fish—most of it coming from coastal waters. This figure does not include the harvest from the U.S. coastal zone by foreign vessels. Fish harvest in the coastal zone and estuaries seems to be approaching the limit of the natural system. Though opinion differs on this point, many believe that it has already reached the limits for sustained harvest. Half of the biological productivity of the world's oceans, in fact, occurs along the coasts, and the estuaries which dot the coastline are the most productive areas known on earth. They exceed the productivity per unit area of most agriculture by a factor of 2 or more. Yet it is the estuaries which are most severely stressed by human urban and industrial activities. Loss of wetlands due to land filling and development and highly concentrated waste sources place irresistible pressures on these highly productive areas. In the past 20 years, California alone has lost 67 percent of its coastal estuarine habitats in the process of coastal development.

As agriculture has increased productivity on land, so aquaculture could increase productivity in the sea. The high productivity of the coastal waters offers the potential of greatly increased food production. Already experiments in Florida have indicated that shrimp farming on a large scale is economically feasible. Yields of one million pounds per acre have been achieved under highly controlled conditions. Experiments with other marine organisms also promise a similar hope of success. Such use of our coastal waters could be highly valuable to man, perhaps yielding more of value per acre than any other human activity.

But aquaculture requires high quality water and a high degree of control over the environment. It can tolerate almost no other use in the same water area. Its potential is great, however, and inevitably it will play an important role in man's uses of the coastal zone in the future.

The waters of the coastal zone blanket vast reserves of minerals and oil. As our needs for these resources grow, the economic and engineering constraints on recovering them will no doubt be surmounted. The U.S. Geological Survey estimates that the potential recoverable reserves of oil on the continental margin amount to 200 billion barrels—about 5 times present proven U.S. reserves. For gas, the figures are 900 trillion cubic feet—about 3 times the proven reserves. In economic terms, U.S. offshore petroleum production in 1967 was valued at $1 billion, or one-half of the total dollar value of all coastal zone resources that year. Gas accounted for another 15 percent of that value. It is clear from these figures that production of oil and gas will continue to be among man's major uses of the coastal zone.

Underwater mining and offshore oil extraction are expected to increase drastically in the next few decades. Some mineral resources of the coastal zone are already being harvested in staggering quantities, nearshore shell deposits are being overexploited, and underwater sand and gravel deposits are under increasing pressure. In 1970, five tons of sand and gravel were extracted for every man, woman and child in the United States. Of this, over 10 percent came from beneath our rivers, bays, and sounds, and this percentage is increasing yearly. The environmental impact of recovering these materials from coastal waters need not be great if care is given to the location of the mining operation. Areas of high sand mobility would be good ones to exploit, for example, since these are known not to be very biologically productive. While sand and gravel represent a resource that we can exploit if we are careful, there are others that may best be left alone. For example, while it is economically and technically feasible to mine the phosphorite deposits along the Atlantic coast, doing this would completely destroy the fragile and valuable salt marshes overlying them.

Man also relies on the coastal zone for recreation. More than half of all Americans vacation on the coasts, and with current population trends, this use of the coasts will no doubt continue to expand. Many reasons for this return to the sea can be suggested, and each of us perhaps has his own. Certainly there is a wide variety of recreational activity taking place along the coastal zone, ranging from swimming, sailing, and sport fishing, to more contemplative activities like wildlife observation or simply sitting along the water's edge.

Attaching a dollar value to these activities is difficult, but we do know that recreation is a booming coastal industry. It is estimated that by 1975, $5.5 billion will be spent yearly for swimming, surfing, skin diving, pleasure boating, and sport fishing. This is more than double the 1967 value of material marine resources garnered from coastal waters.

As man and his activities usurp more and more space on some parts of the earth, and particularly on the coastal zone, the survival of natural habitats and of plant and animal species is threatened. In 20 years, dredging and filling have destroyed over half a million acres of our country's important fish and wildlife habitats. Some coastal preserves have already been established, but as use pressure mounts and we approach total human saturation in some areas, the opportunity to establish adequate additional preserves is fast disappearing. This use of coastal areas must be given high priority since, once lost, these natural habitats cannot be retrieved.

In general, man's uses of the coastal zone can be divided into 6 major categories:

1. Living space and recreation. This is the source of a great deal of pressure on the coastal zone.
2. Industrial and commercial activities. We have only mentioned a few, but this includes power production, mining, and commercial development.
3. Waste disposal. Man uses the coastal zone to dump both industrial and domestic wastes.
4. Food production. As we have seen, this is largely fisheries, but includes aquaculture, a promising future use.
5. Natural preserves.
6. Special government uses. This is a proportionately small area of the coastal zone which includes military and coast guard installations, NASA bases, and government parks and lands.

This array of present uses of the coastal zone and the problems they present is familiar to most of us. It would be interesting to foresee man's future uses—especially their relative importance and impacts. Unfortunately, our crystal ball is cloudy, even after gazing into it for the two weeks of the workshop. We do know that offshore oil extraction and mining are almost certain to increase dramatically. Recreation is increasing rapidly, and space requirements for this use are becoming critical. Human population trends along the coast are more of an enigma. While burgeoning in some areas, population is streaming out of others. Unless we understand and achieve some predictive skill in these population migration and settlement patterns, we will not be able to plan adequately for the living space, waste disposal, and other demands that humans will surely make on the coast.

As we view the uses man makes, and will continue to make, of the coastal zone, we become aware of the constraints put on these uses, both by the environment itself and by their interaction with other uses. For example, fish harvests are ultimately limited by the natural productivity of the coastal waters; in fact, they may have reached that limit already. At the same time, improper waste disposal and the loss of marshes and estuarine wetlands will also reduce the productivity of the system, and hence fish harvests.

If we are to arrive at an effective management scheme for the coastal zone, we will need to sort out these kinds of interrelationships of uses, and recognize the constraints that they impose on each other. We can begin by asking a simple question: Is the use being considered compatible with other uses? If so, which ones, how can they be combined, where, and what limits must be placed on each to ensure that they do not conflict with one another in the long run? This is the familiar "multiple-use" concept. Clearly compatible uses, for example, would be waste disposal, shipping, recreational boating, and urban housing. These are inherently incompatible with others, however, such as scientific preserves and aquaculture, which are examples of exclusive uses.

Some uses of the coastal zone may be displaceable to other locations. As long as use pressures on coastal areas remain low, there is no need for this approach. With increasing demands on the finite and inherently linear coastal environment, we must be selective in what we choose to place there. Clearly, some uses, such as fisheries and wildlife preserves, cannot be displaced. Others, such as power plants and high-density human habitation, could be moved inland or, as some have suggested, placed offshore.

In short, there are three strategies in dealing with coastal zone uses:

1. Multiple use,
2. Exclusive use, and
3. Displaceable use.

If compatible or multiple uses are allowed to grow until they exceed the carrying capacity of the environment or begin to interfere with one another, conflicts arise. Examples of such conflicts arise every day. Conflicts in use patterns require difficult management decisions, but, ideally, most could be avoided by proper advance planning. In some cases there may be technological solutions to problems of conflicting uses, but these solutions should not be limited to single-use conflicts. In other cases, two or more "problems" can be converted to an opportunity. An example is the use of nutrient-rich domestic sewage to enhance the production of food by aquaculture or to raise the level of productivity of existing ecosystems. In temperate or cold climates, waste heat from power plants might be used to control spawning and increase growth rates of fish and shellfish or to warm swimming beaches.

Judgment on the economic feasibility of such solutions should not be based solely on the dollar value of the product or direct benefits resulting from the technological system. The benefits of eliminating or reducing damage to the environment or the savings realized by eliminating the need for other direct, and perhaps costly, technical operations should also be included on the plus side of the ledger.

Understanding the uses man makes of the coastal zone and its resources, understanding the impact and constraints of these uses, and developing strategies to deal with these uses are not enough. Managing the coastal zone for human benefit demands more than economic analysis, more than the opinions of experts, more than any technological solution, elaborate planning technique, or political arrangement can provide. Managing the coastal zone for human benefit rests upon an inherent respect for what the coastal zone is for—its uniqueness and wealth.

In the late 1940s, Wisconsin's great naturalist, Aldo Leopold, made a plea for mankind to develop what he called the "land ethic," which, he wrote, "changes the role of *Homo sapiens* from conqueror of the land-community to plain member and citizen of it. It implies respect for his fellow members and also respect for the community as such." Man has not yet achieved this ideal, but perhaps it is now time for man to adopt a "sea ethic," time he reached some feeling of responsibility for the health of the coastal zone and an ecological conscience in its management.

There is an element of self-interest in this philosophy. After all, man relies heavily on the coastal zone for a great variety of things—recreation, aesthetic refreshment, and resources such as fish and petroleum, to name a few. An insult to this coastal system, however minor, will eventually be paid for by decreased value of the system to man in the form of depleted resources or deteriorated water quality. In this light, harmony of man's uses with the physical, chemical, and biological functioning—in short, the total ecological functioning—of the system should be the overriding goal.

If we attach a quantitative value to this harmony, we shall continue to enjoy the wealth of this unique system. If environmental considerations are built into the cost-benefit analysis of any proposed use, for example, we shall eventually get the most return for our investment. If they are not, and we insist on raping the system, we are doomed to take a loss, whatever our short-term gains.

The acceptance of a sea ethic will not be achieved without sacrifice. Man will have to sacrifice the way in which he now attaches values to things. If he persists in valuing the coastal zone only for the uses which he can make of it today, and the resources he can take from it, he will ultimately destroy it. But if he can better understand what he wants from the coastal zone, what impacts his current practices are having on it, and if he can develop and implement rational and ethical strategies to deal with these uses, he will enhance and preserve this unique area and will be rewarded for his efforts.

1.4 Allocating Coastal Resources

Deliberate changes in the way humans utilize coastal resources must be made if negative environmental impacts are to be reduced. Obtaining public action to protect the natural ecology of the coast is not, as some would hope, simply a matter of making information about the problem known. Defining, agreeing upon, and implementing measures that require a large and heterogenous body of people to forego or reduce the level of utilization of valued resources is seldom easy. Yet this is the task that must be accomplished for the wise management of the coastal zone.

Marine biologists, oceanographers, and ecologists have made the complexity and fragility of marine ecosystems understandable. Social scientists must also make it clear that the social, economic, and political processes that exercise important influences over the formulation and administration of governmental policy are also highly complex. Creating another governmental agency or passing new laws may or may not achieve the goals that gave impetus to them.

Several crucial tasks must be undertaken if public institutions and rules are to be designed which will give adequate weight to environmental values in allocating coastal resources. One such task is to identify the way in which social, economic, and political processes mediate between the welter of issues which individuals and groups must resolve by governmental authority and ultimate action. The content of any coastal policy that is adopted will be affected by these processes. The constraints these processes place upon the capacity of various types of governmental agencies or laws to modify existing patterns of human behavior in relation to coastal resources must be understood. The design of public policy concerning the coast and the agencies to carry it out is too serious a matter to be done without taking certain "realities" into account.

Humans place a high value on access to the coast and engage in a variety of activities there. As previously mentioned, more than one-half of the population of the United States is located in close proximity to the coast, and the percentage is expected to increase. No adequate public policy can be formulated without a clear understanding that there are immense social and economic pressures for increasing the amount of coastal area that is devoted to residential, recreational, and industrial use. Consequently, management of the coastal zone must be concerned with ways of assuring human access. Methods to increase the carrying capacity of the coast for human activities in ways that produce a minimum environmental impact may be essential. Humans, particularly in well-organized groups, are quite innovative in devising strategies for circumventing or neutralizing regulations that interfere with the achievement of highly desired values.

Our governmental system is complex and characterized by a great number of points where policy influence can be exercised. There is little likelihood that all authority concerning the allocation of coastal resources could, even if it was desired, be placed solely in special coastal management agencies. Market transactions, voting, bargaining, and ajudication, as well as administrative agencies with overlapping responsibilities, all interact in the allocation of most major resources in this country. The establishment of any coastal zone management agency must be done with awareness that it will be in competition with other allocative mechanisms in some cases and be able to use them for complementary purposes in others.

Finally, decisions about the coast cannot be made in isolation from the broader geographic regions of which the zone is part. For example, the public interest is not served if an activity that is barred from the coast is located at an inland site where it produces greater overall damage. A means must be available to choose between coastal and inland sites or require that the activity be modified or eliminated. Cases also will arise in which the demand for a widely desired use of the coast exceeds the resource. If the carrying capacity of the coast itself cannot be increased through technology or greater management efficiency, it will be necessary to identify and develop inland sites where similar or substitute water-related activities can take place. A coastal zone management agency, then, will not be in a monopoly position for determining the allocation of coastal resources. Neither can it be given a set of final and inflexible rules to administer. The content and interplay of physical, biological, social, psychological, and economic values concerning uses of the coastal zone are not static. The same is true of the number and influence of special interest groups associated with particular values. They will change over time as will their strategies for influencing policy.

Formulating coastal zone policy will not simply be a matter of deciding on rules to preserve marine ecosystems. Whatever the policy outcome of the current debate over coastal legislation, it will be the result of two sets of trade-offs. One will involve broad choices between environmental and social and economic values. Then there will be a need to ration or choose among the possible alternative uses of the resources that are available and to determine who will be able to use them.

To a large extent, the success of efforts to manage coastal areas will depend upon the ability of the overall political process to provide general guidelines and the capacity of the governmental agencies created to administer them to operate in a dynamic social, economic, and political environment. This will mean that methods must be established to provide fair access to the decision process, a socially equitable distribution of the coastal resources that are available for use, and adequate information for determining the social, economic, and environmental effects of the policies and rules that are considered and adopted.

1.5 The Structure of Management and Planning for the Coastal Zone

Our concern over the future of the coastal zone ultimately focuses on man's ability to control the impact of his activities on its resources. This draws us to an inevitable consideration of the policies, programs, and institutions by which man can exercise his power of rational decision making. Historically, the policies, programs, and institutions which deal with the coastal zone have been lumped under the rubric "coastal zone management." The term has been used to describe anything from theoretical studies of how traditional economics might be used to allocate the resources of the coastal zone to detailed programs for establishing operational agencies and procedures to regulate use of the coastal zone.

In a management sense, the seaward limit or boundary defined in state jurisdictions is suitable for most management programs. In cases of national concern, the limits of federal jurisdiction may be needed. On the landward side, overlapping existing jurisdiction, land-use patterns, and special districts produce a variable definition that relates limits of marine influence inland as defined by each state, since the boundary must relate to the problem being considered.

The multiple political interests and power structures in the coastal zone have made it difficult to manage. Overlapping jurisdiction that we cannot ignore causes nonuniform guidelines and regulations. Furthermore in traditional resource management or planning, man has dealt with each resource subsystem individually—the land, the water, and the submerged lands. We need to view the coastal zone as a natural system in order to utilize resources in harmony with the ecological web that characterizes this zone.

Localized impacts may involve the development of individual units, for example, a nuclear power plant or some other industrial complex. In this class of problem, the issues must be resolved primarily by action at local (or regional) levels. The Hilton Head chemical complex provides a good example of the issues in this category. Some localized impacts have national involvement. The current energy crisis provides an excellent example here. The increasing demand for energy and the projections of national demands suggest that quite soon we may need to develop new deep ports or other offshore facilities for handling supertankers. Although local areas will feel the direct impact of such facilities, the need for fuel is nationwide, and the beneficiaries of the new facility would include a broad cross section of the country.

Historical usage presents a separate set of problems. The New York Bight in its role as septic system for the metropolitan area is a classical example for this category. Can such areas be reclaimed? A willingness to break with past traditions and to reexamine past uses of coastal regions to determine possible new thrusts for action is imperative for future coastal zone management.

Coastal zone management is a process designed to achieve a set of stated objectives. In the coastal zone, the stated objective would be to maintain and to improve its usefulness for man by ensuring the quality and extent of the natural system upon which he depends. This should be done for both the present and the future in ways that would be acceptable to our expressed or imputed economic, social, and environmental goals. With such a best-use objective in mind, we propose a basic definition of coastal zone management which should include (1) developing and understanding of the coastal zone as a system, (2) using this knowledge to create a dynamic plan for its best use, and (3) implementing and enforcing that plan.

The functions of management within this structure are based on the principles of allocation of resources, establishment of priorities, and regulation, including positive use of resources through, for example, incentives, pricing, and other strategies for management. The allocation process requires resource inventory and classification as a resource base line from which allocative decisions can be made, and a plan developed for the future use of various coastal zone resources. This process in turn leads to a program of evaluating and monitoring how well the system is functioning.

In order to achieve these goals, it will be necessary to determine what man's desires are in using the coastal zone and what values and priorities need to be assigned to these desires. It will be necessary to determine the capacity of the coastal zone in relationship to these desires and to determine which uses are compatible within a defined area of capacity. If capacities and uses cannot be matched, trade-offs will be necessary, and the public must be informed concerning the losses they must incur for a given use. Complex mechanisms will be necessary to regulate and promote compatible uses and to determine which parts of the coastal zone should be set aside for exclusive uses. For example, industrial development and preservation are not compatible, but both are desirable uses of the coastal zone system.

The scope of the problems and the common resources involved require that government agencies manage the coastal zone. Federal, state, interstate regional groupings, and local levels (a collective term that includes all levels of government below the state level, such as counties, municipalities, towns, villages, and various groupings thereof) must have clearly defined responsibilities.

The central issues are centralization and decentralization, a fundamental management question, not unique to the coastal zone. Generally, higher levels of government lead to centralization, but also to more general perspective, more objectivity, more access to expert talent, and more funds and political impact. Moving toward decentralization, the lower the level of government, the more intimate the knowledge of the problems, the more myopic the outlook, and the greater the likelihood of living with the effects, whether good or bad, of the decisions. Furthermore, if higher government does not limit its own decision-making appetite, it can become hopelessly bogged down in detail at the expense of the perspective it claimed in the first place.

It is appealing to try to seek out a middle ground, ideally one that preserves the recognized unique attributes of the extremes. To distribute authority in the coastal zone among the various levels of government involved, it is necessary to invoke the principle of delegation of authority. Under this principle, coastal decision making is delegated to the lowest level of government consistent with the scope of the problem, but decisions must conform to the goals and restrictions articulated by successively higher levels. The restrictions are generally formulated to ensure that the external effects of the local decisions are kept within tolerable bounds. A workable system incorporating the principle of delegation of authority and its corollary, management by exception, would place decision making at the lowest level commensurate with the anticipated impact of the decision, while prescribing the policy framework and type of external considerations that must be referred to a higher level.

1.6 Legal Aspects of the Coastal Zone

As already stated, the concentration of population along the coast, the increasingly intensive use of coastal resources in some areas, shoreline development, and waste disposal all have effects that call for consideration of environmental consequences. Man's activities are not only producing substantial effects on the environment, but also foreclosing important future options for society.

Our focus was on those decisions that influence man's use of the coastal zone. One primary concern has been decisions about development. The need for assessment of the environmental impact of man's activities is greatest where there is the most pressure for extension of metropolitan-type use and development into areas of the coast which had previously received comparatively little development.

Pressure from metropolitan centers is felt most acutely at the level of local government. Demands to use beach areas are often met with responses to exclude nonresidents from locally owned or operated beaches. Or, sometimes, the use of the coast by outsiders is favored, as in local communities whose economies depend upon the tourist trade. The exclusion of outsiders raises constitutional questions of "equal protection" of citizens, but local policies that permit too intensive use or development often produce environmental impacts that affect the ecology and resources to the detriment of the wider state and national interests. Both reasonable access and use policies, and policies for appropriate protection of the environment, must be applied in coastal decisions.

Jurisdictional authority in the coastal zone is divided, with several important discontinuities vitally affecting coastal zone law. On the land side of the coastal zone, state law and sovereignty, with the associated police powers for the regulation of public health, safety and general welfare, dominate. Actually, zoning authority in most states is delegated to the county or local municipal governments, and this aspect of the police power is not ordinarily exercised at the state level. The federal government typically exercises little authority on the land side of the coastal zone, although the power to regulate interstate commerce and the ownership of extensive areas of the coast in some states give it a basis for substantial jurisdictional and practical power. The salient point is that the bulk of the decisions on the land side of the coastal zone are zoning and other regulatory decisions made at local levels of government.

When we move into the water areas, we encounter the navigation jurisdiction of the federal government. Under the Commerce Clause of the Constitution, the navigation power extends a jurisdiction throughout the navigable waters of the United States. Construction in these navigable waters is regulated by the federal government, and conservation and environmental factors are now being taken into account in the making of these decisions. Additionally, federal water quality legislation provides for federally approved state water quality guidelines and serves as an additional basis for federal jurisdiction. To the extent that federal law does not preempt state law in these areas, the state also has substantial authority in the water areas of the coastal zone. Thus, the fundamental discontinuity of jurisdiction at the water's edge, with the primary influence of local decisions on one side of the line and the direct and substantial jurisdiction of federal and state governments on the other must be emphasized. While natural processes, and the patterns of man's use, cross this land and water line freely, the discontinuity of jurisdictional authority is a major problem for coastal zone management. Until recently, this discontinuity left gaps in the legal regulation of coastal activities. In the past several years, however, with increasing awareness of the environment in decision making, and with a broadened view of the federal authority over navigable waters to include environmental and water quality responsibilities, there is an increased possibility of federal, state, and local conflicts in decision policies.

Another discontinuity of jurisdiction occurs where the water meets the submerged land. The federal jurisdiction over the waters, as described above, meets the state's ownership of the submerged lands from the coastline out to the territorial sea, and in some cases, beyond. Another major problem with the limits of jurisdiction is that the lateral geographic boundaries of the lower governmental units do not correspond with the natural boundaries of coastal features and thus do not meet the needs of effective coastal zone management.

The proposals usually advanced to deal with these problems of distribution of authority advocate solutions by means of cooperation or coordination of the government levels, or by raising the decisions to higher government levels so that an adequate scope is available for management of activities and resources. While elements of these approaches will undoubtedly be found in future coastal zone programs, they are generally not adequate. Coordination and cooperation has its limits when the interests of the affected government entities do not converge. Elevating decisions to state or federal levels in most cases is impractical, and wasteful of available manpower. Until alternate means of funding at local levels of government are found, conflict will be produced by excluding them from the decision concerning conservation and development of local coastal areas. An alternative approach is the use of guidelines at federal and state levels, with more specific controls provided at the lower government level. Guidelines are a technique with both power and flexibility for assisting coastal zone management.

Recent proposals for land-use management have been made as a means of displacing local decision making. The general approach of these legislative proposals, and of the legislation on this subject recently adopted in Florida, is to provide for the selection of decisions which are of "regional" or "environmental" impact. For these decisions on matters of "critical concern," the decision-making power of the lower levels of government is displaced so that the important choices are determined by the state or federal government. Such an approach necessarily either removes a large proportion of the present decision authority of local and state governments or provides for but a small percentage of the decisions to be so affected. In the coastal zone, the construction of even a small pier or dock, or the dredging of even one submerged acre may have a substantial effect on the local environment. At present the development activities in the water areas of the coastal zone involve local, state, and federal jurisdiction. Accommodating the policies for the land and for the water decisions should be a prime objective, but not by selecting a small percentage of the land decisions for state or federal jurisdiction. The technique of establishing guidelines can achieve the objective of securing conformity of the lower levels of government with standards necessary for regulation of the activities in the coastal zone without losing the participation of local government.

1.7 A Systems View of Coastal Zone Management

With the onset of the technological age, we are witness to environmental destruction of unprecedented scope and magnitude. In recognition of the fact that the present trends cannot long continue without great danger to the life-support system itself, the development of a sophisticated society-environment management system, through which a reasonable balance may be achieved and maintained in perpetuity, is needed.

The necessity and feasibility of developing an analytical model capable of handling a great variety of social and environmental data and of displaying such information in a form suitable for decision making should be recognized. The development of such analytical and display capability is considered to be one of the essential ingredients to wise long-range management. Another essential ingredient is the development of the capability of reaching rational decisions, once the information is freely available. Rationality is a relative matter, subject to the basic values of the decision makers. The underlying value adopted by the workshop is the continuing benefit of the coastal zone for the entire population of the United States. Present managerial systems are simply not equipped to handle the widespread environmental problems attendant upon intensive coastal zone use. A better way is urgently needed.

A most powerful approach developed for attacking complex and otherwise intractable problems is the application of systems analysis. Coastal zone management, while characterized by many special features and uncertainties, belongs to a class of problems to which operations research specialists have devoted much attention. Already available are the general conceptual framework; an inventory of management, social, and physical environmental models; some experience in simulation; and fully adequate computer resources which together provide a strong basis for the development of systems analysis approach.

Managerial data requirements are far less demanding than those required for full scientific understanding, with informed judgments and approximations often providing a quite adequate basis for decision making. Truly refined management at the ecological and socioeconomic levels is, at best, a very long-term goal, the progressive introduction of systematic approaches and of machine support for better human decision making can begin now and would certainly result in substantial practical management improvements.

The advent of the systems approach provides a means of summarizing vast bodies of information and of translating them into various levels of perception, analysis, and action. The use of these techniques, however, bears the risk of losing concrete applicability with every stage of translation. Consequently, great precision in terms and procedures must be linked with continuous scrutiny by those expert in each language—for example, substantive specialists in biology, physics, engineering, or sociology—if the potentialities of systems analysis and integration for improved discovery and decision making are to be realized. This results in a viable man-machine rather than a machine-man process. Over the next decade, the generation of computers and software now being built will together provide great margins of capability for any conceivable data or computational requirements of coastal zone management. Directed by society and constrained by the fragile environment, technology is a means, not an end in itself, and can assist man with the solution to his seemingly overwhelming problems.

The total time horizon with which we are dealing is unlimited, yet it must be phased into foreseeable horizons for short-term planning. Therefore, our immediate objective in planning for a rational future is to establish the framework for a realistic pattern of decision making which is responsive to the many conflicts that are inevitable as demands exceed resources. Thus, as we proceed into the future by a successive series of approximately informed and rational decisions, we can periodically adjust coefficients of the use-environmental quality equations for achievement of successively perceived optimum states. These optimum states, in turn, will, we hope, reflect a wiser and better informed consensus of public opinion.

1.8 Workshop Recommendations

The recommendations made by the Coastal Zone Workshop were guided by the recognition that existing research, managerial, regulatory, and action agencies already develop information and make decisions which they consider in the best interest of all involved parties. Regional zoning schemes, complex transportation programs, Chamber of Commerce advertising, taxing policies, and commercial development contribute to the existing policies, goals, and objectives for the use of the coastal zone. The coastal zone should be considered as an environmental bank. We can withdraw the interest on the capital; but when we withdraw the capital itself, we reduce the interest. Our concern must be to utilize the yield of this environment without destroying its ability to renew its resources. The primary aim of all the recommendations is to provide for the evaluation and development of man's use of the ecosystem.

Legislation at the state and federal level have been discussed and considered in the formulation of these recommendations. The recommendations reinforce some which have been presented in previous reports on the coastal zone. The passage of time has allowed a better definition of the needed focus, and the interdisciplinary nature of the participants in the workshop has produced some novel and innovative approaches.

Management of the coastal zone is of great national concern, but lacks a definitive policy or recognized national goal. The absence of definitive national policy or recognized national goals concerning the numbers, distribution, and movement of people, and the distribution and utilization of natural resources including land, air, materials, energy, and recreational areas, make the development of guidelines for specific regions of the coastal zone very difficult.

Within the limits of our combined interdisciplinary capabilities, the Coastal Zone Workshop has examined the general changes required to meet the proposed goals. The recommendations emphasize actions that we feel to be central to each of the several basic concepts relating to man's use of the coastal zone. These recommendations will require general changes in attitudes, activities, legal structures, management arrangements, and information availability. Some are specific and concern man's use which may appear detailed, but the need for their early implementation is considerable.

The workshop brought together a diverse group, and real communication was not always easy. The participants in the workshop worked diligently and with dedication in an effort to establish and maintain the necessary interaction. The following recommendations were discussed at length and represent the professional opinion of a majority of the workshop participants.

General Recommendations

The Coastal Zone Workshop Recommends—

I.

The development of a vigorous and comprehensive *National Coastal Zone Policy* by the federal government in cooperation with the states that will provide for the wise use of the marine, estuarine, wetland, and upland areas bordering the American shores. All future uses of the coastal zone must be designed to maintain the natural ecosystems and to provide for the use of contiguous resources by the people of the United States. Cooperative action by cognizant federal, regional, state, and local governments will be required. The integral element of the National Coastal Zone Policy should be the focus of management responsibility at the state level, with the active participation of local governments, under federal policies that provide grants and set guidelines for creative and effective programs.

II.

The President of the United States request the National Academies of Sciences and Engineering to create a multidisciplinary *Coastal Zone Task Force* to formulate a management program. The Task Force should assist the federal government in designing the national program and evolving model guidelines for state coastal zone management authorities. The Task Force should work with the coastal states, regional agencies, and the federal government in preparing specific plans for coastal regions of the United States.

III.

The development of *legal institutions and procedures* to make coastal zone management more effective. Substantial improvements in the existing types of decision procedures and laws are required and consideration should be given to—

Development of innovative approaches through new coastal land and water use accommodations;

Alternative means for the regulation of coastal development besides the taking of private property;

Improvement of statutes and administrative regulations for land, water, and submerged land activities;

Increased access of individuals, groups, and governmental units to administrative and judicial proceedings;

Establishment by state legislatures of Environmental Review Boards for appeals of local administrative decisions concerning activities that have coastal and environmental impact;

Establishment by Congress of an expert federal Environmental Court with broad jurisdiction over private persons, state and local government agencies, and federal agencies in controversies involving coastal and environmental impact.

IV.
The establishment of regional *Coastal Zone Centers* to develop and coordinate natural science, social science, and legal research and to provide relevant information about the coastal zone to government agencies and the public. These centers should cooperate with existing research organizations to resolve basic questions of the environment in that region, help appraise management techniques, and provide inventories of coastal resources. Coastal Zone Centers should be established in regions corresponding to the major types of coastal environments and may be international in character.

V.
The creation of a national system of *Coastal Area Preserves* for the permanent protection of the basic genetic stocks of plants and animals and the essential components of their environments, which together constitute ecosystems. These Coastal Area Preserves should be severely restricted in use. Some other coastal areas should be developed for recreational usages that are compatible with the natural life of the area.

Specific Recommendations

To Improve Our Knowledge, the Coastal Zone Workshop Recommends—

1. The acceleration and expansion of comparable surveys and complementary inventories of coastal resources, including demographic patterns, ownership and land-use patterns, and other socioeconomic data in addition to new base-line ecological studies of natural and modified coastal systems.

2. The further development of predictive models to aid in understanding the effect of activities and structures upon the coastal zone environment and to improve the management of the coastal zone by evaluating the impact of alternative actions.

3. Improved environmental impact statements should be prepared for each new or additional activity and structure in the coastal zone to determine the extent to which they would have social as well as environmental effects. More stringent requirements for the preparation, detail, and use of such statements in making specific decisions should be developed.

4. Basic biological, chemical, and physical research directed toward the following types of problems in the coastal zone:

a. Transport, dispersion, upwelling, and cycling of nutrient and hazardous chemicals as they affect the functioning and stability of coastal zone ecosystems;
b. Surveillance of input levels of contaminants, especially chlorinated hydrocarbons, petroleum, and heavy metals;
c. Effects of solid waste disposal;
d. Effects of chronic, long-term, sublethal contaminants on organisms and ecosystems;
e. Assimilative capacity of coastal zone for all kinds of wastes;
f. Epidemiologic and virologic studies;
g. Recovery processes in damaged ecosystems;
h. Factors affecting stability, diversity, and productivity of coastal zone ecosystems;
i. Techniques for increasing production of desirable species or systems.

5. Research in the legal, political, economic, and social aspects of the coastal zone should be directed toward the following types of problems:
a. Exercise of property rights in wetlands and shore areas;
b. Administrative and judicial enforcement of codes;
c. Statutory guidelines and their interpretations with respect to shoreline development;
d. The decision-making process for the coastal zone at local, state, and national levels;
e. Group interests and political pressures in coastal zone uses;
f. Value systems that affect management practices in coastal zone activities;
g. Cost-benefit analysis of ultimate uses of the coastal zone, including ecological effects;
h. Economic models for policy guidance in calculating inputs and outputs;
i. Economic factors and mixes in resource evaluation.

6. Research on the environmental social, economic, and legal effects of:
a. Siting, construction, and operation of coastal and offshore power plants and deep ports;
b. Dredging and deposition of spoil.

7. The creation of regional and national monitoring systems to collect continually chemical, physical, and biological data with a capacity to give advanced warning on conditions that may be hazardous to the ecosystem of the coastal zone.

8. A sustained national commitment to education and training of the necessary talent for the management of the resources of the coastal zone. The goal should be a widespread awakening in the public to the importance of maintaining a sound coastal zone environment as well as the preparation of a future generation of natural and social scientists to manage wisely their environmental heritage.

In Order to Allocate Political Responsibilities Efficiently, the Coastal Zone Workshop Recommends—

9. That the federal government establish a national coastal zone management program, which should be vested in one of the existing federal agencies and should coordinate all agencies involved in coastal zone activities. The federal agency should administer grants to state coastal

zone programs and set appropriate guidelines for such programs as well as for the management of federal coastal lands.

10. The federal and state governments, acting together, create regional councils to assist in carrying out the national coastal zone policy. Such councils would work in concert with federal and state agencies in advising on regional problems of national interest and implement appropriate policies where consensus exists between federal and state governments.

11. The state coastal zone authority should be established as an independent agency, with its expertise and primary responsibility exercised in cooperation with other state agencies involved in the coastal zone. Management programs should view the coastal zone as a complete natural system and not be restricted by political boundaries. Incentive policies, as well as regulatory powers, should be used to improve the management of the coastal zone. Local governments should be strongly encouraged to evolve their own local plans and programs within the guidelines of the state coastal zone program, while citizen's advisory boards should be used to gain public participation in the policy-making process.

12. The application of environmental quality standards and performance criteria based upon monitoring or surveys to be evaluated by all government agencies involved in the management of the coastal zone. They should take into account socioeconomic needs of the community, and resort to general regulations, zoning, and other codes only when necessary for compliance.

13. Recognition of the interest of people dwelling outside the coastal zone, but who are directly affected by its environmental conditions or its productivity. The needs of individuals and groups who have limited resources for competing in the political bargaining process in reaching coastal zone policy decisions must be considered.

14. The cooperation of industry, public utilities, state agencies, and the federal government in the development of regional planning and utilization of energy, including fossil, nuclear, or other fuels in the coastal zone so that costs and benefits of alternative sites of development within and outside the coastal zone can be compared. Public authorities should be guided by both the urgency of protecting the environment and the demand for energy in the United States.

15. Public authorities at all levels should consider methods of increasing the carrying capacity of the coastal zone through technical and managerial means, utilizing airspace over land and water as well as submerged areas in order to achieve community goals.

16. The conduct of a comprehensive investigation by the federal government, in concert with state agencies, into the present management of coastal fisheries and an appraisal of the policies and costs of existing programs. The inquiry should include thorough study of the merits of limited entry to fisheries and lead to an effective national fisheries management policy under the aegis of the federal government. Fishery conservation on the high seas beyond national jurisdiction should be vigorously pursued by the federal government and the right of access to coastal resources by domestic fishermen must be preserved.

17. The federal government lead in establishing regional *Coastal Zone Centers*. However, academic institutions, private foundations and enterprises, state governments, and granting agencies of the federal government should greatly increase their support of both fundamental and applied research in the natural sciences, law, and the social sciences in order to feed original information into the regional Coastal Zone Center and/or to public agencies for improving the management of the coastal zone.

18. Adequate funds be provided for activities that have developed information whose results should be analyzed and published. Where useful, raw data exist that have not been subjected to adequate analysis, funds should be provided to complete the analysis and make the results available to users.

To Provide Special Uses, the Coastal Zone Workshop Recommends—

19. The immediate intact preservation of selected natural land and water areas in shoreline and estuary regions of the United States valued for their unique ecological character. Such areas should be severely restricted from any private or public coastal zone activity.

20. The protection from environmental degradation of those coastal wetlands and estuaries that are highly productive habitats, spawning areas, or nurseries for aquatic life or contain rare and endangered species. Only coastal activities that will not markedly degrade the diversity

and productivity of the existing ecological system in these areas should be permitted.

21. The monitoring of activities in the coastal zone not only for their effect upon the near-shore waters, but upon the seas and oceans. Chemicals, airborne and waterborne, from the coastal zone, as well as certain drilling, dredging, and dumping may cause serious harm to the marine environment and should be regulated to avoid serious damage to oceanic ecosystems.

Part II

Constraints on Man's Activities

Man uses the coastal zone for the production of food, plant products, for the preservation of habitats and ecosystems, for industrial and commercial development, for habitation and recreation, and for various governmental purposes. Often, natural processes prevent man from using parts of the coastal zone for one of these activities, and one activity may place severe constraints on other activities. Some uses that are compatible with others under conditions of low activity may be intolerable at intermediate activity levels and become completely incompatible at full use. An estuary can be used for food production; it can be used as a sewage dump; it can be filled and used for a housing development; it can be used for transportation, and many of man's activities produce wastes that require some treatment, but that ultimately reach the sea. Each of these activities constitutes an important use for an estuary, but today's technology does not permit all uses simultaneously.

The following discussion of man's role in the coastal zone considers his use of biological productivity; his habitation and recreation demands; his industrial, commercial, and governmental uses; the contamination resulting from man's activities; and an analysis of the strategies available for management of man's use of the marine environment.

Chapter 2
Living Resources

PREPARED BY
John M. Teal, David L. Jameson, Richard G. Bader

WITH CONTRIBUTIONS FROM
George W. Allen, C. W. Hart, J. L. McHugh, Ruth Patrick, William Perret, A. T. Pruter, Saul B. Saila, H. Dixon Sturr, Durbin C. Tabb, Gordon Thayer

Chapter 3
Nonrenewable Resources

PREPARED BY
Richard G. Bader, Robert A. Ragotzkie

WITH CONTRIBUTIONS FROM
Dean Bumpus, Howard Gould, J. Robert Moore, John Paden, John S. Schlee, John M. Teal, Linda Weimer

Chapter 4
Recreation and Aesthetics

PREPARED BY
David L. Jameson, Robert A. Ragotzkie, Richard G. Bader, John M. Teal

WITH CONTRIBUTIONS FROM
George W. Allen, Brian Bedford, John Clark, Frank Herrmann, John S. Steinhart, Russell L. Stogsdill, Linda Weimer

Chapter 5
Urbanization and Industrial Development

PREPARED BY
Robert A. Ragotzkie, John M. Teal, Richard G. Bader

WITH CONTRIBUTIONS FROM
Brian Bedford, John Clark, Gabriel T. Csanady, William Gaither, Frank Herrmann, Claes Rooth, John S. Steinhart, Russell L. Stogsdill, H. Dixon Sturr, Linda Weimer

Chapter 6
Transportation and Coastline Modifications

PREPARED BY
Richard G. Bader, Robert A. Ragotzkie, John M. Teal

WITH CONTRIBUTIONS FROM
Gerard A. Bertrand, Dean Bumpus, Gabriel T. Csanady, William Gaither, Frank Herrmann, J. Gordon Ogden III, Claes Rooth, Russell L. Stogsdill, Linda Weimer

Chapter 7
Contamination and Coastal Pollution

PREPARED BY
Donald W. Hood, Eleanor Kelley

WITH CONTRIBUTIONS FROM
E. J. Carpenter, Dayton E. Carritt, Gabriel T. Csanady, Edward D. Goldberg, John Hunt, G. Fred Lee, Patrick L. Parker, Ruth Patrick, Theodore R. Rice, Robert W. Risebrough, Anitra Thorhaug, Oliver Zafiriou

2 Living Resources

2.1 Introduction

Man uses the biological productivity of the coastal zone for food, furs, fertilizer, needles, buttons, ornaments, chemicals, fish meal, and various construction materials. Those species not used for human consumption are frequently used to make oils and meals for fertilizers or animal food. Biological productivity is determined by the availability and transfer of energy in the ecosystem, availability and uptake of nutrient elements, the genetic and species diversity, and the stability of the ecosystem. The availability and transfer of energy and nutrients is disturbed by man's other activities in the coastal zone, and this, in turn, can lead to a simplification of coastal ecosystems by a reduction in diversity and stability.

Under stable conditions, great portions of the nutrients and energy in the ecosystem are in equilibrium. Increased efficiency of the ecosystem results in an increase in the total yield of living organisms. The source of this increase in efficiency is genetic diversity of species; the mechanism of increase in efficiency is the evolutionary process. As a result, the ability of man to use the natural ecosystem yield depends on maintaining the diversity of the system, even though all of this production is not of direct value to man.

Seasonal cycles in the availability of solar energy are characteristic of the coastal zones of North America. Solar energy determines both light and temperature cycles and, when coupled with the mechanical energy from winds and tides, provides transport and mixing of nutrients. The result is a variable, but generally high, productivity in the coastal zone. Different mixed layers of adjacent currents produce sharply differing chlorophyll production rates and hence differing primary production and sharply contrasting fishing resources (Holmes, Schaefer, and Shimada, 1957).

Nutrient availability varies with the seasons as a result of current circulations, variations in water solubility, and restrictions (dams), or additions (waste discharge) due to human activities. The amount of rainfall partially controls the rate of river flow, land drainage, and erosion which, in turn, control the rates of mineral and nutrient transport into estuarine and coastal systems. The availability of nutrients, coupled with temperature and light penetration, controls the rates of phytoplankton and macroscopic plant production, and hence the production of the oxygen required for metabolic processes and for the biodegradation of organic material by microbes (Welch, 1968).

All marine organisms undergo seasonal cyclical processes such as metamorphosis, reproduction, and migration. Alteration of the environment by natural or man-induced activities, especially during critical seasons, may irreversibly diminish biological productivity. Sardines and haddock were overfished during the low portion of a possible long-term cycle (Joyner, 1971). The normal growth and reproductive life cycle of seagrass was stopped by thermal input from a power plant (Phillips, 1969; Roessler and Zieman, 1970; Thorhaug, Stearns, and Pepper, 1972).

The cycles produced by the regular flooding and purging by hurricanes in Gulf of Mexico estuaries and lagoons may be essential to the life cycles of many organisms (Odum and Wilson, 1962). This process could be disrupted if attempts to modify hurricanes succeed. Even so, economic losses due to hurricanes are so large that the social need to reduce the destructiveness of these storms may be the overriding concern. In that case, the reduction of wind intensity without major reduction of rainfall might accomplish the desired flushing of lagoons without storm surge flooding. The use of man-induced changes to increase productivity requires full knowledge of the reproductive and biochemical cycles (Hedgpeth and Gonor, 1969).

Embayment systems tend to be nutrient traps having low tidal flushing rates. As a result of increased nutrients from a buildup of duck farms in the area of Moriches and Great South Bay (New York), the waters of Great South Bay have exhibited increased algal populations of those species not subject to grazing, and this has decreased oyster and fish production (Ryther, 1954; Galtsoff, 1956; Barlow, Lorenzen, and Myren, 1963). The optimum conditions for oyster growth include mixed algal populations. The decline of the oyster industry was directly correlated with the increase in a dominant alga which greatly differs from the flora typical of the area. Worms that are capable of effectively utilizing the new and dominant algae have overrun oyster beds periodically and thereby reduced oyster production.

2.2 Simplification of Coastal Ecosystems

Simplification of an ecosystem has two immediate consequences—it alters the food chains, which when left undisturbed are usually complex, and it alters the genetic composition of the ecosystem by eliminating various genotypes and species of organisms, which modifies the evolutionary processes. For example, in the 1930s a basic food link was destroyed by eelgrass disease on both sides of the Atlantic Ocean. The communities subsisting on this organism disappeared from the area and did not return until their food and/or hiding places were reestablished (Allee, 1923; Stauffer, 1937; Dexter, 1950; Phillips, 1969).

Coastal ecosystems are remarkably diverse systems where species, because of their location at the interface between land and sea, are subject to extreme, and often violent, environmental changes. Many species live near their tolerance limits, and environmental changes from man's activities may result in the elimination of species. Changes in species diversity have become a useful tool of gauging the effects of man's impingement on water quality, since increased stress is generally manifested in lowered diversity. In the Virginia wetlands, the simplest system is in the James River from Richmond to Hopewell, where, because of sewage effluents, the previously diverse nursery food webs have been reduced to the simplified food chain of sewage–algae–tubificid worms–catfish–man (Wass and Wright, 1969).

Tsai (1968) noted that fish diversity was reduced directly below the effluent outfalls of the Patuxent River Estuary (Maryland) and increased further downstream. Wilhm and Dorris (1968) noted similar results for benthic invertebrates in a stream receiving domestic oil and refinery effluents.

Natural disasters, like the disease of eelgrass in the 1930s, can also cause a decline of diversity. But whether from natural stresses imposed by the environment, or from man-induced stresses, the end result is the same—the destruction of the grass followed by a reduction in invertebrate and vertebrate diversity, and often complete elimination of species (Taylor and Saloman, 1968; Roessler and Zieman, 1970; Sykes and Hall, 1970; Briggs and O'Connor, 1971).

Rosenzweig (1971) suggests that an unstable ecosystem can be evolving toward stability, but never reach stability because of the inherent environmental instability of the ecosystem. He further notes that enrichment can be beneficial at moderate levels if it is passed up the trophic chain to predators which eventually may increase. Occasionally this increase also is a cause of instability as a result of exploitation and overharvesting.

Living resources, such as fish, are capable of renewing their population so that the rate of increase in population may balance all mortality due to harvesting. When the resource is a particularly popular species such as tuna or halibut, the catch is often made up of primarily less-than-mature-sized individuals; and this could reduce the size of the breeding stock. The relationship between the size of the breeding stock and recruitment is not clearly understood, but it is probable that the continued taking of immature individuals could lead to disaster in some species. It takes tuna three years to reach maturity and Schaefer (1957a,b) has suggested the severe restriction of fishing for three years to assure attainment of maturity and reestablishment of the self-regulating mechanism.

Figure 2.1 shows Schaefer's (1959) analysis in terms of commercial fisheries. Line C_1 represents an unutilized fish stock, where the cost per unit effort is greater than the value per unit of effort at any effort; hence, the fish are not worth catching. Line C_2 represents the cost of a unit of effort when the average return at a level of fishing effort (E_2) is below the maximum sustainable yield. Schaefer notes that such a fishery is underfished, and no restriction on fishing effort can produce a greater average harvest than is being obtained. In the third case (C_3), cost per unit of effort is equal to the average return at a level of effort (E_3) above that corresponding to maximum sustainable yield. Such a fishery is

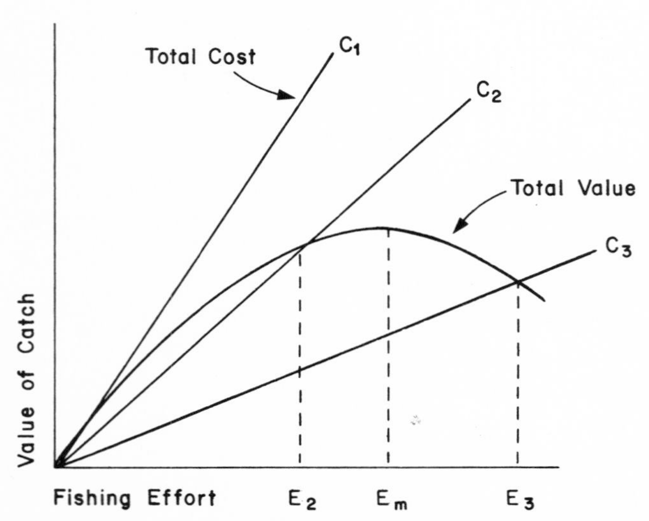

Figure 2.1 Total cost and value of a commercial fishery catch relative to expanded fishing effort. Source: Schaefer (1959).

termed "overfished" since sustainable yield can be increased by restricting fishing effort.

Trade-offs may be one acceptable means of system simplification. Williams and Murdoch (1970) suggest reducing the abundance of "useless" species like jellyfish and stingrays. These and others compete with food species, are not used in the present fisheries, and do not comprise links on food chains to "useful" species. Trading greater efficiency for lower stability by shortening the major coastal food chains could only be achieved by artificially maintaining a delicate balance between herbivore production and carnivore needs. This may not be practical in most cases because of the extra energy that must be supplied to maintain the system.

Man's harvesting of food species leads to uniformity of age classes and often to the reduction of breeding stock and species diversity, and to the development of less complex communities (Schaefer and Beverton, 1962). While some human activities reduce the productivity and harvest of food resources, other human activities, in early stages or small amounts, may stimulate resource availability or growth. Small doses of waste provide nutrients; slight increases in temperature from power plant cooling water can increase production; a few oil derricks, docks, and wharfs increase the diversity of the habitat; military zones provide refuges from overexploitation. But overuse by these and almost any other human activity will lead to deterioration of the coastal zone as a source for human food.

2.3 Commercial Fisheries

From the estuaries and continental shelf of the United States, the nations of the world harvested over ten billion pounds of commercial fish and shellfish in 1970 (Pruter, 1972). In addition, recreational fishing extracts a large crop. The coastal zone, and especially the estuarine areas, support the egg and larval stages of many important species. In some geographical regions of the United States 90 percent or more of the commercial catch of finfish has been shown to consist of estuarine-dependent species (Sykes, 1968). These food and sport fisheries are highly visible yields from the living resources of the zone, but less obvious biological processes are also important. Phytoplankton and benthic communities in open water and wetlands synthesize the organic foods that support useful species, and a wide variety of herbivores and carnivores provide essential links in the food web. Bacteria and fungi and some invertebrates decompose organic materials to permit recycling. If this system is healthy, the coastal zone not only yields food but preserves the aesthetic quality of coastal waters.

Although commercial production from U.S. coastal waters by domestic fishermen has been relatively constant over the last two decades, production by foreign fleets during the same period has grown rapidly, until it now exceeds domestic production. Types of resources and fisheries reflect geographical differences in habitat and extent of continental shelf and estuarine areas. In the Bering Sea the continental shelf is extremely broad, extending several hundred miles from shore; in the central Gulf of Alaska it is some 50 miles in width. The broad continental shelf, with many bays and inland marine waters and temperate water regime, favors the development of large herring, groundfish, crab, and penaeid shrimp resources. The groundfish resource in the eastern Bering Sea ranks among the largest in the world and is being increasingly exploited by commercial fishermen. However, among the extensive groundfish resources off Alaska, only the Pacific halibut is now the object of a substantial fishery by U.S. vessels.

The large river and lake systems in Alaska coupled with favorable coastal environments have resulted in the development of large fisheries for salmon, which in 1967–1968 accounted for 58 percent of the value of all fish and shellfish landed in Alaska. In that year, salmon, crab, shrimp and halibut together accounted for 98 percent of the value of all species landed (Lyles, 1969; National Marine Fisheries Service, 1971). The fishery for shrimp has increased from 7½ million pounds in 1960 to 94 million pounds in 1971.

Fisheries resources and habitats from central California to Washington are similar to those in Alaska. The existence of a continental shelf, at most 20 miles wide, and less extensive river and lake systems results in considerably smaller populations than in Alaska. Current and wind patterns, however, are seasonally conducive to strong upwelling of nutrient-rich waters which support moderately large populations of fishes and invertebrates. As in Alaska, the estuaries are an important habitat, particularly for salmon, oysters, crabs, and clams. Salmon is the most valuable resource in this region; in 1967–1968, it made up about 40 percent of the landed value of all commercial species (Lyles, 1969). Various species of groundfish, including flounders, rockfish, and "cods," support valuable trawl fisheries. Crabs and oysters are the most valuable shellfish. Most of the oysters are harvested in the state of Washington. Pacific hake, which appears to be one of the largest resources in the region from central California to Washington, has been harvested primarily by Soviet fishermen.

Off Southern California, the continental shelf is narrow, but hydrographical conditions are favorable to the development of large pelagic fish resources. In the 1930s and early 1940s, the Pacific sardine supported the largest fishery in the Western Hemisphere, accounting for over one billion pounds taken from California waters in some years. Poor recruitment of young fish and intense fishing was followed by a collapse of the Pacific sardine resource. Anchovy and jack mackerel now support important fisheries, and these pelagic resources are capable of supporting greatly expanded fisheries. Rockfish, flounders, and sea bass are important resources. Most valuable among the shellfish are abalone, squid, and spiny lobster. Tunas are the most valuable fishes landed in California; but most are caught off the coasts of other nations. The offshore kelp beds provide food and habitat for many fishes sought for sport and commerce. The kelp also is a resource of direct economic value.

Although Hawaii is strategically located as a base for intercepting tunas and other high-seas pelagic fishes, it lacks the shallow rich banks that characterize most of the world's great fishing areas. The marine life of Hawaii represents the Northeasternmost extension of the great Indo-Pacific faunal region (Manar, 1969).

A principal commercial fishery is for tunas that are harvested beyond the coastal waters. Among the inshore fishes, the bigeye scad is important for local use. Other inshore fishes serve as bait for the tuna fisheries. Reef fishes appear to be more important for their recreational than their commercial uses. Snappers are among the most valuable of the groundfish and are harvested with handlines. Shellfish include limited quantities of crabs, limpets, spiny lobsters, and shrimp.

Within the Gulf of Mexico, the continental shelf is generally broad, ranging from 20 miles in width off the Mississippi delta to more than 100 miles off the south Texas and west Florida coasts, and the fauna are semitropical. Species of fishes and invertebrates are numerous and diverse and have shorter life spans and more rapid growth rates than those inhabiting more temperate waters. Fishing in the Gulf of Mexico presently harvests over 60 species of fishes and 20 species of invertebrates, and surveys have shown there are other potential resources. The large estuarine areas play an important role in the maintenance of our most valuable fishery, the penaeid shrimps, and our largest volume fishery, the menhaden. The most valuable of the fishes are menhaden, snappers, mullets, mixed groundfish species used for animal feed, sea trouts (weakfishes), and groupers in that order during the period 1967–1968. Oysters and crabs are the important shellfish.

The fauna off the South Atlantic states (Cape Hatteras to southern Florida) is similar in many ways to the Gulf of Mexico and includes both semitropical and temperate species. The continental shelf is relatively broad, often extending 60 miles seaward, and supports substantial concentrations of fish and shellfish. Shrimp are the most valuable shellfish and menhaden the most valuable fish. Other important fisheries are for crabs, lobsters, striped bass, shad, and mullet. A large calico scallop resource, which was recently delineated off the east coast of Florida, appears to be the largest latent shellfish resource in the South Atlantic region.

Commercially important oyster species occur along the east coast of North America from the Gulf of St. Lawrence to the Gulf of Mexico and along the west coast from British Columbia to California. Most important are the American Oyster, *Crassostrea virginica,* and the Giant Pacific Oyster, *C. gigas.*

Oysters grow best in shallow bays or estuaries where suitable temperatures, currents, salinities, suspended matter, food, bottom types, and associated organism densities prevail. The regimens are relatively restricting even before the trespass of man is considered, and oysters (and other sedentary forms such as clams and mussels) exist in a precarious natural milieu.

Oyster production from New York to Texas has increased in some areas between 1956 and 1965, but has suffered an overall decline in that period slightly in excess of $2,000,000. In New York the harvest declined to one-third its former value, and the pollution of the bays of Long Island is cited as the most important cause. Formerly productive beds in Delaware Bay, Rehoboth Bay, and Indian Bay have largely disappeared, and the Chesapeake Bay production has also declined. Only Florida and some Gulf of Mexico areas have shown consistent gains.

Chesapeake Bay is our largest estuarine system. It is also one of the oldest and richest fishing grounds along the shores of the United States and is an important nursery ground for fishes, including the highly prized striped bass. Although landings of oyster meats declined from a level of about 100 million pounds annually in the period 1880–1900 to 26 million pounds in 1967, the oyster continues to dominate the fisheries of the bay in value (McHugh, 1969). Other important shellfish are crabs and clams. Together these made up over 70 percent of the value of the aquatic harvest of Chesapeake Bay in 1967–1968. Most valuable of the fishes are menhaden, striped bass, alewives, flounders, scup, and shad. Alewives and shad are anadromous species that journey through the estuary on their spawning migrations from coastal waters to the rivers and streams. Many of the other fishes found in the bay perform an annual migration into the bay in summer and out to the ocean in winter.

North of Cape Hatteras the continental shelf extends some 60 to 100 miles seaward, and conditions favor the existence of large and important fisheries resources. Commercial fishing has gone on longer in this region than in any other coastal waters of the United States. In recent years, these resources have been subjected to increasing pressure from foreign fishing fleets. Commercial fisheries have included those for haddock, flounders, ocean perch, sea herring, whiting, hakes, pollock, cod, menhaden, and scup. Most valuble of the shellfishes have been the highly prized lobster, scallops, clams, and oysters.

Domestic production in 1970 of fish and shellfish from coastal and estuarine waters of the United States was approximately 4⅓ billion pounds, worth over $500 million to the fisherman.* To this production must be added the catches of foreign fishermen, which in 1970 were estimated to amount to about 6 billion pounds (Pruter, 1972). Competition from foreign fishermen is particularly acute off Alaska, the Pacific Northwest, and New England. It seems only a question of time before similar competition will occur elsewhere off the United States.

Although domestic landings of fish have been relatively stable during the last two decades, consumption of fishery products in the United States has increased. Part of the increase in consumption has been food for people, and the remainder can be attributed to the growing use of fish meal as animal feed. Total consumption of fishery products in the United States amounts to over 60 pounds (live weight) per person per

* Totals for catches from waters off the United States and on high seas off foreign coasts were 4.9 billion pounds and $602 million. Totals here have been adjusted downward to reflect only domestic catches off the United States.

year, of which 49 pounds are used mainly as animal feed (National Academy of Engineering, 1972). Direct per capita consumption of seafood in the United States has been close to 11 pounds (live weight) per person per year for many years. Now, however, use of such products as frozen fish sticks and fish portions has increased at the expense of fresh fish (National Academy of Engineering, 1972). Domestic trawling fleets, the major suppliers of fresh fish, have been particularly affected by this shift, which results in an increasing portion of our domestic food fish market being supplied by imported products. The U.S. production of these industrial fishery products has also remained steady over the past two or more decades, and imports have more than doubled to support increased consumption.

"Institutional constraints" is a broad term referring to laws, regulations, and customs that restrict the operation of fishermen. These constraints are among the major obstacles to increasing the U.S. fish catch and raising the economic status of fishermen. A myriad of conflicting national, state, and local restrictions have been established piecemeal with little consideration of their overall effect on the consumer, the fishing industry, recreation, or conservation (National Academy of Engineering, 1972). The policy of permitting unlimited entry of participants into fisheries contributes to economic inefficiency and to depletion of resources. An additional very important constraint is the general inability of our fishermen to compete effectively with foreign fishermen for resources beyond the present U.S. jurisdictional control, 12 miles from shore.

The future of our commercial fisheries is dependent upon several important actions. We must improve our capacity to manage our domestic fisheries and vigorously seek an internationally acceptable solution to the problem caused by the massive fishing by foreign fleets. Failure to manage adequately the fisheries will result in the addition of more and more species to the already extensive list of overfished resources and thereby deny us the yields the fisheries are capable of providing on a sustained basis (Stroud, 1970).

2.4 Sport Fisheries

Sport or recreational fishing holds an important place among the multiple uses of the coastal zone in terms of the numbers of people involved and the amount of business activities generated. It has been estimated (U.S. Senate Report No. 92–753) that sport fishing currently is an ac-

The Coastal Zone Workshop recommends
The conduct of a comprehensive investigation by the federal government, in concert with state agencies, into the present management of coastal fisheries and an appraisal of the policies and costs of existing programs. The inquiry should include thorough study of the merits of limited entry to fisheries and lead to an effective national fisheries management policy under the aegis of the federal government. Fishery conservation on the high seas beyond national jurisdiction should be vigorously pursued by the federal government, and the right of access to coastal resources by domestic fishermen must be preserved.

tivity of about 11 million people in coastal areas, and that by 1975 this number will have increased to 16 million. Stroud (1969) reported that about $800 million of gross business activity was generated in 1965 by saltwater fishing. He stated that over one-half of the marine angling activity occurs on the Atlantic coast, about one-quarter on the Gulf coast, and the other quarter on the Pacific coast. King et al. (1962) predict that growth in sport fishing activities will increase at a substantially greater rate than human population growth in the United States for the rest of this century.

Sport fishing as presently conducted has less impact on the welfare of living resources in the coastal zone than other activities. Much of the sport fishing interest utilizes species that are exploited by both sport and commercial fishermen, and about 75 percent of these are dependent upon the coastal zone for some part or all of their life history (Stroud, 1970). For example, the anadromous sea basses, striped bass and white perch, spend virtually all of their life histories in the coastal zone. Both species are of sport as well as commercial value, and the striped bass is a prized sport fish on both coasts.

The impact of increased fishing activities in the coastal zone must be considered in terms of individual species and specific areas. For example, there are specialized developmental requirements for fertilized eggs and larvae of striped bass near the heads of estuaries. Recent increases in the relative frequency of dominant year-classes of this species have been hypothesized by Mansueti (1961) to be related to natural and artificial enrichment of these areas. Furthermore, Dovel and Edmunds (1971) have reported that the total area in the Chesapeake Bay region suitable for reproduction and fishing of striped bass increased as an unanticipated result of hydrographic changes due to the construction of the Chesapeake Bay and Delaware Canal. On the other hand, it is clear that severe reductions in populations of Atlantic salmon have resulted from such things as pollution and dams in coastal areas. All species of salmon (Atlantic and Pacific) as well as some trout and char depend upon the coastal zone for a part of their life histories. Some herrings, for example the American shad, have coastal zone requirements similar to the salmon. Of the more than 80 commercially important species of vertebrates and invertebrates dependent upon estuaries and the coastal zone for all or important periods of their lives (Ludwigson, 1970), many contribute substantially to the sport fishery.

It is believed that the increased demands of future sport fisheries can be met with negligible reductions in the present catches of sport fishermen. This prediction assumes that the coastal zone will not be grossly damaged as a spawning and nursery ground, that species are not blocked in their migration patterns, and that an enlightened program of applied research and development for habitat improvement will be put into effect. Research should also be directed toward deciding which coastal species might best be managed solely for the sport fishery. It would appear that the coastal zone is inherently capable of absorbing considerably more fishing pressure than it currently does without serious impact; however, regulation of catches may become a necessity as this pressure increases. It is important that the productivity, diversity, and magnitude of the coastal zone habitat be preserved for its relatively easy sport fishing access, secure refuge areas for fishermen, and abundant fauna.

2.5 Aquaculture in the Coastal Zone

Aquaculture, which includes the culture of marine and brackish water species, has a long history of development and application in parts of Europe and Asia and considerable global potential (Betz, 1971; Bardach and Ryther, 1968). In recent years Japanese scientists have succeeded in culturing pearl oysters, shrimp, and edible algae on a commercial scale. Americans have produced large commercial enterprises around goldfish, bait minnows, catfish, trout, salmon, oysters, and clams. Increased production of oysters grown on rafts over those grown on level bottoms in a conventional fashion gives a spectacular example of the progress being made. Similarly, the cage culture of trout, catfish, coho, and chinook salmon at various inland and coastal experiment stations has provided vastly increased production over conventional pond-rearing techniques.

Although aquaculture in the coastal zone of the United States is beginning to develop in a manner analagous to the land-based livestock industry, there are signs that use of large tracts of naturally productive waters will meet with serious legal constraints (Kane, 1969) because it is necessary to fence large areas of the most productive water where tank-reared young are released to fend for themselves or to build artificial impoundments on uplands adjacent to the bays at the expense of other uses. In general, it seems clear that future developments in aquaculture will be linked to species that command high prices and have ready markets. These species will also be ones that can be grown in a truly intensive fashion, thus placing minimal strain on land and water of the coastal zone. Coastal zone organisms such as the northern lobster, the tropical spiny lobsters, penaeid shrimps, prawns of the genus *Macrobrachium,* scallops, oysters, clams, salmonid fishes, and pompano fall into the high value category. Some are technically easy to rear but have not yet been cultured economically according to Gates and Matthiessen (1971), Scott (1970), and Anderson and Tabb (1970). Others, including the lobsters, spiny lobsters, and pompano have very complex and extended larval life histories which require excessively long rearing times before they are of marketable size.

Although present use of the coastal zone for aquaculture is small, this form of animal husbandry is expected to expand significantly in the near future. Part of this expansion will involve high-density culture such as raft culture for oysters and will probably create little competition for coastal water space. The Olympia oyster is now being cultured in the shallow water beds in Puget Sound. Shrimp culture, on the other hand, is likely to require extensive acreage for many years. One shrimp farming area may cover nearly 3,000 acres of bay and adjacent marshland.

Inasmuch as there is a trend toward high-density closed or semienclosed recirculating systems that restrict acreage requirements, there will be fewer conflicts. There are some indications that aquaculture may be combined with power production to utilize waste heat and flowing water to support dense populations of animals (Gaucher, 1970).

The future of aquaculture in the coastal zone will probably be linked closely to advances in bioengineering and life history research. The elements of these programs will include genetic, nutritional, maturational, and waste disposal studies applied with investigations of physiochemical and space relationships as well as disease and predation studies of intensively cultured organisms. The multiplicity of requirements of modern aquaculture raises some possibilities for conflict with other users of the coastal zone.

Large tracts of water and land, certainly in the thousands of acres, will be required for each major shrimp culture venture. Fencing of the leased waters will restrict free passage, recreational boating, and fishing within the lease. Use of upland acres for pond construction will impinge on other uses, including the functions of marshlands as natural producers of organic material, which contributes to the feeding cycle of wild coastal zone organisms.

Since culture practices must exclude predators and competitors, the fencing of large bay areas will mean taking all such areas out of production for coastal fishes and shellfish. The only known effective way of excluding the undesirable animals is by use of toxic chemicals such as rotenone, followed by recovery of the ecosystem and restocking with the desired species. Feeding large populations of aquatic animals can lead to enrichment of the water as well as to accumulations of excrement. Predation by diving and wading birds such as cormorants, mergansers, grebes, herons, and pelicans will be a serious problem that must be controlled in some acceptable way. Culturists and public agencies should work together to minimize such side effects of aquaculture.

Aquaculture has made significant progress during the past half century and is likely to improve even more in the coastal bays, estuaries, and adjacent uplands. Aquaculture is not likely to compete quantitatively with catches of the same species in the wild; but the product may be of higher quality, and, because of confinement, can be marketed at the optimal stage of the growth cycle or at times when supplies of the wild product are scarce.

Future planning for use of the coastal zone should determine if aquaculture is a legitimate use of a region, and if so, how it can best be accommodated to minimize social conflict and maximize returns from the effort.

To the extent that technological and bioengineering advances do not resolve the problems of incompatibility of aquaculture with other uses, the legal system will be the arena for resolving such conflicts. Institutional constraints result from the uncertainty of property rights in

---

The Coastal Zone Workshop recommends
Research in the legal, political, economic, and social aspects of the coastal zone should be directed toward the following types of problems:
a. Exercise of property rights in wetlands and shore areas;
b. Administrative and judicial enforcement of codes;
c. Statutory guidelines and their interpretations with respect to shoreline development;
d. The decision-making process for the coastal zone at local, state, and national levels;
e. Group interests and political pressures in coastal zone uses;
f. Value systems that affect management practices in coastal zone activities;
g. Cost-benefit analysis of ultimate uses of the coastal zone, including ecological effects;
h. Economic models for policy guidance in calculating inputs and outputs;
i. Economic factors and mixes in resource evaluation.

---

coastal areas. Conventional oyster farmers need exclusive rights to a submerged land area to justify the expense of husbandry, and shrimp or finfish culturists must be able to have some degree of exclusive use of the water column.

The powers of state and local governments to authorize exclusive rights to resources considered to be held as public trust must be clarified before investment will occur. Protection of the public interest in navigation and fishing is considered paramount. Private rights for aquacultural enterprises should not be permitted in areas of high commercial, sport fishing, or recreational value to an extent inconsistent with those uses. Nor should such enterprises be allowed to obstruct the rights of riparian or upland owners of access to water areas. Aquaculture must be subject to the environmental quality standards and regulations imposed at various levels of government. To these ends, legislation such as that pioneered by Florida should be adapted to allow the creation of the necessary rights, to institutionalize planning, and to rationalize the balancing of public uses and the exclusive use of water areas for aquaculture.

2.6 Coastal Marshes and Islands

Coastal zone wetlands and marshes extend all along the Atlantic and Gulf Coasts, though they are less common on the West Coast. Their total area in the coastal region, both fresh and salt wetlands, was about 9.3×10^6 acres in 1956 (Shaw and Fredine, 1956). Along most of the coast, grassy marshes are most common, but in the southernmost areas mangrove swamps dominate. Salt marshes and swamps are intertidal, and though parts are flooded only by spring tides, other parts are covered by every tide. Many also receive freshwater input and thus have both water and salinity gradients.

Coastal wetlands are highly productive both where the higher plants grow and on the mud and sand flats where only algae grow, so that marshes are among the most productive natural systems known (Teal and Teal, 1969). Much of the production becomes detritus which is a food source for organisms both in the marshes and in the estuaries and bays into which they drain. Oysters, clams, and mussels form beds in these tidal areas (Teal, 1962; Odum and De La Cruz, 1967; Steever, 1972). Because of the importance of coastal wetlands and because of the pressure they are under, it is important that they be conserved, since coastal marshlands are one of this country's most valuable types of land (Niering, 1970). A complete discussion of coastal preserves is found in Section 4.3.

The importance of such areas for nursery and feeding grounds has been well documented (McHugh, 1967). These coastal marshes are also utilized by crabs as feeding grounds. Young crabs in the Delaware marshes spend a considerable part of their lives in these areas during the summer months. All three of the commercial penaeid shrimps depend on inshore estuarine areas for the development of their young, as do many species of commercial fin and shellfish. Of the 800 million pounds of fishery products landed in New England in 1960, 500 million pounds were directly dependent upon tidal marshes (Niering, 1972). For the coastal fisheries to continue it is imperative that these estuarine and inshore areas be retained in a viable state.

Such regions of the coastal zone must also be conserved for use by migratory birds. They are important as flyways, nesting and overwintering areas for many species of birds which constitute a valuable resource. If these species are to continue to exist and be used for such recreational purposes as hunting and bird watching, safe resting, feeding, and nesting areas must be provided. Some mammals are also characteristic of the wetlands, particularly muskrats, mink, and otters (Wilson, 1968). Alligators and turtles are regularly found there within their ranges.

The Coastal Zone Workshop recommends
The protection from environmental degradation of those coastal wetlands and estuaries that are highly productive habitats, spawning areas, or nurseries for aquatic life or contain rare and endangered species. Only coastal activities that will not markedly degrade the diversity and productivity of the existing ecological system in these areas should be permitted.

Coastal wetlands can also serve as buffer areas to prevent flood damage, and they moderate the force of waves on upland areas during storms. They are sinks for sediments because of the action of plants in slowing water flow. They absorb nutrients from the water, thus improving water quality in areas where sewage effluents are a problem (Grant and Patrick, 1969; Valiela and Teal, 1972).

Coastal wetlands cannot stand up to the full onslaught of waves, however, so an important aspect of their conservation is the preservation of coastal islands behind which much of our wetland area lies. For the most part these islands seldom exceed ten or twelve feet above sea level, are covered with natural vegetation, and have a pronounced dune-line on the seaward side. They vary in size from a few acres to islands that may be twenty miles long and a mile in width. Historically these islands have been among the first to be settled by colonists, as with Raleigh's disastrous attempt at Roanoke Island in 1587. Some of the smaller islands are uninhabited and have recently been placed in the National Seashore program whereby they have been ensured a preserved status.

There are two basic types of movement of bottom sand at the coastline; onshore/offshore and alongshore. Wave action moves sand toward the beaches and back to the sea, and wind action moves sand seaward or landward. There is a definite seasonal variation to this movement. During the summer, waves are generally long and low and tend to build the beaches seaward. During the winter, storm waves are short and high and beaches erode. Waves generally approach the shoreline at an angle; as a result they generate a current parallel to the shore. This current is usually not strong enough to erode the bottom sand, but it is capable of moving that sand suspended by wave action. This littoral current changes direction in response to different wave conditions, and thus moves sand along the beach in both directions. Net annual sand movements as large as 500,000 cubic yards have been measured, whereas in other areas the annual wave patterns vary such that there is essentially no net sand movement (Caldwell and Lockett, 1965).

The rate of this alteration can be appreciated best by an examination of charts that show that in 1892 the pass between Dauphin Island, Alabama, and Petit Bois Island, Mississippi, was 1.1 miles, whereas in 1962, the distance was 4.4 miles and at the present time is 5.5 miles. The opening between Horn Island and Ship Island is now 6 miles wide, as opposed to 4 miles in 1910.

The result of these increased openings to the mainland is obvious. Formerly productive areas of marshland, now exposed to greater salinity intrusion and wave action, have developed into a more saline ecosystem. At the upper end of the areas, urban development is moving toward the sea, and as a result, many of the marshlands are experiencing intrusion on both the inland and seaward sides.

While urban control may be realized on the upper reaches of the estuaries, the problem of coastal exposure control is less tractable. Granted that major navigational channels will have to be maintained, the continuation of island development by accretion will not be possible across such channels. Solutions to the problem have ranged from the placing of old "moth-ball" ships across the expanded openings to the building of spoil islands with appropriate structural protection on the eastern end, with natural accretion deposits rebuilding the islands to the westward. It is essential that all such proposed modifications be thoroughly investigated prior to implementation.

2.7 Marine Mammals

Important marine mammals of the U.S. coastal zone are the seals, sea lions, sea otters, walruses, manatees, porpoises, and whales. Except for the manatee and baleen whales, they generally occupy the highest trophic level in the marine ecosystem. Many breed on land, and at such times they are particularly vulnerable to man's activities. Others frequent bays or intertidal zones; and one, the Florida manatee, occurs in coastal waterways and fresh waters.

Several species of marine mammals, such as the sea otters, once were seriously reduced in abundance, but with the application of protective measures they are increasing. Some of the seals are now uncommon in local waters, and harassment of the Florida manatee has reduced its numbers and confined it to a few river systems of Florida and the Everglades National Park. On the other hand, the Pribilof Island fur seal herds have been restored to vigor from a severely overharvested state by the adoption of internationally accepted conservation controls. Proper management and protection of all marine mammals is needed to ensure that they provide optimum benefits (Walford, 1958; Magnuson, 1970).

Support for a moratorium on killing marine mammals is strong in the United States. The people who advocate such measures are motivated by the highest ideals and humane considerations, yet few of them understand the full implications of their actions. For example, the decline of the fur seal herds in the late 19th and early 20th centuries was caused not only by overharvesting on the island breeding grounds, but also by unrestricted killing at sea, which was very wasteful indeed, because the kill included females as well as males, and because many of the animals killed were not recovered and used. The present management program on the Pribilof Islands, in which the United States government controls the harvest according to scientific principles, was developed under international agreement with Canada, Japan, and the Soviet Union. This has been considered a model management program, certainly the most successful of all marine resource management schemes. A unilateral decision by the United States to ban killing of all marine mammals could be interpreted as abrogation of the treaty, and if so, the International Fur Seal Commission could collapse. The alternatives then available could destroy the benefits of decades of careful scientific research and management. One action might be a decision by other nations to resume harvesting on the high seas, with results entirely opposite to those intended.

The decision by the United Nations Conference on the Human Environment, held in Stockholm in June 1972, to adopt a resolution of full protection of marine mammals probably has little force in international law. It is unlikely that all nations will be willing to give up the privilege of taking marine mammals. Indeed, it is difficult to understand why such a position is logical in a world that has a need for natural resources, both living and nonliving. Such a position would be inconsistent with the general theme of the Coastal Zone Workshop; that the objective should be rational use, taking into consideration the requirements of various users and the need to maintain resources and their environment in a condition to be of maximum sustainable benefit to man. The United States should reconsider its position on the proposed moratorium, with the recognition that many nations will not accept such a position, and with the realization that the United States, as a leader in the world conservation movement, can exert its leadership most effectively by accepting the principle of rational use and continue to play an active role in world councils.

A great emotional interest in whales and whaling has developed in the United States in the last few years. This concern is not without foundation, for the great era of modern whaling which began about 1925 has until recently concentrated on the once vast Antarctic resource of great whales. The stocks of blue and humpback whales in that region have been reduced to very low levels, and the fin whale resources have been substantially overharvested. Yet conservationists who have advocated a moratorium on all commercial whaling have failed to recognize that the much maligned International Whaling Commission has been neither as impotent nor as indifferent to the condition of the resource as most people think (Gambell and Brown, 1971). The Whaling Commission accomplished little until 1965, and as a consequence the last large reserves of commercial baleen whales in the southern seas and in the North Pacific were reduced well below the level of maximum sustainable yield. All species and stocks, however, were not affected equally.

Since the annual meeting of the Whaling Commission in 1964, when it came close to collapse because of failure to set catch limits for the Antarctic, the commission has slowly improved its performance. Having recommended a ban on killing blue and humpback whales in the Southern Hemisphere, it sharply reduced the combined quota for fin and sei whales in the Antarctic and by 1967 had reduced the Antarctic quota to a level somewhat below the best scientific estimate of the sustainable yield.

Turning its attention then to the North Pacific, where whaling operations with factory-ship fleets had more than doubled in intensity as the Antarctic quota was reduced, the commission quickly recommended a global prohibition on killing blue and humpback whales, a prohibition that was accepted by the member nations. It then proceeded to set quotas by species for fin and sei whales in the North Pacific and to reduce those quotas annually as the scientific stock assessments improved (Gambell, 1972).

The commission has been slow to address itself to two other important matters, elimination of the blue whale unit (BWU) as a method of setting quotas in the Antarctic and adoption of an International Observer Scheme. At the 23rd meeting in 1971, it was agreed that these matters should be resolved either before or during the 24th meeting. Encouraging progress has been made on the observer scheme, which is now in operation in parts of the world ocean and for which agreements have been completed between member nations for all whaling areas. The BWU question was also on the agenda for the 1972 meeting, which decided to abolish the blue whale unit as a means of setting quotas. Harvest quotas are now set for individual species (International Commission on Whaling, 1973).

This substantial progress has gone unnoticed because the commission has neglected to assume a public relations role, a task with which it was never charged. As far as its member nations are concerned, there is now in effect a global moratorium on killing five kinds of whale, the right whales, bowhead, gray, humpback, and blue whale. Insofar as the scientific evidence has clearly established a need, quotas are in effect on stocks of fin, sei, and sperm whales, and the commission has demonstrated its willingness to adjust these quotas downward and to establish new quotas, as the scientific evidence warrants. These steps have not been achieved without bitter battles, and they have been opposed strenuously by the whaling interests. That the Whaling Commission has been able to surmount such difficulties is a sign that it should be encouraged to do better.

When the Secretary of the Interior in 1970 placed eight commercially valuable species of whale on the endangered species list, he went one step further than the Whaling Commission. By recommending bans on killing blue and humpback whales, the Commission had by 1967, in effect, declared that five of those eight were endangered species. In adding the fin, sei, and sperm whale to the United States endangered species list, the secretary acknowledged that it was not necessary to wait until a species was in danger of extinction. While it is perfectly within his right to exercise any statutory authority within his power to prevent destruction of a living resource, as a practical matter, this action has had limited effect on world whaling and on the attitudes of other nations. So far, the other whaling nations have taken the view that each species or stock must be considered separately and that as long as each can yield a sustained catch under adequate scientific control, it is not rational to stop whaling. There is some reason to doubt that this argument applies to the fin whale, for it has been substantially overharvested in the Southern Hemisphere and moderately overharvested

The Coastal Zone Workshop recommends
The development of a vigorous and comprehensive *National Coastal Zone Policy* by the federal government in cooperation with the states that will provide for the wise use of the marine, estuarine, wetland, and upland areas bordering the American shores. All future uses of the coastal zone must be designed to maintain the natural ecosystems and to provide for the use of contiguous resources by the people of the United States. Cooperative action by cognizant federal, regional, state, and local governments will be required. The integral element of the National Coastal Zone Policy should be the focus of management responsibility at the state level, with the active participation of local governments, under federal policies that provide grants and set guidelines for creative and effective programs.

in the North Pacific. But it is difficult to accept the argument advanced that there is an urgent need to apply such drastic measures to protect the stocks of sei and sperm whales.

In the rich waters of the Gulf of Alaska and the Bering Sea and along the coast of Southern California, whales are still important resources of the coastal zone, as they once were along most parts of our coasts. Until appropriate controls can be applied to whalers of all nations, the proper policy is one of caution. The United States should reconsider its present position on whaling in favor of a more positive policy of scientifically controlled harvesting, coupled with support and strengthening of the International Whaling Commission.

References

Allee, W. C., 1923. Studies in marine ecology: III. Some physical factors related to the distribution of littoral invertebrates, *Biological Bulletin* (*Woods Hole*), *44:* 205–253.

Anderson, L. G., and Tabb, D.C., 1970. Some economic aspects of pink shrimp farming in Florida, *Proc. 23rd Annual Session Gulf Caribbean Fish. Inst.* (Miami: Rosenstiel School of Marine and Atmospheric Science, Univ. of Miami), contrib. no. 1334, pp. 114–124.

Bardach, J. F., and Ryther, J. H., 1968. *The Status and Potential of Aquaculture* (Washington, D.C.: Amer. Inst. of Biol. Science), vol. 2, part 3 (Fish culture), publication no. PB. 177 768 Clearinghouse, Springfield, Va. 22151, 275 pp.

Barlow, J. P., Lorenzen, C. J., and Myren, R. T., 1963. Eutrophication of a tidal estuary, *Limnology and Oceanography, 8:* 251–262.

Barnes, H. (ed.), 1966. *Some Contemporary Studies in Marine Science* (London: George Allen and Unwin, Ltd.).

Betz, F., Jr., 1971. Bibliography of aquaculture (Wilmington, North Carolina: Coastal Plains Center for Marine Development Services), publ. 71–74, 245 pp.

Briggs, P. T., and O'Connor, J. S., 1971. Comparison of shore-zone fishes over naturally vegetated and sand-filled bottoms in Great South Bay, *New York Fish and Game Journal, 18:* 15–41.

Caldwell, J. M., and Lockett, J. B., 1965. Effects of littoral processes on tidewater navigation channels, *Evaluation of Present State of Knowledge of Factors Affecting Tidal Hydraulics and Related Phenomena* (Vicksburg, Miss.: U.S. Army Engineer Committee on Tidal Hydraulics), report no. 3.

Dexter, R. W., 1950. Restoration of the *Zostera* faciata at Cape Ann, Massachusetts, *Ecology, 31:* 286–288.

Dovel, W. L., and Edmunds, J. R., IV, 1971. Recent changes in striped bass (*Morone axatilis*) spawning sites and commercial fishing areas in upper Chesapeake Bay; possible influencing factors, *Chesapeake Science 12*(1): 33–39.

Galtsoff, P. S., 1956. Ecological changes affecting the productivity of oyster grounds, *North American Wildlife Conference Transactions, 21:* 408–419.

Gambell, R., 1972. Why all the fuss about whales? *New Scientist 54:* 674–676.

Gambell, R., and Brown, S., 1971. Status and conservation of the great whales. *IUCNN Bulletin,* new series *2*(21): 185–189.

Gates, J. M., and Matthiessen, G. C., 1971. An economic restrictive, *Aquaculture: A New England Perspective,* edited by T. A. Gaucher (Portland, Me.: Research Institution of the Gulf of Maine), pp. 22–50.

Gaucher, T. A., 1970. Thermal enrichment and marine aquaculture, *Marine Aquaculture,* edited by W. J. McNeil (Corvallis: Oregon State Univ. Press), pp. 141–156.

Gloyne, E. F., and Eckenfelder, W. W., 1968. *Advances in Water Quality Improvement, Water Resources Symposium I* (Austin: University of Texas Press).

Grant, R. R., Jr., and Patrick, R., 1969. Tinnisum Marsh as a water purifier, *Two Studies of Tinnisum Marsh, Delaware and Philadelphia Counties* (Washington, D.C.: Pennsylvania Conservation Foundation), pp. 105–123.

Hedgpeth, J. W., and Gonor, J., 1969. Aspects of the potential effects of thermal alteration on marine and estuarine benthos, *Biological Aspects of Thermal Pollution, Proceedings National Symposium on Thermal Pollution,* edited by P. A. Krenkel and F. L. Parker (Portland, Ore.: F.W.P.C.B. and Vanderbilt University), pp. 80–118.

Holmes, R. W., Schaefer, M. B., and Shimada, B. M., 1957. Primary production, chlorophyll, and zooplankton volumes in the tropical eastern Pacific Ocean, *Bulletin Inter-American Tropical Tuna Commission, 2:* 127–156.

International Commission on Whaling, 1973. Twenty-third report of the commission (covering the 23rd fiscal year, 1971–1972; London, in press).

Joyner, T., 1971. Resource exploitation—living, *Impingement of Man on the Oceans,* edited by D. W. Hood (New York: Wiley-Interscience), pp. 529–551.

Kane, T. E., 1969. Aquaculture and the law (Coral Gables: Univ. of Miami), thesis for LL.M. in Ocean Law, 98 pp.

King, W., Swartz, A., Hemthill, J., and Stutzman, K., 1962. Sport fishing—today and tomorrow (Washington, D.C.: U.S. Govt. Printing Office), ORRC study report 7, 84 pp.

Ludwigson, J. W., 1970. Managing the environment in the coastal zone, *Environment Reporter Management, 3*(1): 22.

Lyles, C. H., 1969. Fishery statistics of the United States 1967 (Washington, D.C.. U.S. Dept. of Interior, Bureau of Commercial Fisheries), statistical digest 61, 490 pp.

McHugh, J. L., 1967. Estuarine nekton, *Estuaries*, edited by G. H. Lauff (Washington, D.C.: A.A.A.S.), pp. 581–620.

McHugh, J. L., 1969. Fisheries of Chesapeake Bay, *Proc. Governor's Conference on Chesapeake Bay, September 12–13, 1968* (Baltimore, Md.). ii 135–ii 160.

Magnuson, W. G., 1970. Treaties and other international agreements on oceanographic resources, fisheries, and wildlife to which the United States is a party. Committee print, 91st Congress, 2nd session, 672 pp.

Manar, T. A., 1969. Hawaiian fisheries, *The Encyclopedia of Marine Resources* (New York: Van Nostrand Reinhold), pp. 295–298.

Mansueti, R. J., 1961. Effects of civilization on striped bass and other estuarine biota in Chesapeake Bay and tributaries, *Proc. Gulf Caribbean Fish. Inst. 4th Annual Session*, pp. 110–134.

National Academy of Engineering, 1972. Toward fulfillment of a national ocean commitment (Washington, D.C.: Marine Board, National Academy of Engineering).

National Marine Fisheries Service, 1971. Fishery statistics of the United States (Washington, D.C.: U.S. Dept. of Commerce, National Marine Fisheries Service), statistical digest 62, 578 pp.

Niering, W. A., 1970. The dilemma of the coastal wetlands: conflict of local, national and world priorities, *The Environmental Crisis*, edited by H. W. Helfrich, Jr. (New Haven: Yale University Press), pp. 143–156.

Niering, W. A., 1972. The wetlands, *Ecology Today, 2*(2): 32–33.

Odum, E. P., 1959. *Fundamentals of Ecology* (Philadelphia: Saunders), 2nd edition.

Odum, E. P., and De La Cruz, A. A., 1967. Particulate organic detritus in a Georgia salt marsh-estuarine ecosystem, *Estuaries*, edited by G. Lauff (Washington, D.C.: A.A.A.S.), publ. no. 83.

Odum, H. T., and Wilson, R. F., 1962. Further studies on respiration and metabolism of Texas bays, 1958–1960, *Publications of the Institute of Marine Science (Texas), 8:* 23–55.

Phillips, R. C., 1969. Temperate grass flats, *Coastal Ecological Systems of the United States. A Source Book for Estuarine Planning*, edited by H. T. Odum, B. J. Copeland, and E. A. McMahan (Morehead City: Institute of Marine Science, University of North Carolina), vol. 2, report to Federal Water Pollution Control Administration, contract R.F.P. 68-128, pp. 737–773.

Pruter, A. T., 1972. Foreign and domestic fisheries for groundfish, herring, and shellfish in continental shelf waters off Oregon, Washington and Alaska (Woods Hole, Mass.: Workshop on Critical Problems of the Coastal Zone, unpublished working paper), 16 pp.

Roessler, M. A., and Zieman, J. C., Jr., 1970. The effects of thermal additions on the biota of southern Biscayne Bay, Florida, *Proc. Gulf Caribbean Fish. Inst., 22nd Annual Session*, pp. 136–145; *Sport Fishing Institute Bulletin*, 232 (March 1972), p. 7.

Rosenzweig, M. L., 1971. Paradox or enrichment: destabilization of exploitation of ecosystems in ecological time, *Science, 171:* 385–387.

Ryther, J. H., 1954. The ecology of phytoplankton blooms in Moriches Bay and Great South Bay, Long Island, New York, *Biological Bulletin (Woods Hole), 196:* 198–209.

Schaefer, M. B., 1957a. A study of the dynamics of the fishing for yellowfin tuna in the Eastern Tropical Pacific Ocean, *Bulletin Inter-American Tropical Tuna Commission, 2:* 245–285.

Schaefer, M. B., 1957b. Some considerations of population dynamics and economics in relation to the management of the commercial marine fisheries, *Journal Fisheries Research Board Canada, 14*(5): 669–681.

Schaefer, M. B., 1959. Biological and economic aspects of the management of commercial marine fisheries, *Transactions American Fisheries Society, 88:* 100–104.

Schaefer, M. B., and Beverton, R. J., 1962. Fishery dynamics—their analysis and interpretation, *The Sea,* vol. 2, The composition of sea-water, comparative and descriptive oceanography, edited by M. M. Hill (New York: Wiley-Interscience), pp. 469–483.

Scott, A., 1970. Economic obstacles to marine development, *Marine Aquaculture,* edited by W. J. McNeil (Corvallis: Oregon State Univ. Press), pp. 153–167.

Shaw, S. P., and Fredine, C. G., 1956. Wetlands of the U.S.—their extent and their value to waterfowl and other wildlife (Washington, D.C.: Fish and Wildlife Service), circular 39.

Stauffer, R. C., 1937. Changes in the invertebrate community of a lagoon after disappearance of the eelgrass, *Ecology, 18:* 427–431.

Steever, E. Z., 1972. Productivity and vegetation studies of a tidal marsh in Stonington, Connecticut: Cottrell Marsh (New Haven: Connecticut College), Masters thesis.

Stroud, R. H., 1969. Sport fishing, *The Encyclopedia of Marine Resources,* edited by F. E. Firth (New York: Van Nostrand Reinhold Co.), pp. 671–674.

Stroud, R. H., 1970. Future of fisheries management in North America, *A Century of Fisheries in North America,* edited by N. G. Benson (Washington, D.C.: American Fisheries Society), spec. pub. no. 7, pp. 291–308.

Sykes, J. E., 1968. *Proc. Marsh and Estuary Management Symposium* edited by J. Newson (Baton Rouge: Louisiana State University), 250 pp.

Sykes, J. E., and Hall, J. R., 1970. Comparative distribution of molluscs in dredged and undredged portions of an estuary, with a systematic list of species, *Fishery Bulletin, 68:* 299–306.

Taylor, J. L., and Saloman, C. H., 1968. Some effects of hydraulic dredging and coastal development in Boca Grega Bay, Florida, *Fisheries Bulletin, 67:* 213–241.

Teal, J. M., 1962. Energy flow in the salt marsh ecosystem of Georgia, *Ecology, 43:* 614–624.

Teal, J. M., and Teal, M., 1969. *Life and Death of the Salt Marsh* (New York: Audubon/Ballantine).

Thorhaug, A., Stearns, R.D., and Pepper, S., 1972. Effects of heat on *Thalassia testudium* in Biscayne Bay, *Proc. Florida Academy of Sciences* (in press).

Tsai, C. F., 1968. Effects of chlorinated sewage effluents on fishes in upper Patuxent River, Maryland, *Chesapeake Science, 9:* 83–93.

Valiela, I., and Teal, J. M., 1972. Nutrient and sewage sludge enrichments in a salt marsh ecosystem, *Proc. Intcol. Symposium Physiological Ecology of Plants and Animals in Extreme Environments* (in press).

Walford, L. A., 1958. *Living Resources of the Sea* (New York: The Ronald Press), 321 pp.

Wass, M. L., and Wright, T. D., 1969. Coastal wetlands of Virginia, interim report to the Governor and General Assembly (Gloucester Point: Virginia Institute of Marine Science), special report in applied marine science and ocean engineering, no. 10, 154 pp.

Welch, E. B., 1968. Phytoplankton and related water-quality conditions in an enriched estuary, *Journal Water Pollution Contributions Federal, 40:* 1711–1727.

Wilhm, J. L., and Dorris, T. C., 1968. Biological parameters for water quality criteria, *BioScience, 18:* 477–481.

Williams, R. B., and Murdoch, M. B., 1970. A general evaluation of fishery production and trophic structure in estuaries near Beaufort, North Carolina (Beaufort: Center for Estuarine and Menhaden Research), annual report to A.E.C.

Wilson, K. A., 1968. Fur production on southeastern coastal marshes, *Proc. Marsh and Estuary Management Symposium* (Baton Rouge: Louisiana State Univ. Inc.).

3 Nonrenewable Resources

3.1 Introduction

The coastal zone is a distinct mineral province. Some of its mineral resources, such as shell and phosphorites, have been formed in place, while placer and beach minerals are concentrated there by current or wave sorting. There are four types of mineral deposits found in the coastal zone, namely, *marine placers,* or those deposits that are accumulated by the winnowing of light minerals away from heavy metal-bearing minerals in response to waves and currents; *lodes,* or those deposits that contain metal-bearing minerals in preexisting rocks and that are not related to present marine processes but do occur in the present coastal zone; *chemically precipitated mineral deposits,* which for the coastal zone are largely confined to ferromanganese nodules in the upper Great Lakes; the *buried stream placers* which were formed during the last Ice Age in stream valleys when the sea level was lower but are now covered by the ocean at its present level; and minerals recovered from solution in sea water. The marine placer deposits are the only ones whose origin and exploitation are critically related to coastal processes.

Because they are at or near the surface, these minerals are extracted primarily by dredging and on-site concentration operations. Floating dredges and concentrating equipment, operating in relatively shallow water, are subject to a variety of natural constraints. They must withstand wind and tidal currents and must avoid ice in northern regions. The size and seaworthiness of the dredge depends on its degree of exposure to waves and swell and the likelihood of severe storms and hurricanes in its location. If severe storms are frequent, the floating equipment must be sufficiently mobile to reach shelter in a day or two. In northern regions, ice restricts operations to the summer and fall months and imposes additional economic and operational constraints on the success of the enterprise.

Location of the resource is another constraint on mining operations. High-value placer minerals such as platinum, gold, and silver are likely to be worth mining wherever they can be found. High-volume, low-value materials such as sand, gravel, and shell, however, must be located near the market in order for their recovery to be economically feasible.

Dredging for surface or near-surface minerals disturbs the bottom and results in increased turbidity and siltation, which can adversely affect the natural bottom communities both at the mining site and nearby.

Increased turbidity also reduces light penetration and thereby reduces photosynthesis and the biological productivity of the area. On the other hand, it is possible that increased nutrients released by the dredging operations could have the opposite effect and stimulate plant growth. The net effect, though probably negative, would have to be determined in each individual case. Interference with biological production, both in the water and on the bottom, represents an environmental cost and must be considered a constraint on the operation. Destruction of valuable shellfish beds and recreational beaches, for example, would probably be too high a price to pay for the minerals extracted, and these deposit locations can be bypassed by the mining company.

In assessing the environmental costs and reaching an equitable decision on the exploitation of offshore minerals, it is important to remember that dredge mining of placer and beach minerals is an "in-and-out" operation. The disruption of the biological community is usually temporary and is frequently followed by recovery of the system (O'Neal, 1971).

On the other hand, mining of phosphorite, which usually underlies salt marshes, involves a more permanent cost. Stripping off marshes to recover the phosphorite permanently destroys their productivity and nursery ground function—a cost which would in all probability be an unacceptable consequence of the recovery of this resource.

Oil and gas are extracted on fixed platforms which are less subject to the environmental constraints of waves and currents. Severe storms and hurricanes continue to be a problem, however, and several of these platforms have been lost to storms. With present technology, oil extraction operations are limited to a water depth of 600 feet (CMSE&R, 1969). Siting of these platforms must also take into account navigational requirements and leave room for free passage of ship traffic. Undersea pipelines, which are also necessary to these mining operations, must be arranged so that there is minimal interference with fish and shrimp trawling operations.

The mining company that seeks to exploit offshore minerals must deal with a myriad of legal, political, and social problems as well as the technological ones. One of the most difficult handicaps facing the underwater miner is the lack of legal protection of his exploration rights. Although all coastal states have established offices for supervising oil and gas exploration in state waters, very few have developed the legal means to grant offshore prospecting permits for hard-minerals exploration. Alaska, a notable exception, has initiated a system of legislation to promote and administer mineral exploration in its waters.

Several states do have mineral laws sufficiently broad to cover initial exploration—for instance, Hawaii, Texas, Florida, Louisiana, and California—but most coastal states, including those along the Great Lakes, lack specific laws. Some, like Wisconsin, have no offshore mineral laws whatsoever. Those states that do have laws covering exploration often have no modern statutes to cover environmental quality or transfer of permits to leases.

In short, an underwater mineral law, the prime requisite for orderly mineral exploration and exploitation, is lacking in the majority of coastal states. Without firm legal protection, valid permits, and statutory guides for vessel operations, safety, antipollution measures, and royalties, mining companies will not venture into the coastal zone of a state.

Any mining company wishing to explore in coastal waters must meet the requirements of the Corps of Engineers, Coast Guard, Environmental Protection Agency, and Bureau of Mines, as well as those of a myriad of state regulatory and administrative agencies. Where the operation includes beach property, county and city agencies also must be consulted. Since the prime authority is not yet established in this new field of exploration, the company must make its way through a jungle of bureaus, offices, and agencies on all levels. Coastal management agencies must streamline these procedures and provide for a more orderly and rational decision-making process with regard to mineral resource exploitation.

Quantitative assessment of the environmental impact of mining operations is an essential factor in any rational decision involving mining operations in coastal waters. Such an assessment requires comprehensive and accurate information on the state of the environment prior to the mining operation. Premining surveys have not been systematically conducted in the past; in fact, guidelines for such surveys do not even exist. In order to meet the requirements and responsibilities placed on them by coastal zone management agencies, mining companies must have clear guidance on what these premining surveys ought to include and how they should be conducted.

Beyond a general description of the geological setting and a specific definition of the resource to be exploited, a premining survey should include a quantitative and qualitative ecological survey of the area, a

The Coastal Zone Workshop recommends

The development of *legal institutions and procedures* to make coastal zone management more effective. Substantial improvements in the existing types of decision procedures and laws are required, and consideration should be given to—

Development of innovative approaches through new coastal land and water use accommodations;

Alternative means for the regulation of coastal development besides the taking of private property;

Improvement of statutes and administrative regulations for land, water, and submerged land activities;

Increased access of individuals, groups, and governmental units to administrative and judicial proceedings;

Establishment by state legislatures of Environmental Review Boards for appeals of local administrative decisions concerning activities that have coastal and environmental impact;

Establishment by Congress of an expert federal Environmental Court with broad jurisdiction over private persons, state and local government agencies, and federal agencies in controversies involving coastal and environmental impact.

physical and chemical description of the nearby beach or coast, an economic analysis of the present uses of the area, and an estimate of the hydrological changes produced by the mining operation. Given this kind of base-line information, the effects of the mining operation can then be accurately assessed, and costs for damages, if any, can be equitably distributed.

3.2 Petroleum Resources

3.2.1 Oil and Natural Gas

Oil and natural gas are the most valuable of all nonrenewable resources within the nation's coastal zone. Most of the proved reserves and production in the United States are in Texas, Louisiana, California, and Alaska. A large fraction occurs in the coastal zone, both on land and offshore. Today, offshore resources alone account for 18 percent of the nation's oil production and 15 percent of its gas supply (Bureau of Mines, 1971).

The potential petroleum resources beneath the nation's submerged continental margin have only begun to be tapped. The U.S. Geological Survey (McKelvey, et al., 1968) indicates that the total potential resource base of the nation's continental margin may range from 1.3 to 1.58×10^{12} barrels of oil and 3230 to 4450×10^{12} cubic feet of natural gas. Of this potential, 200×10^{9} barrels of oil and 900×10^{12} cubic feet of natural gas are recoverable with present technology and economics. This is more than twice the amount of all oil and gas produced in the United States to date and exceeds by five times our present proved reserves of oil and by more than three times our reserves of natural gas (American Gas Association, 1971).

3.2.2 Importance of Offshore Petroleum in U.S. Energy Outlook

During the period from 1950 to 1970, the nation's energy consumption doubled, and in the next 15 years, 1970 to 1985, it is expected to almost double again (see also Section 5.3.1). In the past we have met our energy needs from domestic supplies, and consumers have been able to choose among a variety of low-priced energy sources. This is no longer the case. We are now importing nearly one-fourth of our oil, and a shortage of natural gas already exists. The extent to which various energy sources can be expected to contribute to the nation's energy requirements from 1970 to 1985 has been studied by industry advisory groups (National Petroleum Council, 1971), petroleum and mining industrial organizations (American Petroleum Institute, 1972a,b), and individual companies (Shell Oil Company, 1972). One set of results, which is typical of others, is shown in Figure 3.1 (Humble Oil & Refining Company, 1972). According to this projection, nonpetroleum fuels—nuclear power, hydroelectric power, coal, and synthetics—were expected to supply 37 percent of the U.S. energy demand in 1985, as compared to 22 percent in 1970. But because of unforeseen problems, these energy sources have not grown to the extent forecast only a few years ago.

With less than expected increases in the nonpetroleum sources, a larger portion of our energy will have to come from oil and natural gas. They are expected to supply 63 percent of our energy in 1985. However, the limited supplies of natural gas can be expected to supply only 19 percent of that, as compared to 33 percent in 1970. Oil will continue to supply almost 45 percent of the demand in 1985, the same proportion as in 1970. Because of expanded total energy needs over this 15-year period, industry figures indicate that the volume of oil re-

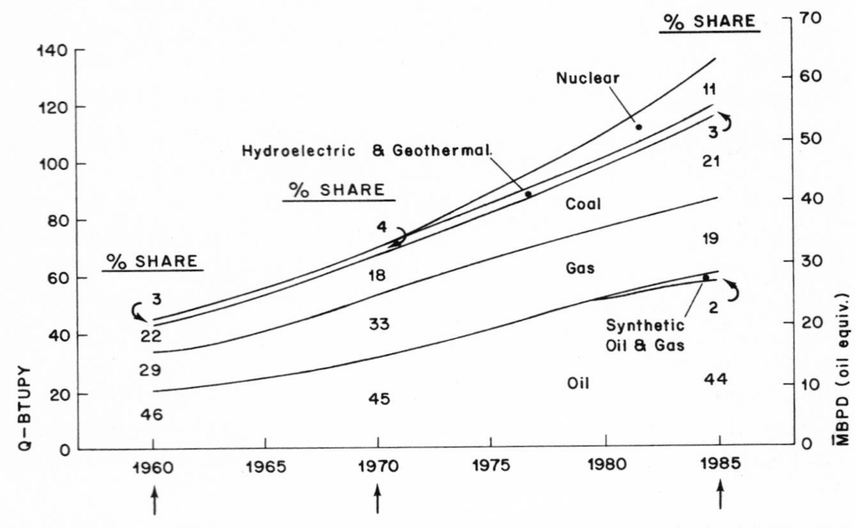

Figure 3.1 United States energy demand by fuel source. $\overline{\text{M}}$BPD refers to millions of barrels of oil per day equivalent; Q-BTUPY refers to quadrillion British thermal units per year.

quired is anticipated to increase to 28 million barrels per day in 1985, nearly twice the consumption in 1970.

In the face of declining reserves and little spare producing capacity, we have poor prospects of supplying these quantities of oil from domestic sources under present conditions. Unless there is a change in national policy to permit and encourage intensive exploration and development of new domestic supplies by 1985, the United States will have to import about 62 percent of its petroleum supplies, mostly from the Eastern Hemisphere (Figure 3.2; Gammelgard, 1971).

Because of the difficulties in predicting the extent to which we can depend on foreign supplies, it seems almost inevitable that it will become national policy to develop our offshore resources more intensively. While Alaska will be an important source, it is the U.S. offshore continental margin that holds the greatest promise for major new domestic petroleum supplies (National Petroleum Council, 1969). However, because of existing government policies stemming partly from public environmental concerns, exploration and development of the nation's offshore petroleum resources have been restricted or halted in many areas. This and low gas prices are partly responsible for our current shortage of domestic petroleum supplies. Examples of federal restrictive actions are the moratorium imposed on Santa Barbara Channel leases, the reduction in frequency and size of federal offshore lease sales, and repeated delays of such sales.

If these restrictive measures and other factors impeding offshore activities are removed, we can expect rapid expansion in exploration and development of the nation's offshore petroleum resources. This will necessarily require increased utilization of the coastal zone. In addition to providing access to shore facilities in support of offshore operations, the coastal areas must furnish the space for receiving greatly increased quantities of oil, whether carried by pipelines or tankers, as well as additional refineries. All of this discussion assumes that neither will there be effective effort to reduce per capita or total energy demands in the country nor will there be any effort to conserve significant reserves of petroleum for the future. It should be apparent that a realistic national energy policy must contain both of these efforts.

3.3 Sand, Gravel, and Shell

Much of the sand and gravel veneering the continental shelf adjacent to the United States was brought there by agents other than these acting today. With lowered sea level during the Pleistocene, rivers flowed

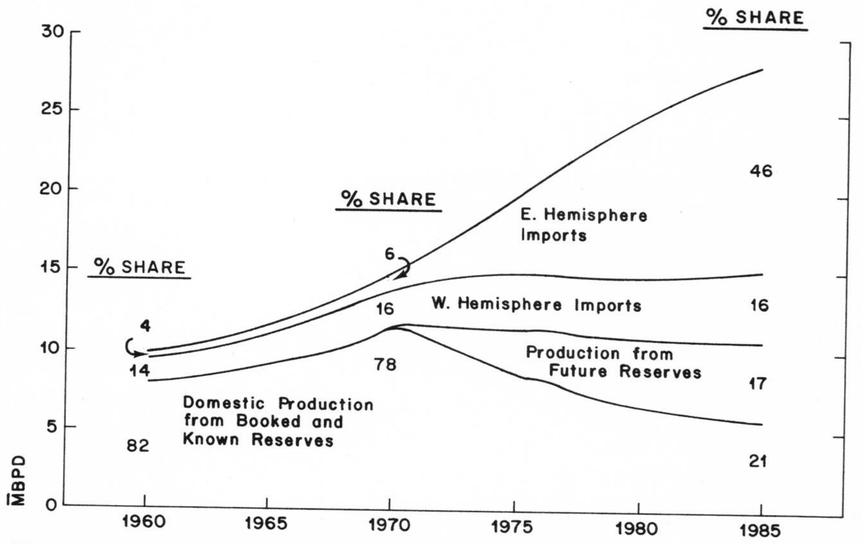

Figure 3.2 United States liquid petroleum supply and demand.

The Coastal Zone Workshop recommends
The cooperation of industry, public utilities, state agencies, and the federal government in the development of regional planning and utilization of energy, including fossil, nuclear, or other fuels in the coastal zone so that costs and benefits of alternative sites of development within and outside the coastal zone can be compared. Public authorities should be guided by both the urgency of protecting the environment and the demand for energy in the United States.

---

across the shelf depositing sediment there and on the continental slope. The ice sheets left extensive glacial deposits on the shelf off the northeastern United States and Alaska, and in and around the Great Lakes. Beginning with the rising sea level and continuing to the present day, tidal and storm currents have reworked the shelf and shaped the sands into a subaqueous dune topography.

The dredging of sand and gravel in the bays and rivers of the coastal zone accounts for about 10 percent of the U.S. demand. Except for pilot beach replenishment programs by the Corps of Engineers, offshore mining has not been conducted. This is a temporary condition since existing land supplies close to population centers are becoming scarce because of the exhaustion or rezoning of potential sites. As land deposits farther from cities are exploited, price rises due to transportation costs lead to the use of offshore sand and gravel. The National Atmospheric and Oceanographic Administration in conjunction with a number of organizations and governmental agencies has begun a three- to four-year study off Massachusetts on the feasibility of offshore dredging and its ecological effect.

The continental shelf near some large metropolitan areas, such as New York and Boston, contains a large supply of sand and gravel (Schlee, 1964, 1968; Schlee, Folger, and O'Mara, 1971). In view of its availability in large quantities, the increase in demand, and the rising prices, it is not surprising that mining companies have begun to look toward offshore United States as a source, just as has been done in European waters (Hess, 1971).

United States sand and gravel production in 1970 was approximately 944 million tons with a value of more than a billion dollars at the site of production. This output is in addition to the 1 billion tons of crushed stone that competes with sand and gravel. The output is 9 million tons higher than the previous peak year of 1966, but it is only about one-half the yearly needs projected for 1980 (Cotter, 1965). The present production figure represents a fourfold increase over the past 20 years with a yearly growth rate of approximately 7 percent from 1950–1965 and a 1 percent growth rate more recently.

Table 3.1. Uses and Value of Sand and Gravel Production in the United States in 1970

	Quantity (10^6 tons)		Value[a] Million Dollars	
Operation	sand	gravel	sand	gravel
Construction				
Building	166	130	198	191
Paving	139	364	149	384
Fill	33	43	21	27
Railroad ballast	1.4	2.8	0.73	2.4
Other	14.4	12.4	16.0	14.6
Subtotal	906		1003	
Total industrial use, ground and unground	29		101	
Total, all purposes	944		1116	

Source: Walter Pajalich, 1972, U.S. Bureau of Mines, oral communication; to be published in Minerals Yearbook for 1970, U.S. Department of Interior, Bureau of Mines, in preparation.
[a] Value at point of production (less transportation).

Table 3.2. Base Prices per Short Ton for Sand and Gravel at Major East Coast Centers[a]

City	Sand[b]	Gravel
Boston	$2.00	$3.00
New York	0.75 to 1.50	3.00
Philadelphia	1.35[c]	3.50[d]
Washington, D.C.	2.75	3.50 truckloads 3.00 large quantities

Source: Manheim, 1972.
[a] Value is for plant, and some sand and gravel pits lie at considerable distance from the metropolitan area. Data obtained by telephone inquiry, April 1972, but vary relatively little from 1971 prices.
[b] Concrete sand.
[c] $3.50 delivered; transport cost about $.04/ton mile.
[d] Crushed stone.

About 90 percent of the sand and gravel is used for building and paving; lesser amounts are used for fill and railroad ballast (see Table 3.1). Wholesale market prices for sand and gravel are given in Table 3.2 for a few large metropolitan centers. Lesser amounts of sand and gravel are dredged for constructional purposes, beach replenishment and fill. Figures in Table 3.1 are not broken down to show how much of the sand and gravel is obtained from the coastal zone, but a tabulation of the production figures from the coastal states shows that the coastal zone accounts for about 73 percent of the total production.

The cost at the pit is a little more than one dollar per ton, although with added processing and transportation costs, the wholesale market price can double or triple (Table 3.2). According to Economic Associates Incorporated, most of the nation's 6000 sand and gravel pits are within 25 miles of their market in an effort to keep land transportation costs down, although aggregate is being brought in from pits as far as 120 miles away for one large construction project in the Boston area.

3.3.1 Utilization of Sand, Gravel, and Shell in the Coastal Zone

Sand, gravel, and shell are important resources of the coastal zone, and the advantages of such underwater deposits lie in cheap transportation and in the availability of water for grading and washing. The price is held down by the fact that the distance between the source of supply and the area of use is very short. Furthermore, in most cases gravel deposits are covered by heavy overburden of sand that must be used if the extraction of gravel is to be commercially feasible.

While most sand is used as an ingredient in concrete, the use of offshore deposits as well as estuarine deposits for coastal beach replenishment should not be ignored. Beach erosion and littoral drift combined with increasing value and decreasing availability of beachfront properties lead to an increasing desire to use submerged sands for beach replenishment and development. Not considering the value of sands used for beach replenishment, it is estimated that the value of underwater sand and gravel is between 20 to 30 million dollars annually.

A relatively unknown consideration in sand extraction from the offshore areas, and to a degree the estuarine areas, is the effect of the dredging upon the organisms of the immediate area. Although some investigations are now underway to determine such effects, there are not yet enough data to predict the rates of bottom-organism recovery or reestablishment of natural bottom configuration (Chesapeake Biological Laboratory, 1970). Deep holes remaining from mining activities may become stagnant pockets void of normal fauna, they may interfere with the migration of benthic organisms, and they could alter natural current patterns. The future use of coastal underwater sands and gravels will probably increase. Land around urban areas is too valuable to be reserved for sand and gravel which bring a low return; however, an area could be mined of some of its sand and gravel before development proceeds.

The extraction of oyster shell from the estuarine bottom is a submerged strip-mining process. This process involves the removal of overburden, the extraction of the resources formed from old oyster shell, the refining of the raw product, and the treatment or care of the area once the material has been removed. The products derived from shell are ingredients of chicken feed, pet food, livestock supplement, cement, and material used in refractory processes. In its original form, or broken up, the shell is used for road fill, highway aggregate surfaces, construction blocks, and as a replacement for gravel.

In certain areas the shell dredging industry also extracts clam shell. Whereas the extraction of oyster shells calls for the use of hydraulic dredges with rotary cutter heads, the extraction or removal of clam shells can be done by suction alone. The density of clam shell is such that processing this material into a form that can be used by the cement industry would call for much higher refractory temperatures than are necessary for oyster shell. Because of this, clam shell is only used directly as roadway surface material.

In 1969 Louisiana initiated a study to determine the effects of suction dredging operations on the habitat of *Rangia* clams. Before and after industrial clam dredging operations, hydrology, plankton, trawl, and dredge samples indicated little variation in any of the measured parameters, except turbidity and suspended solids, although trawl samples taken immediately after the passing of the dredge produced fewer species and individuals.

Disruption and dislocation of bottom-dwelling or more immobile organisms is a certainty, but total mortality is not. Samples from the discharge pipes show that microorganisms are present in numbers nearly equal to amounts found in the bottom prior to the extraction process. Macroorganisms are seldom found, for the rate of progress of the cutter head is such that most mobile organisms have adequate time to escape. Commercial fishermen have found shrimp can be caught in the cuts immediately behind the dredge.

The greatest concern about shell dredging is smothering effect of the suspended silt upon oysters in the area. Investigations have shown that such smothering effects have not occurred beyond 1000 feet from the point of waste discharge, and that oysters placed within 800 feet of the dredge have not suffered mortality. For the purposes of management, it has been generally accepted that 1200 feet is an adequate distance to ensure that siltation from dredge activities will not have adverse effects upon a live oyster reef. However, further work is essential in order to obtain more precise information.

3.3.2 Offshore and Coastal Zone Surveys

Numerous sediment surveys of the nation's shelves have been made, but most are restricted in area and sample only the surface of the sea floor. One of the largest of these was begun in 1962 by the U.S. Geological Survey and the Woods Hole Oceanographic Institution for the Atlantic Continental Margin (U.S. Geological Survey Prof. Paper 529). Other organizations, particularly coastal laboratories, have made similar studies. Hence, much of the general areal pattern of sea floor sediment distribution is known for the U.S. continental shelf. What is far less known is the thickness and shape of the deposits and their areal distribution. A notable exception to this has been provided by the coring connected with the Sand Inventory Program by the Army Corps of Engineers (Duane, 1969). The program has surveyed over 7300 miles of geophysical lines off Florida, New Jersey, Virginia, Long Island, and New England. Their survey, conducted out to a maximum depth of 100 feet, showed sand volumes in excess of 4 billion cubic yards. Manheim (1972) came up with an estimate of 450 billion tons of sand off the northeastern United States, assuming a shelf area of 112,000 square kilometers and an average thickness of 10 feet for the sand gravel cover.

The pressure for sand and gravel exploration has been less off other coasts such as California and the Gulf States because of substantial deposits on land (river terraces and alluvial fans) and a thinner sediment veneer offshore.

3.3.3 British Sand Industry

For the past 40 years, Britain has been mining offshore seabed deposits. According to Hess (1971), 32 companies are presently operating 75 dredges at 80 different sites in 6 principal areas—mainly off the south and east coasts of Britain. They produced approximately 20 million tons in 1971, one-quarter of which was sold to Holland, Belgium,

France, and Germany. The remainder was sold at ports in Britain, where it accounted for 14 percent of the United Kingdom production of sand and gravel.

If this type of an operation is to be conducted in the United States, the British licensing methods and performance should be examined. Standards for the sea-dredged aggregates have been developed by the Greater London Council for limits of sodium chloride content and shell content. Licensing for exploration and exploitation is done by the Crown Estates Commission. Leases are granted on a noncompetitive bid basis to the company that claims the acreage, proves the deposit, and has the capability to mine it. Leasing is granted after consultation with the Coastal Sedimentation Unit, the Hydraulics Research Station, Fisheries Research Group, bordering coastal towns, and the Department of Transportation. The impact of offshore dredging has not been evaluated, but the National Environmental Research Council intends to conduct the necessary studies.

3.3.4 Environmental Impact

In summary, assuming that conventional dragline, bucket-ladder, and cutter-head (hydraulic) dredging systems will continue to be used in mining sand, gravel, and shell, we may expect that local changes in water depth may cause changes in the erosional/depositional patterns of inshore tidal currents. Indeed, remapping of dredged areas should be mandatory subsequent to dredging, particularly in coastal waters frequented by small craft. The benthic fauna of a dredged area will obviously be disrupted, dissolved oxygen will be temporarily lowered, and additional nutrients or heavy metals may be added to the local water body (see Section 7.3.8). Should an unduly large volume of silt be deposited over a sand bottom, a localized mortality of any fauna requiring a sand substrate, for example, flounder and other flatfish, may be expected. Where old oyster reefs are dredged, a paucity of hard (shell) substrate for the new oyster spat could readily develop. Thus, in oyster dredging some shell patches must be left behind.

Extracting sand and gravel from the sea floor or lake bed will remove pressure on aggregate suppliers to open new pits on land. As the low cost-to-volume ratio requires extraction near the use site, underwater production is a more desirable approach to meeting demand, since potential pit areas in the coastal zone will thus be saved for home sites, playgrounds, and parks; also degradation of the landscape will be reduced. Areas to be intensively developed, such as industrial parks, however, might well serve as sand and gravel sources before construction begins. The extrapolated growth of the sand industry suggests that its labor needs will rise sharply during the decade 1970–1980. In large cities, these additional jobs will be important, particularly as semi-skilled labor will be that most needed by the operators. As sand and gravel resource sites are widely distributed, it will not be necessary to mine pleasure beaches or fishing grounds, thus protecting coastal amenities.

3.4 Minerals

The hard minerals, in contrast to oil and gas, are mined chiefly for the metallic elements they contain. Some examples of the metals in this group of deposits are platinum, gold, titanium, zirconium, tin, chromium, and the rare earths. While there are proven reserves of these metals on land, tapping such terrestrial deposits is frequently not without serious political, environmental, and economic problems. Indeed, the environmental problems associated with inland mining operations may exceed in number those found in the coastal zone.

The United States demand for both base and noble metals is expected to increase threefold by the end of this century. The older mines may soon prove uneconomical, and since the ore grade is at a marginal level in several major copper mines already, we must turn to underwater mineral resources or to imports in order to bridge the ever-widening gap between demand and supply (Moore, 1972). Moreover, increased interest by the extractive industries in underwater mining, particularly in the shallow coastal zone, is promulgated, in part, by the adverse political climate in some foreign countries where American-owned mines are operated. Even where the foreign sources of metals have been operated by foreign companies or agencies, there is no assurance that the present supplies will continue. In order to ensure a continuing supply of metals, to avoid political confrontations abroad, to circumvent foreign suppliers from raising prices, and to stem the outflow of American dollars, this nation must soon begin extracting those minerals present in the sovereign waters of the coastal zone, since the metal needs of this nation cannot be met by domestic production from terrestrial deposits alone. Coastal underwater mining has the desirable attributes of proximity, safe political control, and favorable exploration/exploitation potential. One principal area of conflict is the environmental impact of mining on the coastal zone.

3.4.1 Description of Present Coastal Zone Mines and Resource Deposits

Some of the presently operating coastal zone mines and some of the mineral resource deposits that may be worked in the decade ahead are briefly described as examples of these deposit types and to establish the viable nature of this infant industry. For each, we report the real or probable impact on the coastal zone. Plans are being made to mine some mineral resources as soon as certain political, legal, or operational constraints are removed, or when the market becomes more favorable.

PENOBSCOT BAY COPPER-ZINC MINE. This mine is an open-pit, lode sulfide deposit of copper and zinc, with minor amounts of associated metals including silver. It has been mined by damming a small inlet, pumping the water out, and then extracting ore by conventional open-pit methods. In conducting this operation, Callahan Mining Corporation has taken steps to return the mine to as near the original environmental condition as is possible. A problem in this case may be that of sealing the exposed walls so that the sulfide mineral present on pit walls will not come in contact with the seawater when the dam is removed. Callahan has received help from a local Sea Grant investigator who is devising an oyster farm that will attempt commercial operation over the mine site when mining is terminated. Early studies suggest that this mine will not adversely affect the recolonization of local fauna or flora after water has again covered the site. One purely scientific advantage has developed as a spin-off: that of having a coastal zone "testtube" in this area in the form of a deep hole that can be used for chemical, biological, and physical experimentation.

TOM'S RIVER, NEW JERSEY, TITANIUM MINE. This property of the American Smelting and Refining Company is located in the New Jersey coastal zone and should go into production shortly. A placerlike deposit will be worked for ilmenite, a titanium ore. The constituent ore and associated minerals are geochemically stable, and thus no adverse mineral-chemical-water reactions are expected. Nevertheless, the company has initiated several precautionary steps in its scheduled mining operations which, it is hoped, will ensure that the site is returned to a desirable environmental state subsequent to mining.

TRAIL RIDGE HEAVY MINERALS MINE. Located on the southeast U.S. Atlantic Coast, this sporadic operation has been existent for several years and has produced profitable amounts of heavy minerals from the older beach sands. The mine is located in the coastal zone because the sources of heavy minerals and the waves, winds, and currents have combined to produce the ore body. Since less than 3 percent of the sand is removed, the problems normally associated with large volume sand removal do not exist.

HEAVY MINERAL DEPOSITS OFF TEXAS. Although marine geologists of the U.S. Geological Survey announced in the late 1960s the discovery and mapping of large, low-grade, heavy mineral deposits along part of the Texas coast, it appears that these deposits will not be worked within the next few years. Nevertheless, as the metal demand grows the potential of these deposits will be enhanced and some mining is possible by the mid-1980s. The delay should permit coastal zone planners sufficient lead time to prepare for any conflicts that may arise between the mining operators and the fishing and recreation interest groups, including the users of Padre Island National Park.

GOLD DEPOSITS OFF CALIFORNIA. Gold has been identified in the coastal sands off Eureka, California, and adjacent areas, but its recovery has been delayed, not because of environmental concern, but rather because of the engineering problems associated with separating such fine gold particles from the substrate. As the open price of gold now exceeds $50/troy ounce and continues to rise, these California coastal deposits may ultimately be profitable to mine. Coastal zone planners again have ample time and opportunity to specify precisely the rules to ensure a minimum of disturbance to the coastal environment and to avoid conflicts.

CASTLE ISLAND BARITE MINE. This is the only underwater lode mine on the Pacific coast, and its location, off Castle Island, southeast Alaska, makes it economically important to the growing petroleum industry in Alaska. Barite, the chief mineral produced at this mine, is used in powder form as an additive to oil well drilling fluids. The operators of the mine, Alaska Barite Company, a subsidiary of Inlet Oil Corporation, have cooperated with both state and federal agencies to ensure an environmentally "clean" operation. At the present time this mine satisfies about 20 percent of the domestic U.S. barite demand and provides local employment in a depressed area.

GOODNEWS BAY PLATINUM DEPOSITS. Within the past three years, the Inlet Oil Corporation has confirmed the presence of particulate platinum in the sands and muds flooring Goodnews Bay, Alaska. Since the United States faces increasing industrial demand and only modest domestic production, most platinum metal is imported from South Africa and from Eastern Europe. The demand for platinum as a catalyst in antipollution devices of automobiles is expected to rise sharply in 1974–1975 when federal law requires all new vehicles to be so equipped. Meeting this demand without further endangering the trade dollar deficit requires that we make an effort to encourage early domestic production of these deposits. At the present time the available devices for separating the fine platinum from the equally fine mud are not economically feasible. However, a new beneficiation system is being tested and does appear promising. A coastal zone conflict may arise with the local fishermen, but the operators have proposed a plan to schedule alternate fishing and mining periods. Such mining could provide jobs in an area that is economically depressed, and it should be considered.

GOLD, OFFSHORE NOME. Through an intensive coring program, the American Smelting and Refining Company has confirmed that economically extractive deposits of gold are present within three miles of the Nome, Alaska, coast. At this time an engineering plan is being devised to permit year-round mining, even though constrained by winter ice cover. To ensure environmental cleanliness, the company has initiated a base-line survey of premining conditions in the coastal zone. By thorough monitoring now and during actual mining, any environmental harm in the coastal zone can be recognized and corrected. Mining this deposit might provide significant economic aid in the form of jobs for local people. Moreover, the development of offshore mining at Nome would undoubtedly encourage other mining developments at nearby Bluff and farther west on the Seaward Peninsula coast. As the region currently is sparsely populated, no conflicts with recreational users or developmental interest groups are anticipated. Given this situation, Alaskan planners have an excellent opportunity to plan for long-term multiple use rather than for merely short-term economic gain.

MANGANESE DEPOSITS, LAKE MICHIGAN. Although the presence of manganese in lake deposits has been vaguely known for some time, it has only been since 1968 that this mineral resource has been properly described quantitatively mapped, and the reserves confirmed. However, no legal machinery is yet available to regulate extraction or to protect the operator who seeks offshore prospecting permits. Furthermore, no body of law is presently available to provide for environmental monitoring of mining, antipollution regulations for mining, or bistate regulations for mining of this single deposit which crosses underwater state boundaries. Two of the planned uses of the Lake Michigan (Green Bay) manganese nodules are as noxious fume absorbers and as catalysts. It is hoped that coastal states will take notice of these constraints on mineral resource development and enact adequate regulatory legislation to protect both the environment and the operator.

3.4.2 Impact of Mining on the Coastal Zone

As the number of mining operations in the coastal zone of the United States is so small, we have no appreciable basis for assessing contemporary effects on the environment. Present coastal mines that have been monitored and inspected on a regular schedule by environmental agencies since mining began have shown no adverse effects that were recognized by the monitoring agencies, although very few base-line studies have been undertaken prior to mining. Alterations, particularly in shallow bays, are possible. Removal of large volumes of ore rock at lode mines operated as "pits" will result in holes or depressions on the bottom. Since most of these mine sites are in coastal sediment dispersal zones, we may expect underwater holes to be gradually filled with sand and silt. There is no harm to coastal navigation, as the bottom is made deeper, not shallower, but the benthic ecology will be changed. In the case of dredging operations at coastal placer sites, where conventional bucket-ladder or hydraulic cutter-head dredges are employed, a local increase in turbidity and a local decrease in dissolved oxygen can be expected. Brown and Clark (1968) reported that dissolved oxygen in Raritan Bay was reduced 16 to 83 percent below normal during dredging. This appears to be the result of interaction among authigenic sulfide particles and organic matter and oxygen when the bottom is stirred up. Once the agitation stopped, oxygen values returned to "normal."

Large increases in phosphate and nitrogen in the water around a dredge have been reported by Biggs (1967) for a silt dredging operation in Chesapeake Bay. Since the fine-grained sediments of Chesapeake Bay resemble the matrix sediment of potential low-energy marine placer sites elsewhere, it is probable that conventional dredging at mine sites will also cause temporally nutrient increases during actual mining. It must be remembered, however, that not all sediments are similarly enriched, and that navigation channel dredging is a removal process, not a separation process, as would occur in mining.

The long-term effect of dredging on bottom fauna is not known, although in any case where the substrate was drastically altered, for example, from sand to mud, we would expect a change in biota. However, coastal currents are usually sufficient to remove fine muds from a normally developed sand bottom, and a return to a given current–grain size equilibrium on the bottom could be expected.

The socioeconomic impact on local coastal areas can be summed up by saying that operations would be reasonably short term and possibly beneficial to local economy. Marine placers, while rich, are usually small and can thus be worked in periods ranging from several months to four or five years. Such an operation would provide employment and would require local services, both of potential economic advantage to the coastal community.

The geographic distribution of coastal zone mining is limited to relatively restricted areas, largely because of natural geological and environmental patterns. Thus, impact on the coastal zone will be pronounced in areas where lode and placer deposits occur. These are the coast of Maine; parts of the Massachusetts coast; the coast of North Carolina and small parts of South Carolina, Georgia, and Florida; the coastal bend of Texas; the coast of Northern California and part of Oregon; all of coastal Alaska; and the southern shores of Lake Superior, Green Bay, and such waters of the other Great Lakes as are adjacent to crystalline bedrock.

From the foregoing list, it may be assumed at this time that most other coastal areas will not be amenable to mineral mining, although sand and gravel mining near large cities is probable.

Minerals are nonrenewable resources, and once mined, the area is unlikely ever to be worked again. This should be kept in mind by coastal management agencies, for if mining is to be permitted at all, it should be carried out early in any overall scheme in which coastal resources are to be developed. Once extracted, a mining site is from then on available solely for development of the renewable resources. It is prudent management to do the mining first, if it is to be done at all.

3.4.3 Minerals Recovered from Seawater

If dissolved substances in seawater could be extracted economically, they would provide an almost inexhaustible resource for most of man's needs. However, today only four substances—salt, magnesium, bromine, and fresh water—are being recovered commercially from seawater. Recovery of these substances is accomplished by processing plants, all of which are located in the coastal zone.

The total annual value of U.S. production of minerals and fresh water recovered from seawater amounted to $117 million in 1967 and is expected to more than double by 1985 (CMSE&R, 1969). At the end of

1967, there were 286 freshwater conversion plants either in operation or under construction with a total capacity of 39.5 million gallons per day. As the coastal zone becomes more heavily populated and industrialized, the demand for desalted water will grow and thus add to the competition for space.

Since few studies have been made, there is little evidence concerning the effect of these extraction operations on the environment or its ecological systems. However, this is a matter which deserves study, particularly in the case of large freshwater conversion plants projected for the future, which will release an appreciable flow of thermal brine. On the basis of model studies by the Army Corps of Engineers, it was determined that significant effects could be minimized by proper site location (U.S. Army Engineer Waterways Experiment Station, 1971).

References

American Gas Association, 1971. Reserves of crude oil natural gas liquids and natural gas in the United States and Canada and the U.S. productive capacity as of December 31, 1970 (Washington, D.C.: Amer. Petroleum Inst.), Amer. Gass Assn., Amer. Petroleum Inst., and Canadian Petroleum Assn., vol. 25.

American Petroleum Institute, 1972a. The energy-gap (Washington, D.C.: Amer. Petroleum Inst.), 12 pp.

American Petroleum Institute, 1972b. Petroleum and energy (Washington, D.C.: Amer. Petroleum Inst.), 19 pp.

Biggs, R., 1967. Overboard soil disposal I. Interior report on environmental effects, *Proc. National Symposium on Estuarine Pollution* (Palo Alto: Stanford Univ.).

Brown, C. L., and Clark, R., 1968. Observations on dredging and dissolved oxygen in a tidal waterway, *Water Resources Research, 4*(6): 1381–1384.

Bureau of Mines, 1971. Outer Continental Shelf oil, gas, sulfur, and salt, leasing, drilling, production, income, and related statistics, 1971 (Washington, D.C.: U.S. Department of the Interior, Geological Survey), annual petroleum statements, Dec. 23, 1971, Bureau of Mines Annual Statistical Review, A.P.I.

Chesapeake Biological Laboratory, 1970. Gross physical-biological effects of overboard spoil disposal in upper Chesapeake Bay, Solomons, Md.: Final report to U.S. Bureau of Sport Fisheries and Wildlife.

Commission on Marine Science, Engineering and Resources (CMSE&R), 1969. *Our Nation and the Sea* (Washington, D.C.: U.S. Govt. Printing Office), including panel reports, vol. 1, 2, and 3.

Cotter, P. G., 1965. Sand and Gravel. U.S. Bureau of Mines, Mineral Facts and Problems, U.S. Bureau of Mines Bulletin 630.

Duane, J., 1969. A study of New Jersey and New England costal waters: *Shore and Beach, 37:* 12–16.

Gammelgard, P. N., 1971. Oil and environment: the challenges of our times, *Petroleum Today, 12*(2): 6–17.

Hess, H. D., 1971. Sand and gravel dredging reaches major proportions, Presentation to WODA, Pacific chapter meetings 2 March– 6 April, San Francisco, California.

Humble Oil & Refining Company, 1972. Energy: a look ahead (Houston: Humble Oil & Refining Co.).

McKelvey, V. E., Wang, F. H., Schweinfurth, S. P., and Overstreet, W. C., 1968. U.S. Geological Survey, 1968—Study of Outer Continental Shelf Lands of the U.S.: Public Land Law Review Commission Vol. II. (Washington, D.C.: U.S. Govt. Printing Office), P.5AI.

Manheim, F. T., 1962. Marine minerals off the northeast coast of the United States. U.S. Geological Survey circular (in press).

Moore, J. R., 1972. Exploitation of ocean mineral resources—perspectives and prediction, *Royal Society of Edinburgh, Proceedings 1971* (in press).

National Petroleum Council, 1969. Petroleum resources under the ocean floor (Washington, D.C.: National Petroleum Council).

National Petroleum Council, 1971. *U.S. Energy Outlook—an Initial Appraisal 1971–1985,* vols. 1 and 2 (Washington, D.C.: National Petroleum Council).

O'Neal, T., 1971. 14th biennial report (Baton Rouge: Louisiana Wildlife and Fisheries Commission), pp. 188–197.

Schlee, J., 1964. N.J. offshore deposits, *Pit and Quarry, 57:* 80, 81, 95.

Schlee, J., 1968. Sand and gravel on the continental shelf off the N.E. U.S. (Washington, D.C.: U.S. Govt. Printing Office), USGS circular no. 602, 9 pp.

Schlee, J., Folger, D., and O'Mara, C., 1971. Bottom sediments of the continental shelf of the N.E. U.S.: Cape Cod to Cape Ann, Mass. (Washington, D.C.: U.S. Govt. Printing Office), USGS open file report and misc. geol. inves. report no. 1-746, in press.

Shell Oil Company, 1972. The national energy position (Houston: Shell Oil Company), 22 pp.

U.S. Army Engineer Waterways Experiment Station, 1971. Model studies of outfall systems for desalination plants, part 2, Tests for effluent dispersion in selected estuary models (Vicksburg, Miss.: U.S. Army Engineer Waterways Experiment Station), research report H-71-2.

4 Recreation and Aesthetics

4.1 Introduction

Recreation is one of the largest and fastest-growing uses of the coastal zone. Studies conducted by the U.S. Outdoor Recreation Resources Review Commission in 1960 revealed that 44 percent of those engaged in outdoor recreation preferred water-based activities over all others (Ditton, 1971). This broad category encompasses a variety of activities, ranging from the most vigorous—swimming, water skiing, sailing, and diving—to the most contemplative—painting, wildlife observation, sun bathing, or just admiring the view. Recreation has, in fact, become a major economic force in coastal zone management. Annual revenue derived from these activities has reached major proportions. In 1968, for instance, an estimated 112 million people spent about $14 billion seeking recreation in the coastal zone (Winslow and Bigler, 1969). Table 4.1 shows the projected level of some particular recreational activities for 1975.

The natural constraints on use of coastal areas for recreation are primarily the limitations of space and access to that space to pursue a particular recreational activity. The greatest number of constraints are posed by other uses that preempt or interfere with recreation along the coast.

Climate is also a limiting natural factor. Water-based recreation is a mainly seasonal activity in the north, with the exception of limited use for ice fishing on many lakes. By the same token, the recreational need for a proper climate has played a large role in the population explosion of southern coastal areas. Other limiting factors depend largely on the particular recreational activity being considered, as

Table 4.1. Projected Level of Recreational Activities, 1975

Activity	Participation (in millions of people)	Revenue (billions of dollars)
Swimming	40	2.0
Surfing	4	0.2
Skin Diving	3	0.9
Pleasure Boating	14	1.0
Sport Fishing	16	1.3
Total	77	5.4

Source: NEMRIP (1969).

well as on the public's perception of proper conditions and pleasant surroundings.

Swimming, for example, demands water of a certain temperature and high quality from a public health and aesthetic point of view. Human behavioral patterns play no small role in determining constraints on such recreational activities. People generally prefer to swim in a place with a sandy beach and bottom and in water that is clear and odorless. A study of the public's perception of water quality revealed that, in general, the presence of algae was the most important indication of pollution to most people, while dark, murky water ranked second (David, 1971). People often prefer to swim at a beach that is visibly clean but contaminated with salmonella, rather than swim in safe, but turbid water.

Other activities, such as recreational boating, are confronted with different constraints. While water quality and access to the shoreline have less effect on the recreational boater, rough weather and proximity to a safe harbor are important limiting factors. Overcrowding could conceivably become a limiting factor in the future. Of all recreational activities, boating is projected to undergo the greatest popularity growth in the coming decade. The Bureau of Outdoor Recreation estimates that between 1965 and 1980, boating will increase 76 percent while the population will increase 29 percent. In this light, lack of launching access and harbor development will become real constraints before very long (Ditton, 1971, 1972). Most limiting factors are a result of man's other coastal activities. The greatest limiting factor, overall, is access to the shore. Of the 21,724 miles of the U.S. coast classified as recreational shoreline, 19,934 miles or 92 percent was privately owned in 1962 (NEMRIP, 1969). Although some of this has been developed by the private recreational industry, the overwhelming bulk of this land is not available for recreation. Only 6 percent or 1209 miles of shoreland is publicly owned recreational area. If it is assumed that 10 percent of the population will use 150 square feet of beach per person, this public area would serve only 21 million people, and this estimate is made with the assumption that all 1209 miles of shoreline is beach, which, of course, it is not (NEMRIP, 1969).

In terms of access, another limiting factor is proximity to the desired recreation area. This is a problem, especially for the poor who cannot afford to travel to coastal resort areas to enjoy water recreation. Artificial islands and embayments to increase ocean frontage have been

The Coastal Zone Workshop recommends
Recognition of the interest of people dwelling outside the coastal zone, but who are directly affected by its environmental conditions or its productivity. The needs of individuals and groups who have limited resources for competing in the political bargaining process in reaching coastal zone policy decisions must be considered.

among the suggested ways of increasing recreational shoreline in urban areas to overcome this problem (CMSE&R, 1969).

The living resources of the coastal zone pose yet another constraint on recreation. Many sports, for example, fishing, hunting, and wildlife observation, are totally dependent on the natural fauna and flora of the coastal zone. Thus, we can argue for the preservation of natural habitats from a purely selfish point of view. If the native organisms are disturbed by man's activities, he will inadvertently deprive himself of a sizable recreational activity. The U.S. Bureau of Sport Fisheries has calculated that 8,305,000 saltwater anglers spent $800 million on this sport during 1965 (U.S. Bureau of Sport Fisheries and Wildlife, 1965).

Other activities in the coastal zone limit its use for recreation in a variety of different ways. As we have noted, there is marked competition for space. In the cities, there is competition with commercial and shipping interests and industrial plants; in the suburban and rural areas, the competition is usually with private residential developments and, to a limited extent, industry. Recreational activities also compete with the use of shoreline for wildlife refuges and preserves.

Pollution stemming from industrial and waste-disposal uses of the coastal zone seriously limits or precludes certain recreational activities. Further, there are conflicting demands among the recreation seekers themselves—sailors, swimmers, fishermen, scuba divers, power boaters, and water skiiers are not always good neighbors. Zoning and restricted-use policies are becoming more and more necessary as the use pressure increases.

The growth in recreational use is a threat to itself as well as to the natural ecosystem of the coastal zone. It is a beautiful notion to camp by a lonely stretch of beach. It is quite another to camp in an instant rural slum with hordes of like-minded vacationers. Not only is the latter situation aesthetically less desirable, but sanitation and waste-disposal requirements quickly impose further limits on the carrying capacity of an area.

Recreation at the water's edge occurs in an ecological system that must be protected if recreational uses are to continue. It also occurs in an economic matrix that must be viable if continual degradation is to be avoided. A recreational industry must support the resident population or there will be pressure for nonrecreational industrial and commercial development which in turn will reduce the recreational potential. Unplanned recreational developments, such as have characterized the past, do not result in an attractive coastal zone, satisfaction for the vacationing public, or a viable recreation-based economy.

4.2 Coastal Tourism and Recreation

Virtually the entire coast of the United States, including Hawaii, Alaska, and U.S. possessions, has resources necessary for use in the field of tourism, recreation, and aesthetic amenities. Some sites possess greater strength than others and many already provide for high social and economic input. At the same time, extensive coastal areas are undeveloped.

Those natural and cultural resources along the coast that form the base for tourism and recreation include the following major categories:

1. Water and water life;
2. Vegetative cover;
3. Topography;
4. Geology and soils;
5. Climate;
6. Historic characteristics;
7. Legend, lore, and ethnic characteristics;
8. Housing;
9. Industry and institutions.

The degree to which these may be important for tourism and recreation depends upon the planning, development, and management of coastal parks, preserves and historic sites, and the influence of three major enhancement factors: (1) markets; (2) infrastructure; and (3) transportation and access (Gunn, 1970a).

These resource elements, their development, and their enhancement may combine in many ways to support a wide range of tourism, recreation, and aesthetic uses, including the following:

- Swimming
- Surfing
- Scuba/skin diving
- Water skiing
- Fishing
- Pleasure boating
- Horseback riding
- Hiking
- Golf
- Court and field games
- Motorcycling/bicycling
- Mountain and dune climbing
- Picknicking
- Pleasure driving
- Camping
- Hunting
- Attending events
- Vacation home use
- Visiting friends and relatives
- Plant tours
- Photography
- Collecting; botany, geology
- Nature study
- Lounging
- Visiting exotic cultures
- Visiting historic sites
- Cruises and excursions

Those areas that are developed and administered by public agencies account for 6 percent of the total recreational shoreline (Council on Environmental Quality, 1970, p. 178) and include 22 operated by the National Park Service—13 national parks and monuments and 9 national seashores and lakeshores. In addition the Service administers 28 historic areas along the coast (NCMR&ED, 1970, p. 46). The National Wildlife Refuge System now includes 91 refuges in the coastal zone containing 20.4 million acres devoted to management of migratory birds and other wildlife, providing a great amount of recreational activity (NCMR&ED, 1970, p. 47). In addition, many states, county and municipal recreational agencies own and operate recreational lands along the coast. Included in the not-for-profit sector are the holdings of quasi-public and nonprofit organizations, such as Mystic Seaport, Connecticut, and Williamsburg, Virginia. In fact, most of the historic redevelopment and restoration is performed by nonprofit organizations. In Hawaii, for example, over 200 nonprofit organizations have specific recreational programs and activities in the coastal zone (Hawaii Department of Planning and Economic Development, 1969, p. 32).

Private development of the coast for tourism and recreation is great and growing. It has been estimated that "over 68 percent of the total recreational property values along the coasts and Great Lakes are accounted for by shorefront homes (Council on Environmental Quality, 1970, p. 177). Commercial private developments are well known especially at areas of concentration, such as Miami Beach, Atlantic City, Los Angeles, and Waikiki. Estimates now identify recreation as the leading economic activity of coasts, greatly exceeding all other consumer expenditures.

In 1968 it is estimated that approximately 112 million people participated in a total of 7.1 billion ocean-oriented occasions and spent about $14 billion. This figure is more than twice all other government and nongovernment expenditures in the oceanographic market, estimated at $6.5 billion in 1968 (Winslow and Bigler, 1969, p. 51). Boating alone, for example, is now engaged in along the coast by 20 million people and is increasing at the rate of 200,000 boats a year (NCMR&ED, 1970, p. 47). Tourism is generally defined as activity involving travel, including the expenditure of money at a location other than where earned, and participation in a pleasurable objective. The tourism-related establishments in Hawaii employ over 14,700 persons, who, in turn, spend $60,280,000, primarily within the local economy. Visitors pay approximately 16 percent of the state's total tax revenue (Hawaii Department of Planning and Economic Development, 1969, p. 33).

Tourism, recreation, and aesthetic uses of the coastal zone take place in what could be described as the following four subzones paralleling the coastline (see Figure 4.1):

1. NERITIC. This ecological "near-shore" marine zone spreads from the

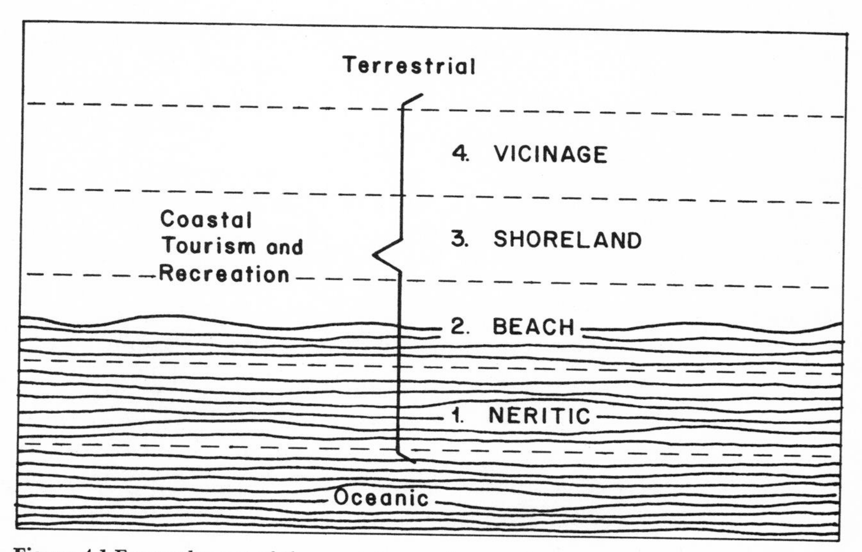

Figure 4.1 Four subzones of the coastal zone. Source: Gunn (1970b, p. 13).

continental shelf to the beach. It is the richest zone for fishing and often contains interesting bars and reefs. It is well suited to cruising and sailing and sometimes is used for travel to nearby islands. Visual contact is dominantly with the sea rather than the land.

2. BEACH. The beach zone reaches both into the water and onto the land and, especially if wide and sandy, supports the most popular of water-based recreation. This zone has always meant relaxing, building sand castles, shore water activities, and beach sports. As most coastal travelers well know, the more passive activities, such as watching and photographing people, the surf, shorebirds, plantlife, and sunsets also are very popular in this zone.

3. SHORELAND. The shoreland zone identifies that ribbon of land lying behind the beach and providing many supporting marine recreations, such as camping, picnicking, and hiking. In some locations it supports hotels and other service businesses. An important aspect of this zone is the visual connection with the sea.

4. VICINAGE. The marine coastal backland generally contains supporting services for recreational activities and is often the locale for vacation homes. Coastal scenic appeal is often enhanced by variations in topography and vegetative cover. This zone depends upon geographic image of nearness rather than visual linkage with the beach and sea.

The seacoast does in fact include a unique mix of resource attributes that contribute to strong leisure appeal and use. Many of these attributes derive primarily from the sea-land interface, but others are due to the abundance of resources that are found in interior regions as well. For example, coastal fishing is special fishing. Fishermen tend to be selective and realize that the fishing conditions of weather, wave action, equipment, and objective species are often quite different on the ocean and along the seashore as compared to other water bodies. Surfing can be done on a pond with mechanically contrived waves but there is no substitute for the lure of the famous Makaha Beach in Hawaii.

Several other recreational activities are peculiarly oriented to coastal resource attributes. Coastal hunting for waterfowl or amateur study of the special coastal flora and fauna cannot be duplicated elsewhere. Skin diving and scuba diving among saltwater shallows and depths as well as among coral reefs are quite different from such activities in inland fresh waters. The challenge and risk involved, as well as the rewards, are much greater. Beachcombing, shell collecting, coastal lounging, swimming, and merely strolling along the water's edge are rewarding experiences for many who seek this special environment. Perhaps the greatest lure of all, but one that defies precise measurement, is that of aesthetic enjoyment. Viewing the deep and unbroken vistas over the sea, watching for ocean-going vessels, feeling the brisk flow of the sea breeze, contemplating the historic past and legends of the sea, and possibly seeing porpoises or whales on the horizon are just a few of the special activities that are extremely meaningful to visitors at the seacoast. These are not possible without the special resource assets of the coastal zone.

Already, considerable investment in recreational activity areas has been made along the seacoast, and much more is planned both on the land and in the sea. Florida has the John Pennekamp Coral Reef State Park at Key Largo, and the National Park Service has underwater parks off the Virgin Islands at St. John and St. Croix (Stalker, 1968, pp. 50–51).

Within this zone can often be found a variety of resources in addition to the sea, such as bays, estuaries, river mouths, and deltas, and mountain scenery backdrops—all adding materially to the physical and psychological appeal for recreational uses. The compounding of many resource features in close juxtaposition adds materially to the strength of attributes taken separately. Therefore, water skiing, canoeing, sailing, boating, swimming, and fishing popularity associated with inland water bodies are many times possible in connection with exclusive sea-oriented activities. Coastal geography usually also allows the opportunity to develop the same typical recreation offerings as elsewhere: hiking, sunning, golfing, horseback riding, picnicking, camping, photography, sketching, painting, and sightseeing, to name a few.

4.2.1 Concentration at Access; Cities

Coastal tourism and recreation is unevenly distributed. Leisure uses are highly concentrated at certain locations and virtually nonexistent at others—and not always because of differences in physical resource qualities. Other factors, especially access and metropolitan concentration, appear to be exerting a great influence on both the provision for and participation in coastal tourism and recreation (Gunn, 1972).

Wherever highway, ship, or air access penetrates to the coast and gives many people access to coastal resources, the demand for recreational activity is high. Along some beaches, wherever public road right-of-way ends at the waterline, mass recreational use is stimulated even though the area may not have been overtly planned for recreation. Conversely, coastal stretches without easy access generally remain untouched, whether privately or publicly owned.

Closely correlated with access is the influence of metropolitan concentration upon recreational uses of coastal resources. Most major cities of the world, including those in the United States, are located along the seacoast. These become focal points for tourism and recreation concentration as well as concentration for all other metropolitan activity, such as housing, trade, industry and cultural opportunity. Generally, along any coast the recreational use is low between the cities and high at the cities. It should not be surprising that the combination of cities and coastal natural resources would produce powerful magnetism for tourism and recreation. Ample testimony is provided by the developments along many coasts such as Miami and Waikiki.

Contrary to the escape theory of recreation—the desire for man to escape the ills of the city—many people are lured to cities because of their special attractions. Mumford (1961) states that welcoming visitors has been one of the prime functions of cities for many years. "Increasingly people sought out the city and became part of it by willing adoption and participation. Whereas one gains membership in a primary group like the village or clan solely by the accident of birth or marriage, the city, probably from the first, offered an opening to strangers and outsiders."

Many of the things that a coastal city wants for itself and its citizens are the same things desired by tourists. For example, San Francisco, for reasons of its own, defends the retention of an old, obsolete cable car system; it tears down old red-light districts and rebuilds them into swank shops; it retains a park on an old promentory where ships were signaled as they entered the harbor; it builds a graceful bridge to link it with suburban communities and it turns its waterfront into world-famous seafood service businesses. Although primarily motivated for local good, these landmarks turn out to be the greatest tourist assets San Francisco has.

Coastal cities are generators of tourists as well as targets for them. As populations along the coasts increase, so do the markets. Evidence continues to support the hypothesis that coastal living begets coastal tourism and recreation. For example, 45 percent of Hawaii's tourism comes from Los Angeles county, a coastal city region with many similar amenities. At any time, one can find a great many visitors from Florida's coasts having a good time at Waikiki Beach. This phenomenon appears to be influenced greatly by a desire to see and enjoy the types of activities that one is already accustomed to.

Finally, as compared to most other uses man makes of the coastal zone, tourism and recreation are low consumers of natural resources. With the exception of the "consumption" of prime beach lands by structures and facilities for recreation, generally the resources upon which tourism and recreation depend are left moderately unimpaired, though "used." Whenever a visitor views the panorama, whenever he uses a boat on the water, whenever he photographs a scene, whenever he runs along the beach, and whenever he studies the historic lore and background of a coastal city, he leaves the resource virtually the same as he found it as far as his principal recreational activity is concerned. The same cannot be said about extractive industries or manufacturing plants whose waste products are of much less value than the original resources. The product of tourism and recreation is the individual experience. As such, it is composed not so much of material goods as of psychological impact. Therefore, what one experienced today may be replicated day after day by thousands more with virtually no decay in the resource.

Obvious exceptions to this argument occur with the construction of facilities and services to accompany recreational experience and with certain recreational activities, such as fishing, that do "consume" natural resources to some degree. Also, the blight of trash and litter, which is too often left by visitors to an area, must be considered an environmental abuse and needs to be planned for. In several instances, however, environmental enhancement is an equal argument. Deer hunting, for example, is often cited by game biologists as a necessary management practice if any balance between the quantity of natural food supply and numbers of deer is to be maintained.

4.2.2 An Evaluation System; Land-Use Units

Regional planning for tourism, if it is to be of greatest value to designers, planners, developers, and resource managers, must be comprehensive enough to allow full development of potential, rather than narrowly reflecting special interests. The potential of an area is dependent upon both its range of resource characteristics and the range of activities desired by the public. Review and observation of present tourism reveals that it consists of a very wide range of user activities. Identifying these activities usually includes a modifying description that relates to the environment where that activity thrives best. The combination of the environment and the activity might be described as an attraction land-use unit, being made up of both man-made development and natural resources.

Attraction land-use units, therefore, are the physical entities that provide for people's leisure-time activities and have a number of characteristics important to regional development. First, they are frequently anchored to specific geographic settings because of heavy dependency upon certain natural or cultural resources. Those land-use unit locations where the least modification of given resources is necessary are most preferred. This is an ecological approach to tourism. Second, land-use units have great economic impact upon local communities and are "successful" if they satisfy nearly all of the people who have interests in the activities supported. Therefore, geographic relationship with population concentrations and markets is important to land-use units. Generally, the greater the numbers of people accommodated, the more completely successful the tourism system. Third, the land-use units are productive only if accessible. Therefore, transportation systems and access are important. Fourth, the fact that an existing service area (a city or other collection of services—health protection and care, police protection, public services, businesses) is within easy reach is an important advantage to any attraction land-use unit. Fifth, developed attraction destinations are already attracting visitors and have established reputations as destinations. Locations near or within such well-known attraction areas are superior to others because less new effort of image creation and promotion is necessary. Finally, the success of attraction land-use units is as dependent upon the quality and functional success of the design, construction, and management of the unit as on the other factors enumerated. To summarize, keys to the planning of a region for tourism are the following factors: natural and cultural resources, nearness to markets and size of markets, transportation and access, infrastructure, ongoing destination reputation, and the design, construction, and management of land development.

The derivation of attraction land-use units can be approached in many ways. One approach is to identify tourism activities. Many lists were studied, including outdoor recreation activities identified by the Bureau of Outdoor Recreation and other governmental agencies working with outdoor recreation. These lists included a wide range of activities that depend upon the out-of-doors. Tourism and travel studies, however, reveal that indoor activities as well as outdoor activities are important to the tourism and recreation markets. Further study showed that activities are frequently identified by the resource involved in the par-

---

The Coastal Zone Workshop recommends
A sustained national commitment to education and training of the necessary talent for the management of the resources of the coastal zone. The goal should be a widespread awakening in the public to the importance of maintaining a sound coastal zone environment as well as the preparation of a future generation of natural and social scientists to manage wisely their environmental heritage.

---

ticipation. For example, the activities of fishing and hunting imply resource areas that contain ample amounts and qualities of available and "catchable" fish and game.

From review of research reports of both tourism and recreation and from observation of the full spectrum of activity in many regions of the United States and from review of resources related to recreation and tourism, a list of attraction land-use units was prepared (Table 4.2). Based upon the relative importance of natural and cultural resource factors, the list is divided into three components: (A) those primarily dependent upon special natural resources, (B) those primarily dependent upon special cultural resources, and (C) those not heavily dependent upon either special natural or cultural resources.

4.2.3 Planning

The attraction land-use units provide a basis for overall regional planning. If a given region can be tested for its potential development of each of the land-use units, an evaluation of its overall tourism potential will result. It does appear, however, that each land-use unit should be carried through all factors, making total regrouping at the final stage rather than earlier. In other words, a region can be described in terms of its support for each attraction land-use unit—assuming, of course, that it would be provided with adequate design, construction, and management of each unit.

Based upon the above rationale, a series of steps appear to be appropriate for evaluating the potential for each of the attraction land-use units. A first step is the identification of characteristics generally associated with the land-use unit in an abstract sense, that is, not associated with a specific area of land. For example, to what extent is climate important for carrying on picnicking or boating? Perhaps a scale of importance could be made among the various factors; then this scale could be applied to a specific region to determine how well (or poorly) the given region meets the qualities needed.

Table 4.2. A Typology of Tourism and Recreation Attraction Land-Use Units

A. Primarily Dependent upon Special Natural Resources

1. Beaches
2. Picnic areas
3. Camping areas, nature
4. General scenic areas
5. Scenic spectaculars (waterfalls, etc.)
6. Rock collecting areas
7. Shell collecting areas
8. Hunting areas
9. Fishing areas
10. Skiing and winter sports areas
11. Snowmobile areas
12. Boating, canoeing, sailing areas
13. Resorts, winter (northern)
14. Resorts, winter (southern)
15. Resorts, summer
16. Camps, organization and group
17. Marinas, harbors, boat launching areas
18. Wilderness
19. Animal observation areas
20. Waterways
21. Vacation home sites
22. Prospecting sites
23. Forest produce collecting areas
24. Trail bike areas
25. Nature trail areas (foot, horse)
26. Bird watching areas
27. Spelunking areas
28. Scuba/submarine exploration areas

B. Primarily Dependent upon Special Cultural Resources

1. Archaeological sites, digs
2. Museums
3. Historic restorations, ghost towns
4. Landmarks, firsts, one-of-a-kind
5. Ethnic cultures, special concentrations
6. Engineering and scientific wonders
7. Manufacturing plants
8. Institutions (outstanding)
9. National shrines
10. Sightseeing tour sites, culturally oriented
11. Dude ranches
12. Legend, lore special areas

C. Not Heavily Dependent upon Either Special Natural or Cultural Resources

1. Concert, drama, pageant areas
2. Craft exhibits
3. Camping areas, urban
4. Spectator sports arenas
5. Golf areas
6. Amusement parks
7. Shopping centers
8. Night clubs
9. Hotels, motels (for tourists, recreationists)
10. Restaurants (for tourists, recreationists)
11. Information centers, rest areas
12. Playfields, playgrounds
13. Residential areas of friends and relatives
14. Festival, parade, derby areas
15. Marine festivals, regattas
16. Convention centers

4.3 Coastal Preserves

Many species have become extinct in North America since the advent of European man, and the list of endangered species is growing rapidly. Since speciation appears to be a rarer event than extinction, we are forced to conclude that this loss of species can only result in fewer spe-

---

The Coastal Zone Workshop recommends
The creation of a national system of *Coastal Area Preserves* for the permanent protection of the basic genetic stocks of plants and animals and the essential components of their environments, which together constitute ecosystems. These Coastal Area Preserves should be severely restricted in use. Some other coastal areas should be developed for recreational usages that are compatible with the natural life of the area.

---

cies and less ecological stability. The effect of specific uses and interactions of multiple uses on the stability of the varying habitats of the coastal zone is meager. Will ecosystems with reduced diversity and stability persist under the continuing changes which normally occur and which are speeded by man's activities? When the results of man-induced stresses on a specific habitat are known, it is difficult to apply this knowledge to other habitats because of the variety of natural environmental stresses on the systems. The relevant ecological information thus forms an inadequate basis for recommendations on how to protect and manage the coastal zone. The final and irretrievable loss of a species, native habitat or native ecosystem reduces the range of genetic and species diversity man can use to respond to natural or man-induced changing conditions.

Coastal area preserves need to be established and maintained to provide natural physical-chemical-biological habitats where the species diversity and stability of the natural community are assured continued persistence; refuges to compensate for overexploitation of resources; nursery grounds for food species and sport species; species populations and communities for restocking decimated or highly stressed systems with assemblages of stable systems; and areas for research to obtain base-line information and data, relative to stressed systems, which can be used to assess the effects of man's acute and chronic (but subtle) stresses on communities, habitats, and ecosystems. The main constraints on the establishment of coastal area preserves are imposed by other human uses, establishing boundaries and size limitations, and determining the number, location, and types of preserves.

Natural coastal area preserves, by their very nature, require that the environment be subject to minimum impingement by man on considerable parts of the preserve. With the possible exception of limited recreational access for such activities as viewing, hiking, fishing, and hunting, the major part of the area should be restricted to preserve maintenance functions.

4.3.1 Distribution of Coastal Zone Preserves

Different portions of the coastal zone are characterized by seasonal variations and by different features, circulation patterns, and species (Table 4.3). Thus, the basic components of biological productivity in the coas-

Table 4.3. Regional Classification of Coastal Zones of North America. Only the outstanding descriptive characteristics of each are given. A variety of coastal area preserves is needed in each region.

1. *Arcadian*—Northeast coast of North America (from the Arctic to Cape Cod)

Characteristics: Rocky, glaciated shoreland and submarine topography; shoreline subject to winter icing; large attached algal species important producers; biota essentially boreal.

2. *Virginian*—Coast of middle Atlantic states (from Cape Cod to Cape Hatteras)

Characteristics: Climate, topography, and biota transitional between Regions I and III; lowland streams, coastal marshes, and muddy bottoms becoming prominent; biota primarily temperate with some boreal species.

3. *Carolinian*—Coast of south Atlantic states (from Cape Hatteras to Cape Kennedy)

Characteristics: Extensive marshes and (cypress) swamps; muddy bottoms very important; waters turbid and highly productive; biota temperate with some subtropical elements.

4. *Louisianian*—Northern coast of Gulf of Mexico (from central Florida to Tuxpan, Mexico)

Characteristics: Quite similar to Region III, but more tropical in environmental conditions and in biotic composition; bottoms mostly terrigenous.

5. *Vera Cruzian*—Eastern coast of Mexico (from Tuxpan to the base of the Yucatan Peninsula)

Characteristics: Diverse shoreland (hills and volcanic mountains grading southward to extensive low plains, marsh, and swampland); bottoms mostly terrigenous; biota distinctly tropical, but with some temperate elements.

6. *West Indian*—Eastern coast of tropical America (southern tip of Florida, Yucatan Peninsula, Caribbean coast of Central America, West Indian islands)

Characteristics: Shoreland low-lying (karst) limestone varying to mountainous, but distinctly calcareous; foreshore and seabed with calcareous marls, sands, and coral reefs; biota tropical.

7. *Columbian*—Northwestern coast of North America (from Arctic to southern California)

Characteristics: Shoreland predominantly mountainous; rocky foreshores prevalent; extensive algal communities, especially offshore kelp beds; biota boreal to temperate.

8. *Californian*—Western coast of North and middle America (from southern California through Mexico and Central America)

Table 4.3 (*Continued*)

Characteristics: Shoreland generally mountainous (often volcanic); rocky coasts with volcanic sand; general absence of marshes, swamps, and calcareous bottoms; biota tropical.

9. *Great Lakes*—Great Lakes of North America

Characteristics: Rocky, glaciated topography, limited wetlands; cold-temperate climate; fresh water only; biota a mixture of boreal and temperate species together with anadromous and marine invaders.

10. *Fiords*—Tidal, glacial and turbid backwash. Alaska

Characteristics: Precipitous mountains, deep estuaries often with glacial moraines.

11. *Subarctic*—Ice-stressed coasts, Bering Sea and Arctic Ocean

Characteristics: Shoreline subject to icing, biota Arctic and subarctic.

12. *Insular*—Hawaii

Characteristics: Precipitous mountains, considerable wave action, endemic tropical and subtropical fauna.

---

tal zone differ from region to region. Different regions have different requirements for the establishment of ecosystem preserves.

4.3.2 Size and Management of Coastal Zone Preserves

Several criteria for establishing the size of habitat preserves can be identified. Since small populations tend to become uniform in genetic diversity (Fisher, 1958) and small communities have a smaller number of species (Preston, 1948; Fisher, Corbet, and Williams, 1943), small numbers of individuals result in reduced genetic and species diversity. This further results in a decrease in the ability of the populations, communities, and ecosystems to remain stabilized.

In response to changing environmental conditions small populations have entirely different evolutionary patterns than do large homogeneous populations (Lerner, 1958; MacArthur, 1965), and Wright (1948) has emphasized that the most stable responses to general environmental variations are found in very large populations variously subdivided into smaller groups with diverse limitations on migrations between groups. Varying sizes of preserves are needed, but in each case the inclusive size must be adequate to maintain the internal physical, chemical, and biological integrity of the habitat. This internal integrity will give the ecosystem preserve the capacity to weather external environmental stresses such as long-term fluctuations in currents, tides, sunlight penetration, and temperature. Since simplification of a habitat or eco-

---

The Coastal Zone Workshop recommends
The immediate intact preservation of selected natural land and water areas in shoreline and estuary regions of the United States valued for their unique ecological character. Such areas should be severely restricted from any private or public coastal zone activity.

---

system has two immediate results—altering of food chains and altering of genetic patterns by loss of various types and species—the size of the preserve must not be so limited that the biological compartments of the habitat will simplify rather than maintain their natural diversity and stability. Simplification eventually leads to instability and destruction of the system that is being preserved. Some areas will need to be thousands of square miles while others may be as small as a hundred acres.

In establishing coastal area preserves, the adjacent areas must also be taken into consideration, since the uses of these boundary regions may be immediately or eventually incompatible with the criteria for preserving the habitat. Thus, where a marsh area is set aside, some form of control over the uses of the watershed and the adjacent estuarine areas is imperative. Activities or uses of the adjacent land area must not encroach on the coastal area preserves. The preserve cannot be bounded by fences, as in aquaculture, nor by an aquarium or zoo, because this would restrict the movement and flow of organisms and energy into and out of the system and change it to a new and different habitat from the one to be preserved.

The inner region of such coastal area preserves would be nonaccess areas with only occasional checks by employees to assure that there are no intruders that might disrupt species interactions. On either side of the protected areas would be regions that would be used as seed stock to restore the managed areas lying outside this zone. These managed areas would, in turn, be used to provide stock and nursery material for the restoration of areas once heavily used by man for spoils, mining strips, and canals but being converted to other uses. Still further from the protected area, hiking, camping, and other light recreation could be allowed, and these activities grade naturally into more intensively used parks, with other human activities. The size of each of these zones would need to be determined individually.

4.3.3 Species Preserves

An expanded program of species preserves also needs to be established and maintained to preserve and protect migratory pathways and endemic, endangered, or overexploited species. The constraints imposed by the establishment of these preserves would not be as stringent as those imposed by habitat preserves. Species preserves can be part of managed habitat preserves or may be managed areas where other activities predominate.

Agricultural utilization of the coastal zone, for instance, is compatible with maintenance of migratory pathways for birds as long as pesticide levels and other persistent potential toxins remain well below biomagnification levels. On the other hand, resting, breeding, and feeding areas for migratory animals often are not compatible with industrial land or harbor use, high-density human activity, or waste disposal. Physical structures for mining and extraction can enhance the diversity if managed to be compatible with species preservation. Some military reservations can be beneficial in that they prevent overharvesting of animals present or migrating into the area. Management of heated effluents from power plants can be used to enhance the growth and development of species. Thus smaller areas with diverse and less restrictive requirements than those of the coastal area preserves, which can be associated with other compatible, but managed, uses, can also be established for species preservation.

To establish, maintain and use coastal zone preserves many studies are needed. New ecosystems created as the result of man's activity need to be compared with unaltered or preserved systems so that the effect of stresses created by man's activity and by general environmental change can be identified and evaluated.

Attempts to move entire ecosystem stocks into areas nearly barren as a result of overuse are needed. Analysis of the subsequent stages of succession where ecosystems have been moved are essential. Studies to provide criteria for the design of adequate managerial and legal structures are essential.

References

Commission on Marine Science, Engineering and Resources (CMSE&R), 1969. *Our Nation and the Sea* (Washington, D.C.: U.S. Govt. Printing Office), including panel reports, vol. 1, 2, and 3.

Council on Environmental Quality, 1970. *Environmental Quality, The First Annual Report* (Washington, D.C.: U.S. Govt. Printing Office).

David, E. L., 1971. Public perceptions of water quality, *Water Resources Research*, 7(3): 453.

Ditton, R. B., 1971. The future of boating on Lake Michigan (Madison: University of Wisconsin Sea Grant Program), WIS-SG-71-106.

Ditton, R. B., 1972. The social and economic significance of recreation activities in the marine environment (Madison: University of Wisconsin Sea Grant Program), WIS-SG-72-211.

Fisher, R. A., 1958. *The Genetical Theory of Natural Selection* (New York: Dover Publications, Inc.).

Fisher, R. A., Corbet, A. S., and Williams, C. B., 1943. The relation between the number of individuals and the number of species in a random sample of an animal population, *Journal of Animal Ecology, 12:* 42–58.

Gunn, C. A., 1970a. Texas marine resources—the leisure view (College Station, Texas: Recreation and Parks Department and Sea Grant Program, Texas A&M University), report of workshop on tourism-recreation development of Texas marine resources.

Gunn, C. A., 1970b. A new approach to coastal tourism development (presented at National Marine Recreation Conference, Long Beach, California), May, 1970.

Gunn, C. A., 1972. Concentrated dispersal, dispersed concentration—a pattern for saving scarce coastlines, *Landscape Architecture, 62*(2): 133-134.

Hawaii Department of Planning and Economic Development, 1969. *Comprehensive Outdoor Recreation Plan* (Honolulu: Department of Planning and Economic Development), prepared by Donald Wolbrink & Associates, Inc., and Arthur D. Little, Inc.

Lerner, M. I., 1958. *Genetic Homeostasis* (New York: John Wiley and Sons, Inc.).

MacArthur, R. H., 1965. Patterns of species diversity, *Biological Reviews, 40:* 510–533.

Mumford, L., 1961. *The City in History* (New York: Harcourt, Brace and World).

National Council on Marine Resources and Engineering Development (NCMR&ED), 1970. *Marine Science Affairs—A Year of Broadened Participation* (Washington, D.C.: U.S. Govt. Printing Office).

New England Marine Resources Information Program (NEMRIP), 1969. Outdoor recreational uses of coastal areas (Kingston: Univ. of Rhode Island, New England Marine Resources Information Program), no. 1, p. 18.

Preston, F. W., 1948. The commonness, and rarity, of species, *Ecology, 29:* 254–283.

Stalker, L. V., 1968. Seacamps—tomorrow's outdoors?, *Field and Stream, 72*(11): 50–51.

U.S. Bureau of Sport Fisheries and Wildlife, 1965. National survey of fishing and hunting (Washington, D.C.: U.S. Govt. Printing Office), resource publication 27.

Winslow, E., and Bigler, A. B., 1969. A new perspective on recreational use of the ocean, *Undersea Technology, 10*(7): 51–55.

Wright, S., 1948. On the roles of directed and random changes in gene frequency in the genetics of population, *Evolution, 2:* 279–294.

5 Urbanization and Industrial Development

5.1 Introduction

Over half of the nation's population lives in the counties bordering the Great Lakes and ocean coasts. According to the 1970 census report, there was a shift during the 1960s from the center of the country to the seacoasts (Taeuber, 1972), and this trend is expected to continue. According to the Commission on Marine Science Engineering and Research (CMSE&R 1969), the 31 coastal and Great Lakes states contain more than 75 percent of the U.S. population. The estuarine regions of the coastal zone encompass only 15 percent of the land but have about 33 percent of the population.

This does not imply equal distribution; indeed quite the opposite is true. The areas between Washington and Boston, Chicago and Detroit, Los Angeles and San Diego, and the cities of Miami, Houston, and San Francisco represent areas of major concentration. This is not unique to the coastal zone; most large cities are targets for people immigrating from rural areas. The seven largest metropolitan areas in the country lie along the coast (CMSE&R, 1969). With these urban areas have come major problems of waste disposal, overcrowding, and overdevelopment. Dense settlement has given rise to heavy industry, demand for more food from the sea, and increasing pressures on natural coastal lands for habitation and recreation.

Throughout history man has always been preferentially attracted toward the sea for a variety of reasons including commerce, transportation, food, and climatic moderation. In recent years, new dimensions have been added: recreation, jobs in industrial complexes, and, as in south Florida, retirement in a warm climate. In the process, man has urbanized the most select sites, generally by the estuaries.

5.2 Housing

One of the greatest pressures on urban areas is the need for housing. The Baltimore Regional Port Shoreline Land Use Study (1967) indicated that residential use comprised 55 percent of the 266 miles of shoreline survey. Industry was second with 16 percent (Trident Engineering Associates, 1968). A similar study conducted in San Francisco (BCDC, 1969) showed that 20 percent of the occupied shoreline was devoted to residential use, with transportation second.

Habitation is one use of the coastal zone in which relatively few con-

The Coastal Zone Workshop recommends
The President of the United States request the National Academies of Science and Engineering to create a multidisciplinary *Coastal Zone Task Force* to formulate a management program. The Task Force should assist the federal government in designing the national program and evolving model guidelines for state coastal zone management authorities. The Task Force should work with the coastal states, regional agencies, and the federal government in preparing specific plans for coastal regions of the United States.

straints have been heeded. Man has become adept at overriding both natural and man-made constraints on coastal zone development, and few things can bar him from filling in swamp and marshlands if he is determined to live there and is willing to accept the consequences. In the past 20 years, dredging and filling have destroyed over half a million acres of important fish and wildlife habitats in the United States (CMSE&R, 1969); and by 1975, housing developments will become the leading cause in the loss of estuarine areas (U.S. Department of the Interior, 1970). Often, man will fill or spray estuarine or marshy areas just to enhance his comfort and eliminate mosquitos and other insect pests.

The Commission on Marine Science, Engineering and Resources (1969) noted that 7 percent of the nation's important estuarine areas has been lost as a result of housing development in the past 20 years. The construction of low-density, single-family residential units required large tracts of land—some adequately planned with reasonable attitudes toward economics and aesthetics, others contributing to urban sprawl. Recently, condominiums have been added to the scene. The shore of southeast Florida has many examples of both adequate and inadequate planning. The shore is often bulkheaded, requiring large amounts of fill usually obtained by dredging, and the natural ecological system has been seriously upset. Drainage from the land is interrupted, the nutrient cycle disturbed, wildlife breeding habitats lost, nursery grounds impaired, oyster beds diminished, and waste disposal intensified. In general, commercial, recreational, and aesthetic assets are deleteriously altered. A primary hazard associated with coastal development is the danger of exposure to violent weather events, winter storms in higher latitudes and hurricanes in mid and low latitudes. Through the action of wind and waves, landowners incur extensive property damage which is too often accompanied by loss of lives. If these events were regular in a particular region, the problem would probably be easier to deal with, either by avoidance or by proper safety precautions. But, because of their irregular occurrence and the inability to predict precisely when they will strike, storms are natural catastrophes which residents of the coastal zone seem willing to risk. Hurricanes are the most tangible illustration of this problem and are of rapidly growing concern. According to hurricane damage statistics, hurricane casualties have dropped, probably because of improved warning, while the amount of property damage has taken a turn upwards, largely because capital investment in land along the coastal zone has increased (Pardue, 1971). While early hurricanes killed tens of thousands, a few hundred fatalities is now considered a substantial death toll. Formerly a few million dollars property damage was typical for a severe storm, but in the last decade two hurricanes have each been responsible for about $1.4 billion worth of damage (Pardue, 1971). Even the cost of emergency storm preparations for major urban areas is substantial. According to the national hurricane center in Miami, it costs Florida's Dade County $2 million for each hurricane alert, whether or not the storm eventually hits (Simpson, 1971).

While not as dynamic or dramatic a process as the coastal storm, shore erosion poses a critical problem to those living along the water's edge. Of the 100,000 miles of U.S. shoreline, about 50,000 miles are judged to be vulnerable to erosion and require attention (CMSE&R, 1969). The most critical areas are along the shores of New Jersey, Florida, Texas, southern California, and southern Lake Erie. While the effects of waves and high water levels can destroy vast and valuable expanses of land in a few years, coastal sedimentation processes can be a boon to other shoreline settlers by increasing their coastal property. Development without attention to the normal sand movement patterns, migration of beaches, and erosion has already brought pressures on the Corps of Engineers and other agencies to interfere with natural processes that have continued for eons. In some cases, attempts to protect unwise construction on the shore seem destined to the same fate as King Canute when he ordered the sea to retreat.

There are other constraints imposed upon the use of the coastal zone for human habitation. Water supply and waste disposal are severe problems in areas of high population density. Shoreline development without adequate water supply and waste treatment facilities quickly leads to soil and surface water pollution by overloaded septic tanks and associated land disposal problems.

Utilization of our coasts as a place to live has both tangible and intangible advantages. Coastal oriented industrialization, commerce, and resource production all require workers, and workers need a place to live and play. Not surprisingly, demographic and socioeconomic indicators all project continued increase in water-oriented housing in coastal areas. Not all coastal areas are increasing in population, however. Many rural coasts along both the oceans and the Great Lakes are losing population. The patterns of human migration and settlement are not well understood, and as these factors affect the coastal zone, they constitute a major information gap in our attempts to devise comprehensive management schemes.

It is probably not realistic to attempt to discourage or systematically control the movement of people. It is essential to improve our understanding and hence our predictive skills of how population shifts will occur in the future. Based on our best estimates of population changes, careful and comprehensive planning for human habitation must be undertaken.

As the migration into the coastal zone continues, more land will be required for housing at the sea's edge and in from the immediate shoreline. The piecemeal, unplanned, uncontrolled, unregulated system that has been used in many areas cannot continue. Growth must be planned for where it is likely to occur, adequate services must be planned in advance, and population densities must be regulated so as not to exceed the natural and service capacity of the area. In areas where population is decreasing, the factors responsible must be determined and decisions made whether to allow the decrease to continue and recover some of the coastline for other purposes such as preserves, or to develop the coast for recreational or industrial uses and thereby reverse the population trend. The long-range plans must determine the types of development suitable for particular areas. Effective use of planning and management techniques can reduce or eliminate uncontrolled urban sprawl, achieve higher aesthetic standards, and maintain economic balance. In addition, regionalization should be given serious consideration in planning the future of an urbanized coastal zone.

5.3 Industrial Development

In general, industries concentrate in the coastal zone because of trans-

The Coastal Zone Workshop recommends
The acceleration and expansion of comparable surveys and complementary inventories of coastal resources, including demographic patterns, ownership and land-use patterns, and other socioeconomic data in addition to new base-line ecological studies of natural and modified coastal systems.

portation advantages, import and export requirements, and the need for large quantities of water. This is demonstrated by the fact that over 40 percent of our industrial complexes are located not only in the coastal zone, but in the estuarine counties which represent only about 15 percent of the land (U.S. Senate Document 91-58).

The concern that has arisen over the condition of our nation's coastal zone can often be traced back to problems stemming from industrial and commercial uses of this area. Though the total area which industry occupies in the coastal zone is not very great, its impact on the coastal system can be severe. Waste disposal, oil spills, and the escape of toxic materials into aquatic ecosystems are all unfortunate by-products of industry which affect the coastal environment.

Industrial uses of the coastal zone vary widely. Some industrial activities, such as oil extraction, mining, and commercial fishing, take place in the waters offshore and are designed to harvest the natural resources of the coastal zone. Other industries, such as power plants and paper mills, locate directly on rivers, estuaries, or along the coast in order to have large quantities of water available for their industrial processes.

In general, the coastal zone is an attractive location for five major types of industry:

1. Industries that benefit from location near low-cost water transportation and inland transportation systems;
2. Industries that derive power from water or use water for process or cooling purposes;
3. Industries that are beneficially located near centers of population but that do not have direct dependence on, or need for, water or water access;
4. Marine transportation industries; and
5. Industries that depend directly on the marine environment for raw material for commercial activity.

The Coastal Zone Workshop recommends
The application of environmental quality standards and performance criteria based upon monitoring or surveys to be evaluated by all government agencies involved in the management of the coastal zone. They should take into account socioeconomic needs of the community, and resort to general regulations, zoning, and other codes only when necessary for compliance.

Each type of industry is unique in the conditions and resources it demands from the coastal zone. Each is also unique in the constraints that the environment and other coastline uses impose on it. Beyond that, it is important when discussing industrial uses, to consider the impact that these activities, in turn, have on the coastal zone. In many cases, impacts lead to further constraints. For example, public concern over the effects of nuclear power plants along the coastal zone has resulted in real political constraints for companies desiring to build new plants.

A study made by the Industrial Economic Research Division at Texas A&M University (Wright and Mathews, 1971) for the Gulf Coast is presented in Table 5.1. The Commission on Marine Science, Engineer-

Table 5.1. Factors Affecting Industrial Location

Firms	Rank of Location Factor
Petrochemical	1. Near raw materials
	2. Good transportation
	3. Skilled labor supply
	4. Land availability
	5. Water supply
Metals and machine tools	1. Near product markets
	2. Good transportation
	3. Unskilled labor supply
	4. Skilled labor supply
	5. Land availability
All others	1. Unskilled labor supply
	2. Near product market
	3. Skilled labor supply
	4. Land availability
	5. Office and plant space availability

Table 5.2. Existing and Future Industries

	Type	Examples
Existing industries	Mature, healthy and growing	Oil and gas, food processing and packaging, chemical extraction from seawater, mining of sand, gravel, sulfur, fisheries, surface marine recreation
	Early stages of growth	Desalination, bulk and container transportation systems and associated terminals, aquaculture, fresh water and estuarine underwater recreation
	Mature but static or declining	Most segments of fishing, shipbuilding, merchant shipping
Future industries	Near-term promising (where near term is less than 15 years)	Mining of placer minerals, oils and gas beyond the continental shelf, continental-shelf sand and gravel
	Long-range	Sub-bottom mining (excluding sulfur), aquaculture, open ocean deepwater mining, power generation from waves, currents, tides, and thermal differences

ing and Resources (1969) has presented the information in Table 5.2 on existing industries in the coastal zone and the likely industries of the future.

In view of the probable expansion of industry with its associated influx of people and housing developments, it is essential that the coastal states in conjunction with the federal government formulate on a bilateral or regional basis an adequate coastal zone plan to provide for appropriate housing, commercial, industrial, and recreational development.

5.3.1 Energy Needs

Basic to any industrial expansion is the need for energy. The source of this energy is oil, gas, coal, flowing water, and the atom. In the future, we can expect that geothermal energy, synthetic oil, and synthetic gas will assist and, hopefully, that solar energy and fusion reactors will become practical.

In 1950, 34×10^{15} Btu (British thermal units) of energy were used in the United States; this increased to 45×10^{15} in 1960 and 68×10^{15} in 1970 (National Academy of Engineering, 1972). The respective annual increases were 2.8 percent and 4.3 percent. Thus, the United States may consume as much as 220×10^{15} Btu by the year 2000. Electric power generation, which has consumed about 25 percent of the total energy for production, shows a more striking trend. Maximum estimates based on a 7 percent rate of growth per year indicate an increase from 1.6×10^{12} kWh (kilowatt hours) in 1970 to 12.5×10^{12} kWh in 2000. The Federal Power Commission has projected a somewhat lower figure, as shown in Table 5.3 (National Academy of Engineering, 1972).

It is difficult to determine immediately just how much electricity will be consumed in the coastal zone, but considering that about 70 percent of the population resides there some 1.0×10^{12} kWh was consumed in 1970, and a predicted 8.7×10^{12} kWh will be used in the year 2000. In 1950 the power demand for the state of California was about 5×10^{6} kW (kilowatts); in 1970 this increased to 31×10^{6} kW; by the year 2000 it is expected to increase over threefold (EQL, 1971).

This increase in the national electrical energy demand will require an investment of about $250 billion in power generating equipment over the next 20 years (EQL, 1971). A large percentage of this money will be spent in the coastal zone and in areas of high population density, which implies the estuarine regions. In 1969 there were over 86 fossil-fuel power plants on the East Coast of the United States pouring heated effluent into the coastal waters (Sorge, 1969), and 32 on the West Coast (Adams, 1969). In addition, nuclear plants are in operation and many more are planned for the future.

The average size of power plants using fossil fuel and taken out of use from 1961 to 1965 was about 20 MW (megawatts). The average fossil unit now in operation is about 300 MW and the evidence indicates that the nuclear units now averaging 700 MW must increase to at least 1000 MW to be economically feasible in the future (Bader, Roessler, and Voss, 1972). The amount of heated effluent discharged into the coastal environment presently poses some serious problems; these will

Table 5.3. Electric Production and Use (10^{9} kWh)

Year	Generation	Loss	Residential	Commercial	Industrial	Other
1950	329	48	70	52	142	17
1960	755	72	196	115	345	27
1970	1,530	139	448	313	572	58
1980	3,075	275	880	650	1,110	160
1990	5,825	528	1,590	1,325	2,015	370
2000	10,000	900	2,640	2,365	3,370	725

increase at an accelerated rate. For example, in Florida, at Turkey Point some 50 miles south of Miami, two fossil-fuel plants are now in operation (Florida Power and Light, 1970), producing about 860 MW and using cooling water at the rate of 1270 cubic feet/second. In the near future two nuclear units are expected to become operative producing an additional 1,520 MW. The total effluent discharge will be about 4250 cubic feet/second at temperatures up to 4°C above ambiant (Bader and Roessler, 1972). Similar operations are proposed for the West Coast and the northeast.

A few years ago it was estimated that the amount of water withdrawn for steam electric cooling was 40 trillion gallons per year, roughly 10 percent of the total water flow in the United States (Parker and Krenkel, 1969). This water is not "consumed"; a major portion or all is returned to the river in a modified condition. If our power demands increase at the predicted rate—an approximate sixfold increase (National Academy of Engineering, 1972)—240 trillion gallons per year would be used for power production in the year 2000. This does not consider increased efficiency. If the prediction is realistic, the water utilization will amount to more than 50 percent of the estimated flow of water in all U.S. rivers and streams. However, much of the water will come from the sea.

Power plants locate along the coast because of the availability of cooling water and the dilution capacity of the adjacent water body for waste heat. The power plants that are encountering the greatest number of constraints in operating along the coastal zone are steam generating plants run by fossil fuel or nuclear energy.

There are some general constraints operating on any power company wanting to build along the coast which are shared by other waterfront industries. All confront problems of zoning, which prescribes restrictive use of the coastal zone for specified purposes; availability of sufficiently large and well-suited locations; reasonably priced land not subject to marine-related catastrophes such as storms and floods; and the capacity of adjacent land or water to accept and assimilate or dilute waste products such as process or cooling water and industrial or domestic wastes.

The primary waste product of power plants is heat, which has traditionally been disposed of by running cooling water through the plant and then diluting this hot water in adjacent water bodies. Discharged cooling water is usually about 10°C (18°F) warmer than the receiving water. This warm water has a variety of ecological effects, depending on the ecosystem and the natural temperature range.

Many investigations have been conducted on the ecological effects of thermal effluents. California has at least 23 power stations discharging into coastal zone waters (West Region Task Force on Generation, 1968). Adams (1969) discusses the plants at Morro Bay, Humboldt Bay, and the Sacramento River. The Patuxent River estuary in Maryland has received considerable attention (Cory and Baumann, 1969; Heinle, 1969). Donis and Snyder (1967), Snyder (1968), and Coutant (1969) concerned themselves with thermal effects on salmon in the Columbia River. An intensive study in tropical to subtropical southeastern Florida has been underway for over four years. Here it has been demonstrated that tropical marine organisms are living precariously close to their upper thermal limit. Actually, they are under natural heat stress during the summer months, and thus are very susceptible to thermal effluents (Thorhaug, Stearns, and Pepper, 1972). Additional work in this estuarine system has demonstrated the sensitivity of tropical organisms to elevated temperatures (Wood and Zieman, 1969; Bader, Roessler, and Thorhaug, 1971; Reeve and Cosper, 1971; Thorhaug, Devany, and Murphy, 1971). Thus in tropical regions, no summer discharge of waste heat into an estuarine or coastal ecosystem can be tolerated by the ecosystem; and this mode of heat disposal for the power plant must be eliminated if the natural ecosystem is to be preserved. In temperate regions, however, discharge of the warmer water may be less damaging, especially if it is done so that there is rapid dispersion and transport of the heat away from the source.

Based on these considerations, the environmental constraints on power plants on tropical waters are much more severe than in temperate or cold regions. Alternative methods of cooling, such as lagoons, canals, and cooling towers, will probably have to be used in the tropics. High air temperatures and high humidity, however, impose additional limits on the efficacy of these methods of direct transfer of heat to the atmosphere.

In temperate or cool regions, waste heat disposal in water is less of a problem, but still requires careful selection of the site. The intake must be located so that warm water from the thermal plume is not short circuited because of prevailing currents and tidal circulation. Maximum dispersion and mixing is necessary to reduce the temperature of the plume as rapidly as possible. Therefore, power plant outfalls into closed bays and lagoons should be avoided. Oscillating currents, as in tidal estuaries, can result in the return of the warm water plume to the source and further heating of the same mass of water.

There are other ecological effects of warm water plumes. For instance, migratory fish can be blocked by thermal barriers. Therefore, power plants must be sited to avoid setting up such barriers in rivers or estuaries.

There may be technological ways to turn the problem of waste heat into a benefit. One of the more attractive possibilities is using waste heat for central heating in cities. This is a feasible alternative. Geothermal heat is used routinely in Iceland for this purpose, and a section of Chicago was heated by waste heat from a coal-fired power plant in the early 1900s.

Nuclear power plants, because of their larger size and lower thermal efficiency, have a higher heat production and hence offer the most attractive possibilities for waste heat utilization. Ideally, this low-grade heat is best used in the cities for space heating of buildings or for snow melting. Yet, because of the risks of nuclear accident, these power plants cannot be sited in urban areas. In short, the potential economic and environmental benefits of waste heat utilization cannot yet be realized because of engineering constraints.

The radioactive hazards associated with nuclear plants are coming under renewed scrutiny by scientists. The questions of low-level radioactivity normally discharged from nuclear power plants, as well as the risks of accident due to cooling water failure, are the subject of major public and scientific controversy (Gillette, 1971a,b). Until these questions are satisfactorily resolved, continued delays on plant siting due to public reaction and legal actions are inevitable.

Power plants, especially nuclear ones, must be as free as possible from the hazards of earthquakes or flooding. The site must be away from known seismic or earthquake centers, and it must be protected from the waves and surges of hurricanes or other severe storms at sea.

Even where the environmental needs of satisfactory land, satisfactory provisions for waste heat disposal, freedom from natural catastrophes, and seclusion from urban areas can be met, the power industries are still faced with a long and involved permit process. They must get permission from various agencies and levels of government before a

The Coastal Zone Workshop recommends
Research on the environmental social, economic, and legal effects of:
a. Siting, construction, and operation of coastal and offshore power plants and deep ports
b. Dredging and deposition of spoil

plant can be built. Following this, they are further faced with possible public reaction and subsequent legal actions. Taken together, these political, legal, and social difficulties represent the greatest constraints acting on the construction and operation of power plants today.

In conclusion, the present demand for electric power exceeds capacity, and the projections are staggering. With little or no population growth over the next 30 years—a nonrealistic assumption—power requirements will be at least 2½ times present consumption. The demand for power in the years ahead and the effect this will have on the entire nation requires that the federal government in conjunction with the states formulate a national energy policy. A plan is essential whereby reasonable power demands can be met, wasteful and unnecessary uses of power can be curtailed, economic development can be planned, and the quality of the environment can not only be maintained but enhanced.

The planning for the development of coastal zone power must involve all levels of government and the public. Guidelines and regional enforcement must come from the federal level; local enforcement from the state; and both the public and the power industry must have a continuous means for expression of needs, requirements, and demands (see Chapter 9).

A system must be devised whereby major research programs can be undertaken. This should be a consortium involving the public, industry, and government. Engineering research on improving the efficiency of the present type plants is essential, as are studies in the utilization of solar and other untapped sources of energy, including fusion reactors. The possibilities of offshore sites should be thoroughly investigated from the viewpoints of engineering, safety, and economics. Extensive investigations on the ecological effects of thermal pollution must be undertaken, both on the acute lethal effects and on those of a sublethal chronic nature, discernible only after years of influence but possibly as important to the survival of species.

5.3.2 Environmental Impact of Petroleum Development

One example of industrial use in the coastal zone is the petroleum industry. Approximately 60 percent of the U.S. refining capacity (7 million barrels per day) is concentrated in the four coastal states of Texas, Louisiana, California, and New Jersey, mostly on the coast. Since the nation's future demand for petroleum must be supplied by offshore production and imports, a considerable expansion in coastal zone refining capacity is anticipated. The facilities for marketing operations are also more prevalent in the coastal zone since they are necessarily located near population centers. The coast is also convenient for waste disposal. (See Chapters 3 and 7 for discussion.)

The environmental risks resulting from tanker operation in the coastal zone relate primarily to the potential discharges of oil due to the intentional or operational pollution resulting from tank cleaning, deballasting, and bilge pumping, or to the accidental or episodic discharges resulting from tanker collision, stranding in the open oceans, and restricted port approaches, or to spills in port due to hose failures, overfill of cargo tanks, and malfunctions of the cargo transfer (see Table 5.4).

After discharging cargo at a refinery, a tanker will take in seawater to facilitate handling at berth, to ensure proper screw immersion, and to provide suitable seakeeping characteristics. This oily ballast must be disposed of prior to arrival at the loading port. Unless properly handled, this operation could result in all of the residue oil from the cleaned

Table 5.4. Estimated Direct Petroleum Hydrocarbon Losses to the Marine Environment (not including airborne hydrocarbons deposited on the sea surface)

	Millions of tons				
	1969	1975		1980	
		Min.	Max.	Min.	Max.
Tankers	0.530	0.056	0.805	0.075	1.062
Other ships	0.500	0.705	0.705	0.940	0.940
Offshore production	0.100	0.160	0.320	0.230	0.460
Refinery operations	0.300	0.200	0.450	0.440	0.650
Oil wastes	0.550	0.825	0.825	1.200	1.200
Accidental spills	0.200	0.300	0.300	0.440	0.440
Total	2.180	2.246	3.405	3.325	4.752

Source: Revelle, Wenk, Ketchum, Corino (1971).

tanks and approximately 15 percent of the residue oil from the tanks which were initially ballasted at the unloading terminal being pumped overboard. The amount of oil pollution that can result from this operation is termed "clingage" and can amount to 0.3 to 0.4 percent of the cargo. Many tankers do not function in this manner but have a "slop tank" which allows for a separation of water and oil, with the water being drawn off the bottom, thus leaving the oil behind to be combined with the next cargo. However, all ships have not adopted this system and should be encouraged to do so. Tankers also contribute to oil pollution via minor leakage, bilge pumping, and spills during bunkering, cargo handling, and accidents.

Oil terminal operations can result in loss of oil to the sea by many means. Some means are available to minimize the occurrence and magnitude of such spills and to effect their clean up. The experience of the Milford Haven Oil Terminal in Britain indicates what a well-equipped, well-operated terminal can accomplish. The spillage rate at Milford Haven has been reduced to 0.00004 percent of the total handling capacity in recent years, although a single spill could increase this average by at least an order of magnitude (Soros Associates, 1972). This figure is down from a 1967 high of 0.0014 percent which was mainly due to spillage from the damaged *Chrissi P. Goulandris* (Dudley, 1970). According to Cowell (1971) surveys reveal no widespread persistent damage to the fauna and flora at Milford Haven due to oil terminal outfalls. However, evidence from Buzzards Bay, Massachusetts, shows that a fuel oil spill had severe and long-lasting effects, and there is evidence of petroleum hydrocarbons in many organisms from the coastal zone (Sanders, 1972; Blumer, 1972; Burns, 1971; IDOE, 1972).

Accidental discharges of quantities of petroleum and petroleum products are environmental risks and spills of crude oil are unsightly on beaches and are detrimental to natural populations. Spills of refinery products may be even more damaging. Some evidence suggests that offshore oil terminals could produce minimum harm to the coastal marine environment if properly designed, equipped, and operated. Navigational safety is one of the most significant aspects in the abatement and control of pollution in the coastal zone from tankers and bulk carriers. Navigational aids, navigating equipment on board, traffic management, and training of personnel are basic areas to be considered in improving navigational safety.

With regard to petroleum development, the environmental impact in offshore areas is somewhat different than in coastal marsh and estuarine areas. In areas of intensive development, such as offshore Louisiana, the physical presence of the more than 1700 platforms (St. Amant, 1971) has an impact, especially on commercial fishing activities. These structures have not preempted commercial fishing, but they have required closer attention to navigation by fishing vessels. Platforms used in offshore producing operations provide artificial reefs which concentrate marine life around the platforms so that sport fishing is enhanced in such areas.

Offshore platforms pose aesthetic problems when located near urban areas or off coasts of scenic beauty and high recreational value. In some places, such as off Long Beach, California, artificial islands have been substituted for platforms and developed so as to shield petroleum operations from view. In some other places, the aesthetic problem has been solved by completing offshore wells entirely underwater on the ocean floor, thus eliminating structures that extend above the ocean surface. However, the hazard to fishing trawlers still exists.

The chief potential hazard of offshore drilling and producing operations is that of major oil spills resulting from blowouts. Two of the most publicized oil spills in recent years were the wreck of the *Torrey Canyon* and the Santa Barbara oil well blowout. Both involved crude oil which reached the beaches in variable amounts some time after release. In the Santa Barbara spill many birds were killed and entire plant and animal communities in the intertidal zone were killed by a layer of encrusting oil which was often 1 or 2 centimeters thick (Holmes, 1969). At locations where the oil film was not so obvious, intertidal organisms were not severely damaged (Foster, Neushul, and Zingmark, 1970). Straughan (1971), in summarizing various biological studies, concluded that little damage to fauna and flora occurred in the Santa Barbara Channel due to the direct toxic effects of the oil.

The most serious effects of the Santa Barbara spill were the damage to birds, patchy destruction of benthic plants and animals in the intertidal zone, and contamination of beach and shore facilities (Straughan, 1971). Recovery of the intertidal forms began a month after the spill (Anderson et al., 1971) and had apparently returned to normal less than a year later though there were no behavioral, physiological, chemical, or biochemical studies. Studies of fish and mammals (Brownell and LeBoeuf, 1971) concluded that they had suffered no observable damage from the oil spill. Nonetheless, oil spills resulting from producing operations are a serious matter, and everything possible should be done to prevent them and to contain and remove the oil before it can reach shore should such accidents occur in the future.

In the case of the wreck of the tanker, *Torrey Canyon,* the deleterious effects have been attributed more to the detergents and dispersants that were used than to the oil itself (Smith, 1968). A large number of intertidal organisms and organisms on the beach were killed but it was impossible to distinguish between the effects of the oil itself and of the methods that had been used in an effort to control the oil pollution.

A relatively small oil spill in West Falmouth, Massachusetts, has been studied by both biologists and chemists since it occurred. On September 16, 1969, an oil barge, the *Florida,* was driven ashore in West Falmouth, Massachusetts, during a stormy night. Between 650 and 700 tons of No. 2 fuel oil were released into the coastal waters. Studies of the biological and chemical effects of this oil spill are still continuing (Hampson and Sanders, 1969; Blumer, Souza, and Sass, 1970).

Many fish, shellfish, worms, crabs, other crustaceans and invertebrates were affected. Bottom-living fish and lobsters were killed and washed ashore. Dredge samples taken in 10 feet of water soon after the spill showed that 95 percent of the animals collected were dead and others were moribund. Much of the evidence of this immediate toxicity disappeared shortly after the spill either because of the breaking up of the soft parts of the organism, burial in the sediments, or dispersal by water currents, so that most of the dead animals had disappeared within a few days. Careful chemical and biological analyses revealed, however, that the damaged area has not only been very slow to recover but the areal extent of the damage has been expanding with time.

The catastrophic ecological effects of the oil spill in West Falmouth appear to be more severe than those reported from other oil spills such as those of the *Torrey Canyon* wreck and the Santa Barbara blowout. In the former case, No. 2 fuel oil was released closer to shore than either the *Torrey Canyon* or the Santa Barbara accidents which released crude oil into the oceans. The differences in the character of the oil as well as the proximity to shore may have contributed to the more dramatic effects of the former. It is becoming clear, however, that any release of oil in the marine environment carries with it the threat of disastrous destruction of the marine ecosystem and constitutes a danger to world fisheries.

In addition to the potential hazard from oil spills, other problems may occur when petroleum development takes place in the shallow inshore coastal zone. In Louisiana, for example, where over 25,000 wells are operated in one of the most productive fish and wildlife habitats in the world, several types of ecological disturbances have been associated with petroleum industry activities (St. Amant, 1971). These are (1) mechanical, physical and navigational problems associated with structures; (2) seismic methods used in oil exploration; (3) direct and indirect mechanical, hydrological, and physical effects of inshore shallow water oil production which result in adverse ecological and environmental effects; and (4) physical and ecological effects of oil production activities in unstable marshlands.

Prior to the 1950s, little was known about the effects of these activities on the envrionment. Since that time, however, marine research programs and certain administrative procedures designed to protect the coastal environment have been developed, but many problems associated with oil development remain to be solved. Much of the research to date has necessarily dealt with short-term effects of industrial activities on coastal ecology and natural resource production. There is need now to consider the effects of chronic low-level exposure.

5.4 Governmental Uses of the Coastal Zone

Governmental use of the coastal zone is not now an urgent problem and is not likely to increase in the future. In some cases, it has been beneficial in preserving large areas of natural habitats. Governmental uses of the coastal zone include military installations; space stations, both NASA and military; and miscellaneous minor uses, such as for immigration or quarantine stations. On the Atlantic coast, all of these uses combined occupy less than 200 of the 4300 miles of coastline (Annex University Center for the Environment and Man, 1971). The reports of the Commission on Marine Sciences, Engineering and Resources (1969) tabulate a total of 21,724 miles of United States shoreline. Access to about 581 miles of this coastline is restricted, 2.7 percent of the total. The states of California, Florida, and Maryland have a combined restricted coastline of 335 miles or 57 percent of the total restricted shoreline in the country.

Military installations in the coastal zone include those within existing commercial harbors, such as navy yards, small training facilities, operational bases, and repair facilities, and some in more remote areas, such as extensive defense installations using shoreline areas for development of weaponry, test ranges, test sites, and operational activities. Berthing of ships and logistics operations have similar impact and are subject to the same constraints as civilian shipping and commerce. Some national defense activities, such as radar sites, anti-ballistic missile sites, and undersea sound surveillance, generally have minimal environmental impact, and by virtue of their security from outside interference, they often provide long-term sanctuaries for natural ecological communities. Test sites and target ranges, however, are exclusive uses, at least at times. Their impact on the environment can be locally destructive as in the case of island bombing and gunnery ranges. The requirements for such uses are that they be isolated from human habitation and other coastal zone activities.

These installations frequently involve vast acreages, and are generally restricted from use by the general public. Elgin Air Force Base near Panama City, Florida, for instance, extends for a distance of about fifty miles along the coast and to a distance of about thirty miles inland. While much of this area is closed to the public, there are large areas within the reservation that are made available for regulated hunting and fishing. Sound forestry practices are sometimes followed in the growth and management of the forest resource, while other areas are held intact and managed as wildlife refuges. Beaches are usually restricted from public use except for certain areas that are specifically developed for public recreation. At some installations used for shore-based radar security systems, coastal defense systems, and anti-ballistic missile sites free access is denied to the public, since it might impair national defense security or endanger the people who enter the area.

The present trend is to reduce the size of military holdings; but though some areas have been transferred to federal, state, and local recreational agencies, others are retained in the restricted-use category. Much of this land has remained in a relatively inviolate state; and in some cases, it is the only remaining segment of undeveloped beach in a long series of high-rise apartments and condominiums. These regions can serve as a source of base-line data and information relative to beach processes for application to investigations of developed areas where no

The Coastal Zone Workshop recommends
That the federal government establish a national coastal zone management program, which should be vested in one of the existing federal agencies and should coordinate all agencies involved in coastal zone activities. The federal agency should administer grants to state coastal zone programs and set appropriate guidelines for such programs as well as for the management of federal coastal lands.
The federal and state governments, acting together, create regional councils to assist in carrying out the national coastal zone policy. Such councils would work in concert with federal and state agencies in advising on regional problems of national interest and implement appropriate policies where consensus exists between federal and state governments.

recognizable natural characteristics remain. They may also serve as areas of refuge for coastal flora and fauna, as well as for animals, such as turtles, which require undisturbed beaches for egg laying. In areas of prime recreational demand, such as the Hawaiian Islands, there are existing recreational beach facilities that are suffering under extreme public-use pressures; whereas only a few miles away, there are many miles of equally suitable beaches within military installations. Under such conditions, the military should be responsive to local and state planning agencies regarding future conversion of this property to other uses when it is not required for military purposes.

References

Adams, J. R., 1969. Ecological investigations around some thermal power stations in California's local waters, *Chesapeake Science, 10*(3): 145–154.

Anderson, E. K., Jones, L. G., Mitchell, C. T., and North, W. J., 1971. Preliminary report on the ecological effects of the Santa Barbara channel oil spill, *Biological and Oceanographic Survey of the Santa Barbara Oil Spill 1967–1970,* vol. 1 (Los Angeles: Allan Hancock Found., Univ. of Southern California), pp. 4–5.

Annex University Center for the Environment and Man, 1971. North Atlantic regional water resources study (Hartford: Annex Univ. Center for the Environment and Man).

Bader, R. G., Roessler, M. A., and Thorhaug, A., 1971. Thermal pollution of a tropical marine estuary, *Report of FAO World Conference on Pollution in the Ocean,*

Dec. 9–18, 1970 (Rome: Food & Agricultural Organization of the U.N.), F.A.O. fisheries report no. 99.

Bader, R. G., and Roessler, M. A., 1972. An ecological study of South Biscayne Bay and Card Sound. U.S. Atomic Energy Commission, annual report.

Bader, R. G., Roessler, M. A., and Voss, G. L., 1972. Power production, heat dispersion and possible solutions, *Symposium on Biological Effects of Electrical Power Generation* (Miami: Florida Academy of Sciences).

BCDC, 1969. San Francisco Bay Conservation and Development Commission, San Francisco Plan, January 1969.

Blumer, M., and Sass, J., 1972. The West Falmouth oil spill II. Chemistry (Woods Hole, Mass.: Woods Hole Oceanographic Inst.), technical report no. 72–19.

Blumer, M., Souza, G., and Sass, J., 1970. Hydrocarbon pollution of edible shellfish by an oil spill, *Marine Biology, 5:* 195–202.

Brownell, R. L., and LeBoeuf, B. J., 1971. California sea lion mortality: natural or artifact? *Biological and Oceanographic Survey of the Santa Barbara Channel Oil Spill 1967–1970,* vol. 1 (Los Angeles: Allan Hancock Found.: Univ. of Southern California), pp. 287–306.

Burns, K. A., and Teal, J. M., 1971. Hydrocarbon incorporation into the salt marsh ecosystem from the West Falmouth oil spill (Woods Hole, Mass.: Woods Hole Oceanographic Inst.), technical report no. 71–69.

Commission on Marine Science, Engineering and Resources (CMSE&R), 1969. *Our Nation and the Sea* (Washington, D.C.: U.S. Govt. Printing Office), including panel reports, vol. 1, 2, and 3.

Cory, R. L. and Baumann, J. W., 1969. Epifauna and thermal additions in the upper Patuxent River Estuary, *Chesapeake Science, 10*(3): 210–217.

Coutant, C. C., 1969. Temperature reproduction and behavior, *Chesapeake Science, 10*(3:) 261–274.

Cowell, E. B., 1971. Some effects of oil pollution in Milford Haven, U.K., *Proc. Joint Conference on Prevention and Control of Oil Spills* (Washington, D.C.: Amer. Petroleum Inst.), pp. 429–436.

Donis, G. E., and Snyder, G. R., 1967. Aspects of thermal pollution that endanger salmonoid fish in the Columbia River (Seattle: U.S. Bureau of Commercial Fisheries).

Dudley, G., 1970. Milford Haven Conservancy Board, Milford Haven, U.K., personal communication.

Environmental Quality Laboratory (EQL), 1971. People, power and pollution (Pasadena: California Institute of Technology), report no. 1.

Florida Power and Light, 1970. Electric load and capacity (Dade County: Florida Power and Light).

Foster, M., Neushel, M., and Zingmark, R. 1970. The Santa Barbara oil spill II. Initial effects on littoral and kelp bed organisms, *Santa Barbara Oil Pollution, 1969.* F.W.P.C.A. Water Pollution Control Research Series, pp. 25–44.

Gillette, R., 1971a. Nuclear reactor safety: a skeleton at the feast, *Science, 172*(3986): 918–919.

Gillette, R., 1971b. Reactor emissions: AEC guidelines move toward critic's position, *Science, 172*(3989): 1215–1216.

Hampson, G. R., and Sanders, H. L. 1969. Local oil spill, *Oceanus, 15:* 8–10.

Heinle, D. R., 1969. Temperature and zooplankton, *Chesapeake Science, 10*(3): 186–209.

Holmes, R. W., 1969. The Santa Barbara oil spill, *Oil on the Sea,* edited by P. P. Hoult (New York: Plenum Press), pp. 15–27.

IDOE, 1972. Baseline studies of pollutants in the marine environment and research recommendations (New York: IDOE Baseline Conference 24–26 May, 1972).

National Academy of Engineering, 1972. Engineering for resolution of the energy environmental dilemma (Washington, D.C.: Marine Board, National Academy of Engineering).

Oil and Gas Industries, 1971. *Environmental Conservation,* vol. 1 (Washington, D.C.: National Petroleum Council), 106 pp.

Pardue, L. G., 1971. The hurricane season of 1970, *Weatherwise, 24*(1): 24.

Parker, F. L. and Krenkel, P. A., 1969. Thermal pollution: status of the art (Nashville: Vanderbilt Univ. Dept. of Environmental Water and Resources Engineering).

Reeve, M. R., and Cosper, E., 1971. Acute effects of power plant intrainment on copepod *Acartia tousa, Report of F.A.O. World Conference on Pollution in the Ocean,* Dec. 9–18, 1970 (Rome: Food and Agricultural Organization of the U.N.), F.A.O. fisheries report no. 99.

Revelle, R., Wenk, E., Ketchum, B. H., and Corino, E. R., 1971. Ocean pollution by petroleum hydrocarbons, *Man's Impact on Terrestrial and Ocean Ecosystems,* edited by W. H. Matthews, F. E. Smith and E. D. Goldberg (Cambridge, Mass.: M.I.T. Press), pp. 279–318

St. Amant, L. S., 1971. Impacts of oil on the Gulf Coast, *Trans. 36th Amer. Wildlife and Natural Resources Conf.,* pp. 206–219.

Sanders, H. L., Grassle, J. F., and Hampson, G. R., 1972. The West Falmouth oil spill I. Biology (Woods Hole, Mass.: Woods Hole Oceanographic Inst.), technical report no. 72–20.

Simpson, R. H., 1971. The hurricane—perennial but increasing threat to our coast, address before Council of State Governments, Atlanta, Ga., Jan. 28, 1971.

Smith, J. E. (ed.), 1968. *Torrey Canyon—Pollution and Marine Life* (New York: Cambridge University Press).

Snyder, G. R., 1968. Nuclear power (Seattle: U.S. Fisheries Bureau of Commercial Fisheries).

Sorge, E. V., 1969. The status of thermal discharge east of the Mississippi River, *Chesapeake Science, 3*(4): 131–138.

Soros Associates, Inc., 1972. Feasibility study on the evaluation of deep water offshore terminals for the maritime administration (Washington, D.C.: U.S. Dept. of Commerce).

Straughan, D., 1971. Biological and oceanographic survey of the Santa Barbara channel oil spill 1967–70 (Los Angeles: Allan Hancock Found., Univ. of Southern California).

Taeuber, C., 1970. Population trends of the 1960s, *Science, 176*(4036): 773–777.

Thorhaug, A., Devany, T., and Murphy, B., 1971. The effects of temperature on shrimp, *Proc. Gulf of Caribbean Fishery Institute.*

Thorhaug, A., Stearns, R. D., and Pepper, S., 1972. Effects of heat on *Thalassia testudinum* in Biscayne Bay, *Proc. Florida Academy of Sciences* (in press).

Trident Engineering Associates, 1968. Chesapeake Bay study. Report prepared for the National Council on Marine Resources and Engineering Development, Annapolis, Md. 117 pp.

U.S. Department of the Interior, 1970. Fish and Wildlife Service, Bureau of Sport Fisheries and Wildlife and Bureau of Commercial Fisheries, National estuary study, vol. 5 (Washington, D.C.: U.S. Govt. Printing Office).

U.S. Senate Document 91–58, 1970. National estuarine pollution study, 91st Congress, 2nd session.

West Region Task Force on Generation, 1968. Future generation patterns of the west region. West Region Advisory Committee of the Federal Power Commission.

Wood, E. J. F., and Zieman, J. C., 1969. The effects of temperature on aquatic plant communities, *Chesapeake Science, 10*(3): 172–174.

Wright, A. L., and Mathews, W. T., 1971. Economic development and factors affecting industrial location on the Texas gulf coast (College Station: Industrial Economics Research Division, Texas A&M University).

6 Transportation and Coastline Modifications

6.1 Introduction

There are several types of construction activities that result in physical modification of the coastline. These include fills for housing or industrial development, ports, navigation channels and jetties, levees, causeways, seawalls, groins, breakwaters, beach and dune stabilization projects, hurricane protection barriers, and bulkheading. All either protect shore-front property or are necessary for marine transportation.

In the natural state, beaches are not stable, but exist in a state of dynamic equilibrium. Because of the high value of shore-front property, man often interferes with the natural beach and dune migrations in order to achieve artificially a "stable" condition. Among the various means used for such stabilization are the construction of groins, breakwaters, seawalls or sand fences, artificial sand nourishment, and planting beach grasses.

Groins are relatively short structures extending seaward from about the high-water line for the purpose of impounding sand and thus eliminating erosion. Once the impoundment capacity of the groin is filled, sand will move past the groin to feed the downdrift beach. Unless the capacity is artificially filled immediately after construction, erosion will occur on the downdrift beach until natural sand movement past the groin occurs. Several groins are usually required to stabilize a significant stretch of beach. Since groins do affect the local movements of bottom material, they have an impact on the ecology of beach life; however, the impact has been reported to be relatively minor (Cronin, Gunter, and Hopkins, 1971). On the other hand, they are usually aesthetically unpleasing and can be hazardous to swimmers.

Seawalls are barriers designed to protect the land from wave attack erosion. For the most part they are relatively expensive, not particularly aesthetic, and not designed to protect the beach itself. If properly constructed they have no great environmental impact.

Artificial beach nourishment is often used to replace beach materials lost by erosion, either natural or due to man's updrift activities. It is also used for the creation of new beaches. Such action, however, does not eliminate the cause for the erosion, and it is necessary to repeat the artificial nourishment periodically. Nourishment material may be dredged from the estuary or lagoon behind the beach, reclaimed from the impoundment of an upstream structure (usually a jetty), dredged from a nearby navigation channel, or removed from the seabed offshore. Such removal of offshore sands will probably have minor ecological effects, since the sediments contain a relatively low biomass and the system can recover in a relatively short time (Cronin, Gunter, and Hopkins, 1971). Effects of dredging materials from estuaries or navigation channels is discussed in the section on construction of navigation channels. The ecological impact of burying an actively eroding beach with new material will be minor, while the impacts on a relatively stable beach may be significant but probably of short duration. If the new material is not quite clean, the new beach may not be satisfactory for recreation for some time. In addition, the method of placement (particularly, of large pipes) may limit public access to the beach. Constructing sand fences or planting appropriate grasses and shrubs in sand dunes reduces the windblown loss of sand from the dunes, thus stabilizing the dunes and limiting the amount of beach erosion during storm periods.

Hurricane surges can cause very extensive property damage. Many municipal and industrial developments are located in areas subject to inundation from such surges, and thus there is often a desire to provide positive protection. Hurricane surge protection barriers may be located along the coastline at various distances from the shore, depending on the location of the property to be protected. These structures may be unsightly and may cut off the ocean view. To provide economic protection for property along the shore of an estuary, it may be necessary to construct a barrier across the estuary. Such a barrier must have openings for navigation and tidal movement to allow normal circulation within the estuary. Care must be taken in designing the tidal openings to ensure that salinity intrusion, water levels, current patterns, sedimentation rates, and flushing of pollutants are not adversely affected by the barrier. Such alterations in the estuary could have a serious impact on the ecology of the system.

Causeways for vehicle traffic are often constructed across coastal areas. Since solid fill construction is usually cheaper than placing the roadway on piers, many solid fill causeways have been constructed. As with hurricane barriers across coastal waterways, these structures can have serious ecological impacts if they significantly alter existing circulation patterns. North Biscayne Bay in the vicinity of Miami, Florida, and North Tampa Bay on the west coast of Florida are monumental examples.

The Coastal Zone Workshop recommends
The further development of predictive models to aid in understanding the effect of activities and structures upon the coastal zone environment and to improve the management of the coastal zone by evaluating the impact of alternative actions.

Bulkheading for land reclamation is another practice used to extend housing and industrial lands. In some instances this can be justified; however, this practice may interrupt land runoff and eliminate nutrient supply from brackish marsh and mangrove areas. One of the most controversial topics is mangrove areas, where in some regions wholesale destruction has occurred. Before land reclamation is permitted, thorough investigations should be undertaken to assure that valuable coastal ecosystems are not upset.

There is no doubt that with man's ever-increasing use of the shoreline, modifications will be necessary. However, since poorly planned alterations can have serious consequences, they must be prevented by adequate planning and consultation at all levels of government, and adequate investigations must be undertaken to determine the impact of such modifications.

6.2 Shipping and Commerce

Commerce has played a major role in encouraging man to establish cities and industrial centers in the coastal zone. Water transport has historically been the only practical, as well as the most economical, means to move goods and people to developing regions. Once started, cities and industrial centers on rivers, bays, and estuaries develop not only as ocean ports but also as focal points for inland transportation systems. Typically, ocean ports are located at the mouths of rivers which serve a broad inland area. Such rivers have often been canalized and developed as major shallow-draft inland transportation systems.

As railroads developed in the nineteenth century, the gradual slopes up the courses of rivers were also recognized as the most practical routes for trains. Thus both canalization of rivers with locks, dams, and dredging, and the building of railroads represented a large investment in fixed facilities which converged at each ocean port city.

As commercial centers grew at ocean port sites and jobs became available, population grew. Thus commerce increased at ocean port cities both because commodities were needed in inland areas and because the population of the port cities, themselves, required consumable commodities. In short, man's place in the coastal zone developed primarily from ocean commerce and the need for a transfer point to inland distribution systems. Industrial activities and related worker populations subsequently developed at port locations.

As a counterconsideration, there are several constraints of the coastal environment on commerce. Since ocean port cities are typically located for functional reasons—as transportation and commercial hubs—they are usually not planned for compatible and nondestructive development of the coastal zone.

Specific constraints on man's use of the coastal environment for commerce include—

1. Water depth and sedimentation rates to maintain navigation channels of adequate depth;
2. Channel dimensions, length and circuity, to allow for an acceptable level of marine traffic density;
3. Suitable terrain for landside construction of handling facilities well-protected from wave damage and flooding;
4. Maintenance of estuarine life and natural habitats in the face of increased need for filled sites and expanding ports; and
5. Political and social constraints which may be inadequately reflected in laws and regulations but which are often expressed by private and group action.

These constraints will pose even more of a problem in the future, and new ones will arise, particularly in relation to port and channel size, if the projected trends in shipping are realized. It is estimated that the 2 billion tons of shipping afloat now will rise to 5 billion tons early in the 1980s (Brockel, 1972). This projected rise is the result of the new shipping and cargo-handling techniques being developed of which two of the most significant are containerization and the supertanker.

Major economical and operational advantages are gained by conversion to container shipping. The percentage of cargo moved in containers, though presently quite small, is increasing each year. It is estimated that 70 percent of nonbulk cargo could be containerized (CMSE&R, 1969, vol. 1). This technique will require expansion of landside facilities, both in terms of areas and of improved rail and highway access to the loading docks. The economic costs of this expansion may be prohibitive in some older ports where adjacent land is already heavily industrialized or developed for other urban uses. Therefore, the establishment of new and specialized ports or the designation of regional ports to serve these specialized needs must be considered. In either case, whether existing harbors are expanded or new ones built, competition for coastal land will be a serious constraint on future port development.

Supertankers are revolutionizing the oil and gas shipping industry. Originally stimulated by the closing of the Suez Canal, the building of larger and larger tankers has proven very economical. Ironically the success of these giant ships has now rendered the Suez Canal permanently obsolete. At the same time, it has also created a whole spectrum of new problems.

6.2.1 Transportation and Ports

A generalized schematic diagram of a transportation system is shown in Figure 6.1. The ocean ports of this system are generally located in the coastal zone (Gaither and Sides, 1963), though some river ports may be far from the coastline.

Typical commodities that form the major part of world ocean trade include liquid bulk products such as petroleum and liquid chemicals; coal; minerals; grain; and general cargo. Inland transportation is needed to bring the commodity, much of which is produced in the hinterland, to the port, or to distribute the landed cargo to the consumers. Processing or refining may take place either at the production area or on the coast. In the latter case, it may require a major industrial site located adjacent to the ocean port. The ocean port is, thus, a link in the transportation system connecting the coastal zone with the hinterland.

Ports may be located in sheltered natural or man-made harbors, or estuaries inland at the head of dredged canals, or offshore. Ship sizes now span a broad range of deadweight tonnage (DWT) capacities. Typically, general cargo ships serving major ports such as New York, New Orleans and San Francisco have a loaded draft of 25–40 feet and a capacity of 10,000 to 25,000 DWT. A variety of bulk carriers typically ranges from 30,000 to 160,000 DWT capacity. Some of the large crude oil carriers are nearly 500,000 DWT with drafts up to 95 feet (Soros Associates, 1972). These large and deep-draft vessels exceed the depth capacity of most ports, and the 700,000 or 1,000,000 ton tankers of the future will need 100–105 foot depths for navigation (MacCutcheon, 1972). Deepening many existing harbors and channels would involve cutting into bedrock (MacCutcheon, 1972). Providing sufficient depth of channels to existing ports for supertankers would involve major extensions in the length of the offshore portion of the channel, especially

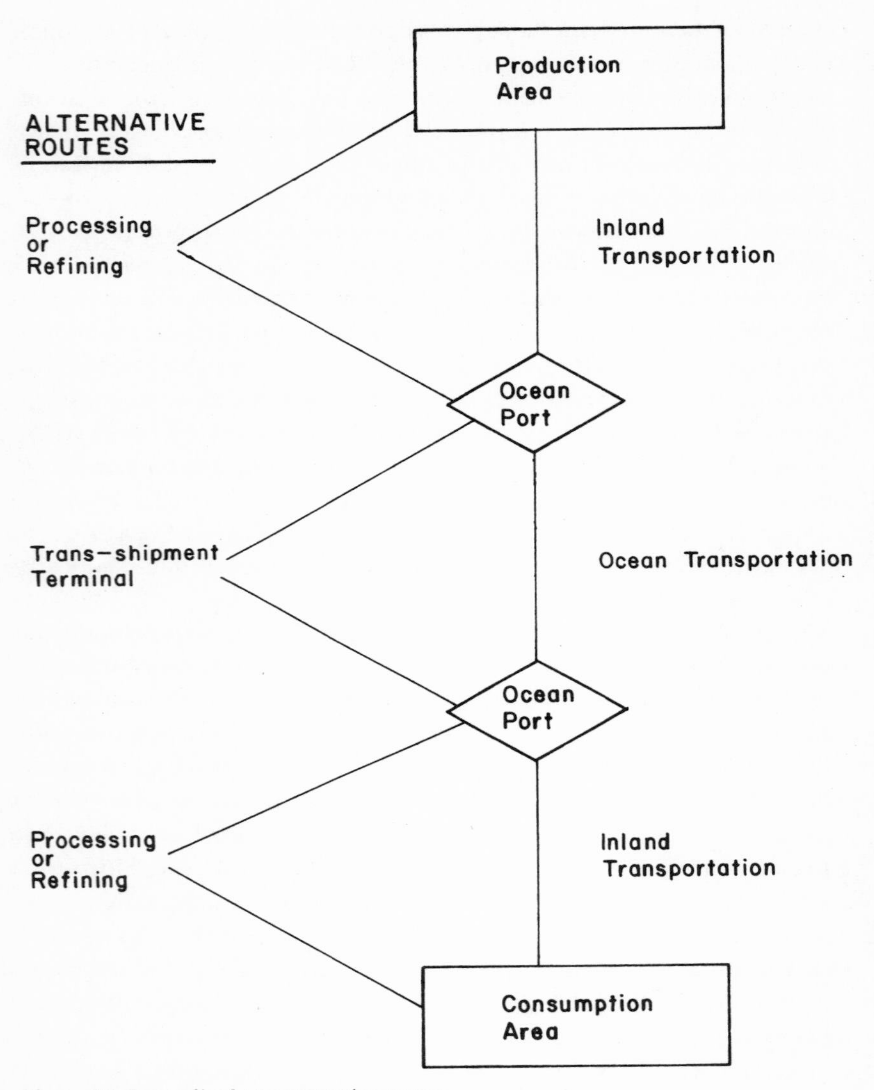

Figure 6.1 Generalized transportation system.

in areas such as the Gulf of Mexico, where the slope of the continental shelf is relatively small. Movement of bed material by offshore currents could result in a high shoaling rate in the channel.

There are other constraints on accommodating these ships in existing harbors, for example, the existence of highway and railway tunnels under harbors or entrance channels, and the need to move or rebuild docks and other directly adjacent landside loading and terminal facilities. The cost of channeling through bedrock or moving tunnels or other structures almost certainly far exceeds the benefits to be gained by giving supertankers access to conventional harbors.

One possible solution to this problem is to build offshore terminals. Though expensive, they are both economically and technically feasible; and their development would not only solve many of the current harbor problems but also would alleviate the competition for coastal land. Large ships can be unloaded or loaded in locations requiring less shelter than smaller ones, and oil and gas could be unloaded offshore and shipped in by pipeline in a largely automated process. By reducing the number of human operations in handling these potentially hazardous materials, the risk of oil spillage or fire is reduced; and thus environmental impact and hazards to human populations are minimized. At present, federal as well as state studies are underway to evaluate alternative East Coast deepwater port sites for crude oil importation. Both the Maritime Administration and the Corps of Engineers are conducting East Coast studies (Gaither and Sides, 1963).

On the high seas, supertankers with 100-foot-draft requirements and weights approaching one million tons constitute potential hazards of major proportions. There are many shallow areas of the ocean, such as the North Sea or the Straits of Malacca, where they cannot travel unless special channels are constructed. Because of their lack of maneuverability and long turning and stopping distance, ordinary navigational hazards such as rocks, reefs, or other ships are of even greater concern. (A 300,000-ton tanker cruising at normal speed requires 5 miles to come to a stop.) A collision or grounding of one of these superships, which may carry 2,225,000 barrels of oil, would far overshadow the effects of the *Torrey Canyon* disaster, where 800,000 barrels of oil were spilled (MacCutcheon, 1972). Navigational precision and improved marking of existing obstacles up to 110 feet below the surface are needed to reduce the chances of such a disaster. The environmental hazards of oil are discussed in Chapter 5.

Some serious political, legal, and social constraints must be considered in addition to those placed on shipping and commerce by the environment and by competitive uses of the coastal zone. For example, though containerization offers major economic and operational advantages, it is vigorously opposed by the longshoremen's labor unions, which see it as a threat to jobs. Loading and sealing these containers

The Coastal Zone Workshop recommends
Public authorities at all levels should consider methods of increasing the carrying capacity of the coastal zone through technical and managerial means, utilizing airspace over land and water as well as submerged areas in order to achieve community goals.

at factories and shipping them to dockside by truck or rail raise further labor questions.

Offshore terminals are also being opposed. Nearby land communities are against them, both because of real or imagined hazards and because of the nearby onshore refining facilities that are needed to process the crude oil. Pipeline easements are difficult to obtain in some states, and it may be necessary to resort to lighters or small tankers to bring the oil ashore. This would be an expensive alternative, both in economic and environmental costs, and in the use of coastal land.

6.2.2 Port Site Selection

Commercial interests seek to optimize efficiency and lower the cost of moving commodities from producer to consumer. When considered only as part of a total transportation system, the location of ocean ports is not an independent decision made on the merits of the coastal site alone. In the case of petroleum transportation systems with long inland pipeline links, the controlling economic factor in realizing the lowest cost is the pipeline length, construction difficulty, and pumping horsepower required (Gaither and Sides, 1963). As a general observation, it can be said that inland transportation is more costly on a unit cost basis than is the ocean voyage link.

In selecting a specific port site, favorable conditions include (1) available real estate; (2) adequate submarine foundations; (3) suitable bottom material for channel dredging, submarine pipeline burial, and minimum sedimentation; (4) anchor-holding capacity of the bottom; (5) adequate shelter; (6) minimum environmental impact; (7) minimum environmental risk to marine and coastal life if accidents occur; and (8) minimum secondary environmental impacts on shore and hinterland areas. An integral part of larger port facilities is the large areas of fill which must be raised to a suitable height above high water to permit continuous operation. Such fill material is dredged from the immediate vicinity of the new construction or is imported from a distant location by barge, scow, or dredge.

For each site, environmental base lines and predictive physical and environmental models are needed. The value of predictive models includes an ability to examine the potential effects of a variety of operational mishaps such as collisions, loading or unloading spills, or failure of storage facilities.

6.2.3 Navigation Channels

The principal types of channel improvements which affect the coastal zone are flood control levees, bank erosion protection works, channel straightening, and lock and dam systems for navigation. These projects can alter salinity, sedimentation, flushing, and dispersion characteristics in the estuary by changing the time distribution of freshwater inflow and the sediment load of the stream (Simmons and Herrmann, 1969). The objective of channel design is to meet the needs of the traffic expected, provide safe depths for the vessels using the waterway with minimum maintenance, and minimize the environmental impact of construction. The depth of the navigation channel must provide for the loaded draft of vessel in salt or fresh water; minimum safe clearance under the keel, including additional draft due to squat, pitching, rolling, or heaving when under way; and the effects of channel depth on ship power requirements. The channel alignment must also permit safe navigation for all vessels.

Natural channels are seldom straight, so that initial channel construction costs for a straight channel would be high, and the channel would probably have a relatively high sedimentation rate and a severe influence on circulation patterns. The combined influences of tidal action, changed current patterns, Coriolis effect, freshwater flow, bank and bottom disturbances, and littoral drift at the entrance will affect the stability and maintenance requirements of the channel. However, straight approaches to the entrance, bridges, locks, barriers, and control structures are required for safety (McAleer, Wicker, and Johnston, 1965).

The federal government has participated in the development of over 500 commercial harbors, and has primary responsibility for the construction and maintenance of the required navigation channels and harbors, including sheltered water areas, but not including terminal facilities. The total federal investment in coastal and Great Lakes harbors and channels through 1966 was about $2.2 billion, which was divided almost equally between construction and maintenance. In addition, there was nonfederal participation in federal projects in the amount of $0.2 billion and nonfederal investments (1946–1965) in port facilities of $2.0 billion (CMSE&R, 1969).

Construction of a navigation channel in the coastal zone can have several effects on the physical and the biological system. If the navigation channel appreciably increases the cross sectional area of a tidal waterway, the upstream extent of saltwater intrusion in the channel will increase. If velocities of flow or volume of the tidal prisms are appreciably altered, the degree of vertical mixing of the fresh and salt waters can be changed, producing changes in the flushing rate and in the distribution pattern of sedimentation. Deepening a navigation channel affects the sedimentation rate (Simmons, 1965) and may permit saltwater seepage into the aquifer and the groundwater supplies of coastal communities. Bottom materials removed from new cuts include every possible substrate, from solid rock through sand and gravel to silts and clays, but maintenance dredging generally involves fine sediments or sand. Channel dredging may remove the original interface between the water and the bottom, which is frequently an area of high biological activity, and deeper substrate materials, which might house burrowing organisms (Sykes and Hall, 1970). The new deep-water areas may affect the distributions of animal and plant populations, and increased upstream intrusion of salt water which can result in upstream migration of desirable or undesirable species. Sediments released into the water may clog the gills of many kinds of animals, reduce photosynthetic production, reduce the buoyancy of eggs, and smother bottom dwellers; and the dissolved or absorbed chemicals released into the water may result in stimulation by nutrients or toxicity by poisons (Cronin, Gunter, and Hopkins, 1971; Davis and Hidu, 1969; Odum, 1963; Sykes and Hall, 1970). These effects vary from ephemeral and insignificant to permanent and highly significant. Although the Corps of Engineers and others are engaged in investigations to identify the ecological impacts of dredging activities, much more work still remains to be done in this field.

6.2.4 Spoil Disposal

Channel dredging necessarily creates spoil to be disposed of, which can be done by barging to sea, disposal on beaches, marshes, and other unconfined areas; or used as fill on land, including diked areas of marshes or wetlands or upland areas. It is estimated that annual open water disposal by the Corps of Engineers in the coastal zone amounts to about 90 million cubic yards of maintenance dredging spoils (CEQ,

1970b). There is an additional annual 50 million cubic yards of new work spoil disposed of in open water, but it is not known how much of this is in the coastal zone. About 67 million cubic yards of maintenance spoil annually is placed in containment areas throughout the nation (Boyd, 1972). The importance of our navigable waterways and harbors requires the disposal of large amounts of channel dredging. However, efforts aimed at reducing the amount of shoaling that occurs in navigation channels would obviously reduce the extent of the problem.

"Open water" disposal may consist of discharging spoil through a floating pipe onto a beach, in a marsh, or in the middle of a tidal waterway; discharging from the bottom of hopper dredges or barges into deeper water; or sidecasting through a boom-supported pipe on the dredge directly into the water body. Each of these procedures will change the bottom or bank-line configuration. If extensive disposal is performed in a given area, current patterns may be altered. If spoil banks or islands are created adjacent to the navigation channel, the altered flow conditions may reduce subsequent sedimentation in the channel. On the other hand, this practice may also adversely affect gross circulation patterns within the estuary; north Biscayne Bay in Florida is an excellent example.

Increased turbidity, smothering bottom organisms, and oxygen depletion are usually of relatively small extent and short term and are considered to be generally insignificant (Cronin, Gunter, and Hopkins, 1971). However, continued use of an area can severely change the morphology and biological values. The deposition of dredge spoil in Laguna Madre covered the grass beds and blue-green algal mats of the shallow flats, thus eliminating the dominant benthic producers. Because of the shallowness of the area and the lack of circulation and flushing, many plankton and fish species disappeared (Odum and Wilson, 1962). In one area of Florida, dredge and fill operations in estuaries alone are estimated to result in biological productivity losses of $1.4 million annually (Taylor and Saloman, 1968).

Since marshes are an important source of food for marine animals and provide nursery areas for many species, covering them reduces productivity of the coastal zone. Destruction of shallow areas may also destroy refuges for birds and mammals. If polluted spoil is released in open water, oxygen depletion can be a serious problem, the deposits may be toxic, and organisms may incorporate heavy metals, pesticides, and other undesirable materials in their tissues (CEQ, 1970b). On the other hand, open water dredging may expose buried shell on which new oyster reefs can develop, and the creation of spoil banks can provide nesting areas for waterfowl. New spoil disposal practices must be investigated. These should include intentional marsh creation and enhancement, planned development and colonization of spoil islands for terrestrial wildlife, and selective spoil disposal to enhance bottom substrate for the improvement of sport and commercial fishing.

Dike disposal areas located within a tidal waterway will obviously alter current patterns, salinity, and sedimentation. Unless the physical changes in the waterway are rather extensive, the significant environmental effects are likely to be the smothering of bottom organisms within the diked area, and destruction of marsh lands. In addition, these areas may be aesthetically unpleasing and produce obnoxious odors.

Upland spoil disposal areas may also destroy valuable wildlife habitats, leach pollutants into the groundwater, and create mosquito breeding grounds. On the other hand, with proper planning, spoil can be used as fill for agriculture or construction. It should be noted that most of the material which will have to be disposed of on land will come from those areas where land disposal sites will be difficult to obtain, such as densely populated harbor areas and ecologically valuable marshes. Future land disposal sites should be selected and developed so that the areas can be used for some beneficial purposes, such as commercial development, recreation area, or wildlife habitat, after they have been filled. The practice of developing confined disposal areas on coastal marshes should be discontinued.

The River and Harbor Act of 1970 authorized the Corps of Engineers to undertake a comprehensive program of research, study, and experimentation to investigate the characteristics of dredge spoil, develop alternative disposal methods, and in cooperation with the Environmental Protection Agency, to study the effects on water quality. This program should be expedited.

6.2.5 Channel Protection

Protective structures may be required adjacent to the navigation channel to reduce wave action and/or sedimentation in the channel. Such structures include jetties, training walls, and contraction dikes. In addition it may be necessary to provide bank protection to prevent erosion by ship waves.

Jetties are constructed in the mouth of a river or the entrance to an estuary to aid in deepening and stabilizing the navigation channel and to provide wave protection. By interrupting the movement of littoral material along the shore, they reduce the sedimentation rate in the navigation channel. At the same time, however, they may reduce the supply of littoral material to the beach on the other side of the entrance and cause an erosion problem on the downdrift beach. Research is underway by the Corps of Engineers to develop improved methods of sand bypassing at tidal inlets to eliminate this latter problem. If jetties are located on both sides of an entrance, too close a spacing may reduce the flow through the entrance, thus reducing the tidal prism of the estuary. Generally, the environmental effects of jetties are minor. They alter the ecology of beach life to a minor degree, but they do serve as places of attachment for many sessile organisms, attract sport fish, and influence the movement of fish and crustaceans into estuaries (Cronin, Gunter, and Hopkins, 1971).

Training walls and contraction dikes are often constructed to alter local flow conditions to reduce sedimentation within the navigation channel. They may thus alter the bottom configuration in the immediate vicinity. Immediately after construction, this may result in areas of localized scour and fill, thus removing or burying bottom habitats. This is usually a short-term effect, but important long-term effects result from the destruction of shallows.

6.3 Upstream Activities

Many upstream activities far from the seacoast have significant effects on the waters of the coastal zone. Pollution problems will be discussed in Chapter 7, but the construction of dams, diversion of river flow from one estuary to another, control of floods, and changes in the character of the landscape such as clearing forested lands will all have indirect effects on the coastal waters.

6.3.1 Effects of Dams

Dams are frequently built for multiple use, including flood control, hydroelectric power generation, water supply, canalization, and recreation. These functions often have secondary effects on the coastal zone, even though the dam is located far upstream (see Figure 6.2). Flood control dams primarily serve to reduce the damage to urban and agricultural areas caused by high-water levels during peak discharge periods, and are often also used to store water for augmentation during low-flow

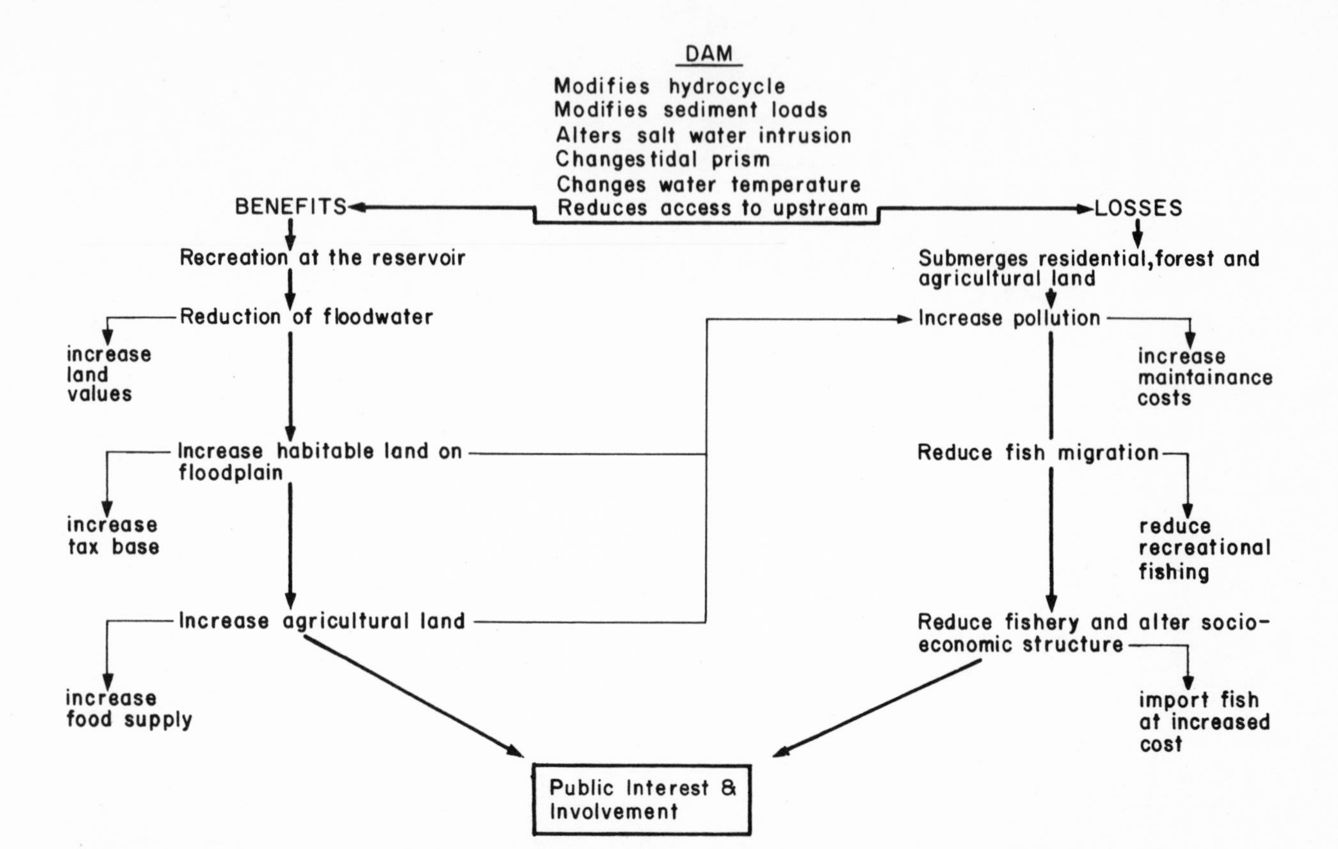

Figure 6.2 Possible primary and secondary impacts of dam construction. Impacts may be detrimental or beneficial but all are interrelated.

periods. The demand for flood protection will continue in the future, although alternatives such as levees and floodplain zoning may be used more extensively. Hydroelectric power generating plants throughout the nation provide only about 4 percent of the nation's energy. Because there are a limited number of suitable sites remaining for future development, the hydroelectric share of the energy market is expected to decline slowly (Humble Oil & Refining Company, 1972). Recreation is usually a secondary function of a dam. Increased leisure time, however, will continue to create demands for reservoirs with related recreation facilities.

When peak freshwater discharges are decreased and minimum flows are increased by flood control structures, the duration of higher than normal discharge may be significantly increased. Since the salinity characteristics of a downstream estuary are a function of freshwater flow, the extent of saltwater intrusion and the degree of mixing of fresh and salt waters will be altered during periods of flow regulation.

A significant portion of the sediments carried by the stream will be deposited in the reservoir behind the dam. This can reduce the sediment supply to the coastal zone, and coastal sedimentation may be reduced. An imbalance between supply and demand may result and promote beach erosion. Sufficient bank and bed erosion may occur downstream of the dam to provide the sediment to the coastal zone, but the nature of the sediments will have been altered.

If water is released from the bottom layers of the reservoir, it will have a different temperature and dissolved oxygen content than the natural stream flow. If the dam is far inland, the effects may not extend to the coastal zone, but it may have a significant effect on the chemical and physical characteristics of downstream areas. The nutrient load and natural flooding of the bank areas which normally supply both nutrients and sediments may be reduced and modify the quality of the water entering the estuary.

Physical model tests of flow regulation at the Tocks Island Dam on salinity conditions in the Delaware River indicated that seasonal maximum salinities in the upstream reaches of the estuary would be significantly reduced, while the seasonal minimum salinities in the lower reaches would be significantly increased. The maximum upstream penetration of seawater containing 2.50 parts per million chlorine* would

* Undiluted seawater contains about 19,000 parts per million chlorine.

be about eight miles farther downstream for the regulated flow than for natural flow conditions (Simmons et al., 1971).

The ecological impacts of a dam are quite varied, and they depend on the nature and degree of the physical changes which occur and the region involved. However, some general statements can be made. Since estuarine organisms are tolerant to salinity and salinity fluctuations in varying degree, some species may be benefited by a change in salinity, while others may be harmed. For example, increased salinity during low-river-flow periods may permit upstream migration of oyster drills, which may reduce the oyster population. The Aswan Dam on the Nile River has reduced the seasonal flow fluctuation to such an extent that organisms that depend on this seasonal cycle are unable to flourish in the river's estuary (Hammerton, 1972). Reduction in the nutrient-rich sediment supply to an estuary reduces the food supply and also causes scouring of bottom habitats. On the other hand, reduced turbidity permits increased photosynthesis and may be beneficial to estuarine biota. Alteration in the composition of the sediment supply may change the bottom composition. A dam may present a barrier to the upstream migration of fish, either by degrading water quality or with the physical structure itself if no provisions have been made for fish passage or if conditions in the reservoir are unsuitable (Cronin, 1967). Among the other coastal zone effects of upstream dams is the increased urban, industrial, and agricultural usage of the floodplain which results from improved protection from flooding. Obviously these new activities would change the quality of the water and thus indirectly cause ecological perturbation in the coastal zone.

FLOW DIVERSION. Diversions of flow may be used to provide municipal water supply, irrigation water, industrial water supply (including cooling water), or flood protection. The diverted water may be subsequently returned to the same stream, diverted to another location within the same estuary, diverted into a different estuary or directly into the sea, or discharged into the atmosphere in cooling towers. For example, major diversions of the Colorado River supply municipal water to Los Angeles and irrigation water to large arid areas in California and Arizona. Mississippi River flood waters are diverted into Lake Pontchartrain and the Atchafalaya River to reduce flooding. Flow may be significantly increased in basins receiving water diverted from another river system. Increased population, industrial usage, and the desire to develop agriculture in arid areas will continue to create a demand for water diversion projects. For example, the Texas Master Plan proposes to divert most of the flow of the Sabine, Neches, Nueces, Trinity, Brazos, and Colorado rivers for irrigation use. There are also plans to divert some of the water of the Sacramento–San Joaquin drainage basin; however, this proposal has been plagued by uncertainty and legal questions.

The physical changes in estuaries caused by diverting flow to another basin are similar to those for reducing peak flow caused by a flood control dam. The effects in this case, however, are permanent rather than seasonal. The reduction in freshwater inflow increases the upstream extent of saltwater intrusion and the degree of vertical mixing of fresh and salt water. Impoundments for irrigation water store water during high-flow periods and may thus cause seasonal salinity fluctuations in the estuary. If the flow is returned to the same estuary through another stream or outfall, salinity patterns within the estuary, the flushing rate, dispersion, and circulation characteristics may be altered.

Charleston Harbor provides an excellent example of the effects of diversion of flow into an estuary (Simmons and Herrmann, 1969). Prior to construction of the Santee-Cooper project of the South Carolina Public Service Authority in 1942, the annual freshwater inflow was 72 cubic feet/second. The annual inflow is now 15,000 cubic feet/second. The system was previously well mixed throughout the entire estuary, and there was essentially no salinity gradient from surface to bottom. Now there is an appreciable salinity stratification, and salt water has been eliminated from the upper reaches. As a result of the density currents which have developed and the increased sediment inflow to the estuary, sedimentation in the navigation channel has increased from about 180,000 to about 10,000,000 cubic yards per year. Solid domestic and industrial wastes are also trapped in the estuary resulting in an increase in harbor pollution.

The permanently altered salinity regime caused by water diversion will create a shift in species composition within the estuary. Lowered peak salinities may permit upstream penetration of undesirable predators or desirable species. Diversion may also disrupt migration patterns because "home-stream" water is not present to stimulate ascent and spawning (Cronin, 1967). The reduced nutrient input may cause a reduction in the feeding grounds of key species and result in an overall reduction of estuarine productivity. As a result of the diversion of Mississippi River flood flow through the Bonnet Carre Spillway into

Lake Pontchartrain in 1937, 1945, and 1950, motile organisms in the lake were temporarily driven out and many sessile forms were killed. Also, large amounts of nutrients were added to the lake; and following return to normal salinities, production of shrimp and other marine life was unusually large. The total effects of these diversions may thus not have been all bad (Gunter, 1953; Cronin, 1967).

As a result of flood control levees along the Mississippi River, runoff is faster and more prone to peaks; increased velocities transport more silt; and flooding of swamps, marshes, and estuaries has virtually ceased. The increase in velocity has increased the rate of coastal zone erosion, and enormous quantities of silt are deposited directly in the Gulf of Mexico. Drainage of nutrients from the land has been reduced, salinity increased and stabilized, and island erosion increased. It is also possible that the bays may be moving inland. The marine organisms appear to be responding to these changes, and some oyster reefs are disappearing (Gunter, 1953, 1957; Cronin, 1967).

6.3.2 Land Use

Land management practices that affect the amount of water lost to the atmosphere by evaporation and transpiration or to groundwater reservoirs will determine the amount of water reaching the river basin as precipitation that will be available for stream flow. Cutting the original forests in colonial days must have greatly modified both the volume, characteristics, and sediment load of most of the major rivers in the country.

The principal effect of clear-cutting forest land is an increase in water flow. Table 6.1 shows the principal drainage systems of the United States together with their annual water yield based on precipitation and runoff data. The humid northeast and coastal Pacific area have more than 50 percent of the precipitation which falls within the watershed areas. Central and southern areas show considerably lower flows, due in part to greater losses by evaporation and transpiration and in part to lower precipitation.

Runoff from urban landscapes varies from 60 to 100 percent of precipitation, while agricultural land yields from 60 to 90 percent of the annual water input. According to Cole (1971), forests cover nearly one-third of the U.S. land area and yield nearly two-thirds of the total water flow. The water yield of forested land is nearly seven times greater than that from nonforested sources. However, runoff from forested land may increase greatly as a result of clear-cutting of forests, which produces rapid runoff in streams and rivers draining clear-cut areas.

Table 6.1. Area, Precipitation, and Average Runoff from Major U.S. Drainage Systems

Drainage system	Area (10^3 square kilometers)	Mean annual precipitation (centimeters)	Mean annual runoff centimeter year	10^9 liters/day	Runoff / Precipitation
New England	153	105	60	255	0.57
Delaware-Hudson	81	100	53	122	0.53
Chesapeake	148	107	47	194	0.44
S. Atlantic	442	123	35	418	0.29
E. Gulf of Mexico	283	130	47	376	0.36
Great Lakes-St. Lawrence	333	97	73	312	0.75
Hudson Bay	156	55	4	17	0.08
Mississippi R.	3224	86	20	1585	0.23
W. Gulf of Mexico	887	67	8	198	0.12
Colorado	671	35	2.7	49	0.08
Great Basin	520	23	2.5	38	0.11
S. Pacific	291	60	30	243	0.50
Pacific NW	668	65	33	604	0.50
Totals	7,857,000 square kilometers		4.41×10^{12} liters/day 1.95×10^{15} liters/year		

Sources: Miller, Garaghty, and Collins (1962); Livingstone (1963).

The total water yield from a deforested landscape may be reduced as a result of increased soil exposure and evaporative losses, but the general response is an increase in runoff and sediment load and a shortening of the response time to precipitation.

Hibbert (1967) showed that forest clearing could increase water yield by 40 percent, but regrowth of vegetation reduced the yield over a twenty-year period. The high yield was reestablished with a subsequent clear-cutting. In this study, there was little increase in turbidity or sediments in the stream draining the area. Large increases in major nutrients in the stream were observed in a similar clear-cutting experiment in New Hampshire (Borman et al., 1968).

Almost all major watersheds in North America show varying degrees of deforestation for man's activities. Urbanization and agricultural uses have the greatest effect on increasing stream flow. As population pressures increase and urbanization activities grow, it can be expected that existing problems of water quantity and quality at the interface between rivers and the coastal zone will be exacerbated. Increased urbanization increases not only the amount of runoff, but the rate of runoff. Since many coastal communities depend upon groundwater for municipal and industrial use, the impact of decreasing upstream groundwater recharge will be most critical in the coastal zone, leading to saltwater intrusion as the downstream hydraulic gradient of groundwater is diminished.

Continued development upstream of the coastal zone will require increased flood control measures. The establishment of impoundments will serve to delay the appearance of runoff peaks at the interface with the coastal zone, and contribute to groundwater recharge downstream from the impoundments, but the quantitative effect of these changes is unknown at present.

References

Borman, F. H., Likens, G. E., Fisher, D. W., and Pierce, R. S., 1968. Nutrient loss accelerated by clearcutting of a forest ecosystem, *Science, 159:* 882–884

Boyd, M. C., 1972. Disposal of dredge spoil, problem identification and assessment and research plan development (Vicksburg, Miss.: U.S. Army Engineer Waterways Experiment Station), unpublished manuscript.

Brockel, H., 1972. The modern challenge to port management (Madison: University of Wisconsin Sea Grant Program), WIS-SG-72-321.

Cole, D. W., 1971. Forest management and agriculture practices, *Impingement of Man on the Ocean,* edited by D. W. Hood (New York: Wiley-Interscience), pp. 503–528.

Commission on Marine Science Engineering and Resources (CMSE&R, 1969. *Our Nation and the Sea* (Washington, D.C.: U.S. Govt. Printing Office), plus panel reports, vols. 1, 2, and 3.

Council on Environmental Quality (CEQ), 1970a. *Environmental Quality, The First Annual Report* (Washington, D.C.: U.S. Govt. Printing Office).

Council on Environmental Quality (CEQ), 1970b. *Ocean Dumping* (Washington, D.C.: U.S. Govt. Printing Office).

Cronin, L. E., 1967. The role of man in estuarine processes, *Estuaries* (Washington, D.C.: American Association for the Advancement of Science), publ. no. 83.

Cronin, L. E., Gunter, G., and Hopkins, S. H., 1971. Effects of engineering activities on coastal ecology (Washington, D.C.: Office of the Chief of Engineers, U.S. Army Corps of Engineers), revised, 48 pp.

Davis, H. C., and Hidu, H., 1969. Effects of turbidity-producing substances in sea water on eggs and larvae of three genera of bivalve mollusks, *Veliger, 11*(4): 316–323.

Gaither, W. S., and Sides, J. P., 1963. Systems approach to petroleum pat site selections, *Proc. IX Conference on Coastal Engineering* (London).

Gunter, G., 1953. The relationship of the Bonnet Carre spillway to oyster beds in Mississippi Sound and the "Louisiana Marsh" with a report on the 1950 opening (Austin: Univ. of Texas), Institute of Marine Science publ. 3, no. 1, pp. 17–71.

Gunter, G., 1957. Wildlife and flood control in the Mississippi Valley, *Trans. 22nd N. Amer. Wildlife Assn. Conf.*, pp. 189–196.

Hammerton, D., 1972. The Nile River—a case history, *International Symposium on River Ecology and the Impact of Man* (in press).

Hibbert, A. R., 1967. Forest treatment effects on water yield, *International Symposium of Forest Hydrology,* edited by W. E. Sooper and W. L. Lull (New York: Pergamon Press), pp. 527–543.

Humble Oil & Refining Co., 1972. Energy: a look ahead (Houston: Humble Oil & Refining Co.).

Livingstone, D. A., 1963. Chemical composition of rivers and lakes (Washington, D.C.: U.S. Geol. Survey), prof. paper 440G.

McAleer, J. B., Wicker, C. F., and Johnston, J. R., 1965. Design of channels for navigation, *Evaluation of Present State of Knowledge of Factors Affecting Tidal Hydraulics and Related Phenomena* (Vicksburg, Miss.: U.S. Army Engineer Committee on Tidal Hydraulics), report no. 3.

MacCutcheon, E., 1972. Traffic and transport needs at the land-sea interface, *Coastal Zone Management: Multiple Use With Conservation,* edited by J. P. Brahtz (New York: John Wiley and Sons), pp. 105–148.

Miller, G. W., Garaghty, J. J., and Collins, R. S., 1962. *Water Atlas of the United States* (Port Washington, N.Y.: Water Information Center, Inc.).

Odum, H. T., 1963. Productivity measurements in Texas turtle grass and the effects of dredging on intercoastal channel, *Publications of the Institute of Marine Science (Texas), 9:* 48–58.

Odum, H. T., and Wilson, R. F., 1962. Further studies on respiration and metabolism of Texas bays, 1958–60, *Publications of the Institute of Marine Science (Texas), 8:* 23–55.

Simmons, H. B., 1965. Channel depth as a factor in estuarine sedimentation. (Vicksburg, Miss.: U.S. Army Engineer Committee on Tidal Hydraulics), technical bulletin no. 8.

Simmons, H. B., and Herrmann, F. A., 1969. Some effects of man-made changes in the hydraulic, salinity, and shoaling regimens of estuaries, *Proc. GSA Symposium on Estuaries* (in preparation).

Simmons, H. B., Harrison, J., and Huval, C. J., 1971. Predicting construction effects by tidal modeling (Vicksburg, Miss.: U.S. Engineer Waterways Experiment Station), miscellaneous paper H-71-6.

Soros Associates, Inc., 1972. Feasibility study on the evaluation of deep water off-shore terminals for the Maritime Administration (Washington, D.C.: U.S. Dept. of Commerce).

Sykes, J. E., and Hall, J. R., 1970. Comparative distribution of molluscs in dredged and undredged portions of an estuary, with a systematic list of species, *Fishery Bulletin, 68:* 299–306.

Taylor, J. L., and Saloman, C. H., 1968. Some effects of hydraulic dredging and coastal development in Boca Grega Bay, Florida, *Fisheries Bulletin, 67:* 213–241.

7 Contamination and Coastal Pollution through Waste Disposal Practices

7.1 Introduction

There is a prevailing opinion in our society that all contaminants introduced into the environment are bad and that waste disposal into the coastal environment is to be avoided. This misleading and unrealistic concept must have clarification in order for society to establish acceptable criteria for management of its resources at the same time it provides for a continuing, healthy environment. Most wastes have a level of concentration which can be reached in the environment without significant alteration to the functional parts of the system; in fact, wastes such as nutrients and, sometimes, heat can be beneficial, if properly interfaced with biological, chemical, and physical features. Wastes added under these conditions should be considered *contaminants,* not pollutants.

Marine pollution has recently been defined by the UNESCO Intergovernmental Oceanographic Commission (IOC) as follows:

Introduction by man, directly or indirectly, of substances into the marine environment (including estuaries) resulting in such deleterious effects as harm to living resources, hazards to human health, hindrance to marine activities (including fishing), impairing the quality for use of seawater and reduction of amenities.

Accepting this definition, we wish to explore the status of our present knowledge of the health of the coastal zone with respect to pollution.

Although the world ocean is a continuous liquid with a long time scale necessary for complete mixing, it is an inhomogeneous solution. The inhomogeneities are caused largely by the energetics of the ocean system, which set up density gradients or layers through which transfer is very slow. These inconsistencies of composition and properties occur when materials are introduced which have a short lifetime in the ocean, such as domestic sewage; which take part in the evaporation and precipitation cycle, such as water and cyclic salts; which have affinity for surfaces, such as polychlorobiphenyls or petroleum; which are particulate and subject to settling, such as plankton and clay minerals; or which exchange with the atmosphere, such as gases. Temperature exchange with the atmosphere also leads to stratification of the water column because of its direct effect on density. Mixing processes in the coastal zone are driven both horizontally and vertically by winds, waves, and freshwater input.

Coastal regions are especially vulnerable to pollution, since much of the waste is discharged directly or indirectly into the shallow coastal waters. Inhomogeneities are frequently great in this area, so that mixing with offshore waters is generally sluggish. The volume of coastal water is only a minor part of the total ocean, but is an important consideration with regard to marine pollution: It is at the shoreline that bathing beaches and other recreational amenities are located; and environmental deterioration of the coastal nursery grounds for many species of commercially important fishes can extend deleterious effects to the fishing industry in larger areas. Some substances of both municipal and industrial origin may cause serious problems if discharged in a nearshore area but if discharged in the open sea might reduce some of the hazards to living resources (Hood, Stevensen, and Jeffrey, 1958). Advantage should be taken, therefore, of the much larger volumes of water available at sea and the fact observed that a substantial portion of our food fish abound on a healthy shelf close to the shore.

7.2 General Concerns about Coastal Pollution

Major concerns about the addition of foreign materials to the coastal waters can be summarized as follows:

1. Concentration of contaminants added to local regions;
2. Rates of mixing and transport from the region;
3. Phase distribution of the contaminants;
4. Specific biology and chemistry of contaminant interaction with the coastal environment;
5. Effect of the contaminant on vital life processes;
6. Rate of biodegradation or geochemical removal;
7. Food-chain concentration factors;
8. Prognosis for amount and kinds of future additions;
9. Other special factors, such as addition of pathogens and aesthetic deterrents.

The total volume of the oceans and the naturally occurring abundance of the contaminant in question must be related to the additions being made by man's activities. Pertinent data on ocean dimensions and rate of additions for land-derived materials are presented in Table 7.1. The amount of ocean available to each living human now on earth is about

Table 7.1. Data Relative to Ocean Dimensions and Natural Addition Rate from Land Sources

Area of oceans	3.6×10^{18} cm^2
Volume of oceans	1.4×10^{24} cm^3
Mean depth of oceans	3.8×10^5 cm
Total discharge to oceans	$\sim 3.6 \times 10^{19}$ cm^3/yr
Total dissolved load supply rate	$\sim 3.6 \times 10^{15}$ g/yr
Discharge per unit area of ocean/1000 years	$\sim 1 \times 10^4$ cm^3/cm^2
Supply of dissolved load per unit area of ocean	~ 1 g/cm^2/1000 yr
Area of ocean/human being on earth	3.6×10^7 cm^2 (0.12 km^2)
Volume of ocean/human being on earth	4.0×10^{14} cm^3 (0.4 km^3)
Area of U.S. continental shelf (to 100 fathoms)	1.8×10^{16} cm^2

Source: Turekian, 1971.

0.12 square kilometers, or 0.4 cubic kilometers. On a total-earth-resource basis this is useful only in lending perspective to the problem of ocean pollution. Full utilization of this resource for disposing of wastes is not practical, except in the sense that mixing and transport processes of the ocean ultimately bring all segments of the ocean toward equal concentration with respect to added materials.

The horizontal and vertical mixing processes in the sea are discussed in detail by Okuba (1971). Mixing results from a multitude of physical processes occurring internally in the sea which tend to produce uniformity in properties. Mixing is essentially irreversible, since without addition of outside energy the mixture cannot return to its original state. Horizontal mixing is rapid relative to the vertical process. Wastes added to the surface waters of the coastal zone will usually be diluted tenfold within 1 or 2 kilometers from the point of addition in a highly stratified system. Much more rapid dilution will occur if the surface layers are well mixed, as is often the case above the thermocline. A knowledge of the rate of mixing and the expected dilution of an added contaminant in solution is fundamental in order to design a system for the discharge of contaminants into the ocean environment. Special consideration must be given to semiclosed estuaries that flush slowly, since high concentrations of pollutants may accumulate. Many contaminants are biodegradable, as long as their concentration remains at nontoxic levels, by indigenous organisms that are able to utilize the contaminant material as a nutrient. Mixing rates and biodegradation rates can then be applied to compute the fate of a given contaminant and its compatibility with a given environment.

The manner in which a contaminant fractionates between phases in the environment has a large bearing on its ultimate fate. Those materials, such as trace metals, that interact with sediments are often removed from the water column and ultimately deposited on the bottom. Materials that are lipophilic in nature, such as PCB (polychlorinated biphenyl) and DDT, may tend to concentrate at the surface of the sea or in fatty portions of organisms. Volatile compounds, such as light hydrocarbons, will partition between the atmosphere and the sea. Gases, such as carbon dioxide, will form a concentration equilibrium between the sea and the atmosphere; whereas atmospheric dusts or added solid materials may interact with the ocean by adhering to the surface, by becoming suspended in the water column, or by settling with sedimentary material. Only those materials that dissolve or form colloids in the ocean are mixed by the same processes as temperature and salinity.

The manner in which the contaminants interact biologically and chemically with environmental components is of importance in their overall effect on the environment. Some compounds are relatively innocuous in themselves, but if easily oxidized they may deplete the environment of dissolved oxygen and thus indirectly stifle the growth of aerobic organisms or even render the system anoxic. Some materials limit photosynthesis by reducing light penetration in the water; other substances may physically cover benthic organisms; and some chemicals stimulate growth of certain unwanted organisms through remineralization at the expense of other more desirable organisms. Overfertilization by sewage or other nutrient sources results in the familiar problem of eutrophication in lakes, streams, rivers, and some estuaries.

Effects of contaminants on vital life processes can often be measured quantitatively, such as inhibition of photosynthesis or respiration, energy conversion, or even decrease in motility. Other deleterious effects are more subtle, more difficult to detect, and if not tested carefully can go unnoticed until a particular species is eliminated. These latter effects may be manifested as reproductive failure, chemotactic interference, or alterations of predator-prey relations. Inhibition of particularly sensitive processes such as photosynthesis or respiration often occurs at contaminant concentration levels below those affecting other vital functions, enabling use of a control technique that sets a maximum tolerable concentration level below which photosynthesis and respiration are unaffected.

Biodegradation and geochemical removal are probably the two most active factors in depleting the oceans of added contaminants. All organic materials are eventually biodegraded, although this process takes place extremely slowly with chemicals such as PCB and other chlorinated hydrocarbons which are not naturally present in the environment. Radioactive nuclides degrade according to their particular nuclear stability in a predictable manner, but they may be removed from the water by geochemical processes in the meantime.

An important index of the potential hazard of a material when released to the ocean is the concentration factor in the food chain. Examples in Table 7.2 show the relative composition of shellfish for certain trace elements as compared to seawater (Brooks and Rumsby, 1965). Toxic elements may be concentrated many thousandfold over the ambient concentration in the environment. This process is well documented for inorganic components and is also important for organic materials, particularly the fat soluble substances. Food-chain buildup of materials added to the environment provides a mechanism to produce hazardous levels of toxic substances that may affect sensitive biota of the ecosystem, including man.

The rapid rate of increase in the production of synthetic organic

Table 7.2. Concentration Factors for the Trace Element Composition of Shellfish Compared with the Marine Environment

		Enrichment Factors		
Element		Scallop	Oyster	Mussel
Silver	Ag	2,300	18,700	330
Cadmium	Cd	2,260,000	318,000	100,000
Chromium	Cr	200,000	60,000	320,000
Copper	Cu	3,000	13,700	3,000
Iron	Fe	291,500	68,200	196,000
Manganese	Mn	55,500	4,000	13,500
Molybdenum	Mo	90	30	60
Nickel	Ni	12,000	4,000	14,000
Lead	Pb	5,300	3,300	4,000
Vanadium	V	4,500	1,500	2,500
Zinc	Zn	28,000	110,300	9,100

Source: Brooks and Rumsby, 1965.

chemicals over the past few decades and their release to the environment by multiple methods of transport make the prognosis for future amounts and types of additions difficult to appraise. Production rates over a 25-year period for one of the better-known environmental contaminants, DDT, is given in Table 7.3. The 1968 production of some commonly used organic chemicals, many of which reach the environment, is shown in Table 7.4. There are hundreds of other synthetic organic compounds produced in varying quantities. Since the quantities of organic compounds produced follow very closely the gross industrial production in the world and since many of the compounds are volatile, it is apparent that the environment will continue to be under ever-increasing stress from additions of man-made organic chemicals.

Table 7.3. Production of DDT in Units of 10^3 Metric Tons/Year (United States only)

Year	DDT
1945	15.1
1950	35.5
1955	59.0
1960	74.6
1965	64.0
1968	63.4

Source: Stanford Research Institute, 1969.

Table 7.4. Production Rate of Some Organic Chemicals

	Annual production (10^6 metric tons)	Boiling point (°C)
Formaldehyde	1.7	−21
Acetaldehyde	0.59	21
Acetone	0.59	56
Dichloro-difluoro methane (freon 12)	0.13	−29
Carbon disulfide	0.34	45
Methylene chloride	0.12	40–41
Diethyl ether	0.050	35
Ethanol	0.86	78.5
Methanol	1.5	65
Cyclohexane	0.86	81
DDT	0.74	—

Source: U.S. Tariff Commission, 1968.

The Coastal Zone Workshop recommends
The creation of regional and national monitoring systems to collect continually chemical, physical, and biological data with a capacity to give advanced warning on conditions that may be hazardous to the ecosystem of the coastal zone.
The monitoring of activities in the coastal zone not only for their effect upon the near-shore waters, but upon the seas and oceans. Chemicals, airborne and waterborne, from the coastal zone, as well as certain drilling, dredging, and dumping, may cause serious harm to the marine environment and should be regulated to avoid serious damage to oceanic ecosystems.

Usage of inorganic chemicals, particularly trace metals and radioactive materials, is also increasing. Most of these compounds are nonvolatile, but they may be transported long distances by winds as particulate matter. Loss to the environment of these less volatile compounds may be more easily controlled, and it is possible that environmental stress from these contaminants can be limited. If a substance is found to cause specific damage, as in the case of DDT, its production and use may be controlled. It is extremely important, however, that the environment be closely monitored to detect damage to an ecosystem before it has become irreversible.

7.3 Specific Pollutants of Particular Concern

7.3.1 Trace Metals

Industrial activity has resulted in increased mobilization of a number of trace elements such as the heavy metals lead, mercury, and cadmium, that are highly toxic to biological systems. Consequently, environmental levels may greatly exceed the concentrations to which organisms are normally exposed in areas where these elements are mined or processed, or where waste materials containing them are discarded. These metals may be released into the atmosphere, resulting in widespread dispersal and fallout in locations remote from the areas of origin and use. The industrial production of potentially harmful elements that reach the environment is given in Table 7.5; of these substances, copper, zinc, and molybdenum are required elements in low concentrations for biological growth but may become toxic in high concentrations. The

Table 7.5. Heavy Metal Production and Potential Ocean Inputs

Substances		Mining production[a] (10^6 metric tons/yr)	Transport by rivers to oceans[b] (10^6 metric tons/yr)	Atmospheric washout[c] (10^6 metric tons/yr)
Lead	Pb	3	0.1	0.3
Copper	Cu	6	0.25	0.2
Vanadium	V	0.02	0.03	0.02
Nickel	Ni	0.5	0.01	0.03
Chromium	Cr	2	0.04	0.02
Tin	Sn	0.2	0.002	0.03
Cadmium	Cd	0.01	0.0005	0.01
Arsenic	As	0.06	0.07	
Mercury	Hg	0.009	0.003	0.08[d]
Zinc	Zn	5	0.7	
Selenium	Se	0.002	0.007	
Silver	Ag	0.01	0.01	
Molybdenum	Mo		0.03	
Antimony	Sb	0.07	0.01	

Source: NAS, 1971a.
[a] U.S. Department of Interior, 1970a,b.
[b] Bertini and Goldberg, 1971.
[c] Estimated from aerosol data of Egorov, Zhigalovskaya, and Malokhov (1970) and Hoffman (1971).
[d] E.D. Goldberg, unpublished data.

other elements vary widely in toxicity, but lead, chromium, cadmium, mercury, selenium, arsenic, and antimony are of greatest concern.

Natural concentrations of some of the trace elements in the oceans are listed in Table 7.6. It is apparent from comparing Tables 7.5 and 7.6 that the total amount in the oceans is substantially greater than that which potentially reaches the ocean annually from man's activities. As indicated earlier, however, the time required for total mixing in the ocean is slow, and local effects, such as contamination of rivers, estuaries, coastal regions, and surface water of the open ocean, may occur.

Another way to evaluate the significance of man's activities in producing heavy-metal pollution is to compare the natural geological rates of mobilization with the man-induced rate. Table 7.7 lists data for twelve chemicals that man cycles more rapidly than nature. Although total amounts of materials differ by several orders of magnitude, the data are shown in order of ratios from man-induced weathering rates to

Table 7.6. Some Heavy Metal Concentrations in the Oceans

Substance	μg/liter	Total in ocean (10^6 metric tons)
Copper	3	4×10^3
Cadmium	0.11	1.5×10^2
Chromium	0.05	7
Iron	10	14×10^3
Lead (estimated natural)	0.03	4
Mercury	0.03	4
Nickel	2	3×10^3
Vanadium	2	3×10^3
Zinc	10	14×10^3

Source: Hood, 1972.

Table 7.7. Rank Order Rates of Mobilization of Materials

Element	Geological rate, *G* (10^3 metric tons)	Man-induced rate, *M* (10^3 metric tons)	Ratio *M/G*
Tin	1.5	166	110.
Phosphorus	180	6,500	36.
Antimony	1.3	40	31.
Lead	180	2,330	13.
Iron	25,000	319,000	13.
Copper	375	4,460	12.
Zinc	370	3,930	11.
Molybdenum	13	57	4.4
Manganese	440	1,600	3.6
Mercury	3	7	2.3
Silver	5	7	1.4
Nickel	300	358	1.1

Sources: Geological rate—Bowen, 1966; man-induced rate—United Nations, 1967.

normal geographical weathering rates, since doubling or quadrupling of an elemental source may have more effect on coastal ecosystems than absolute amounts. In the case of tin, for example, man mines one hundred times the amount that would be cycled naturally, but Table 7.5 shows that only 16 percent of the amount mined reaches the oceans annually.

Increases in heavy metals in river waters have resulted in reduction in salmon stocks in the maritime region of Canada. Merlini (1971) has recorded differences in copper and zinc content of oysters in contaminated and uncontaminated waters in Japan. "Green oysters" contain up to ten times as much copper as oysters from uncontaminated areas. Similarly, three times more zinc is found in oysters from contaminated estuaries than from normal environments.

Increases in the abundance of trace metals in the marine environment are difficult to assess, because so little is known of the natural variations and behavior of many of the elements. This is further complicated by the geochemical dissimilarity of river, estuarine, and coastal regions; by the great organism dependence on trace-metal availability; by the chemical form of the elements in the water column; and by the presence of other antagonistic or synergistic elements. Continued research is needed to work out the mechanisms of heavy metal complexation; the dynamics of interchange between chemical species, the biota, and the sediments; and the effect of one element on the toxicity of another.

Direct input from outfalls or dumping is a major source of trace elements to the coastal areas; however, atmospheric transport is important, especially to the open ocean. Patterson (1971) has shown the concentrations of lead to be ten times higher in surface waters than in the deeper waters of the Pacific Ocean. Most of this transport appears to be atmospheric from aerosols created by automobile exhaust and lead processing industries. It is estimated that atmospheric transport contributes three times more lead to open ocean areas than runoff (NAS, 1971a). If one assumes that other elements are transported by the atmosphere from urban centers in a way similar to lead and are also distributed in the ocean in a similar manner, then the expected increase of specific trace metals in the ocean has been computed as shown in Table 7.8. This analysis allows one to place in perspective the ocean pollution that can result from atmospheric transfer of trace metals from metropolitan areas.

Additional results obtained from broad-spectrum sampling of marine organisms (IDOE, 1972) did not establish increases in levels of lead, cadmium, or mercury in open ocean marine organisms over "natural" levels. In coastal regions, however, the available data suggest that these elements may be substantially more concentrated in areas of severe pollution. The studies are preliminary, however, and it is not always possible to differentiate between the proportion of the metals measured that are man-generated and those naturally present. In almost all areas there is a need for such base-line data against which future accumulation trends might be compared.

Table 7.8. Possible Impact of Atmospheric Pollutants on the Marine Environment

	Concentration[a]				Estimated % increase of trace elements in upper 200 m of ocean[c]	
Element	Open ocean (μg/liter)	U.S. urban air (μg/m³)	Enhancement ratio in air[b]	Ratio air/ocean	High	Most probable
Pb	0.02	1000	2300	50,000	—	—
Al	1	1500	0.5	1500	200	30
Cd	0.02	20	1900	1000	400	20
Sc	0.001	1	1	1000	80	20
Sn	0.02	20	280	1000	—	20
Mn	0.3	200	6	700	60	15
Fe	5	2000	1.00	400	80	8
La	0.01	3	3	300	—	6
V	1	200	42	200	40	4
Zn	3	700	270	200	25	4
Cu	2	200	83	100	100	2
Ag	0.01	2	830	200	—	4
Cr	0.3	40	11	130	20	3
Be	0.005	1	10	200	—	4
Sb	0.2	20	2800	100	10	2
In	0.001	0.1	29	100	—	2
Ti	1	100	0.5	100	4	2
Co	0.03	2	2	70	4	1
Se	0.1	4	2500	40	5	0.8
Hg	0.1	3	1100	30	8	0.6
W	0.1	5	93	50	—	1
Ga	0.02	1	2	50	—	1
Ni	2	30	12	15	4	0.3
Cs	0.3	4	37	10	—	0.3
Ta	0.02	0.2	3	10	—	0.3
As	2	20	310	10	0.8	0.2
Mo	10	10	190	1	—	0.02
U	3	0.1	1	0.03	—	Negligible

Source: IDOE, 1972.
[a] Urban-air values are approximations taken from the compilation of Robertson and Perkins (personal communication, 1972).
[b] The ratio of the concentration in urban particulates of a specific element to iron, divided by the ratio of the same elements in average crustal material.
[c] The percentage increase represents the magnitude by which these trace elements could have been increased in the upper 200 m of the ocean based on the anthropogenic lead concentration in this layer. The high estimates are those which might be expected in oceanic regions adjacent to certain large urban areas (Robertson and Perkins, personal communication, 1972).

Evidence is becoming available that the sink for heavy metals reaching the ocean lies in the estuaries and near-shore sediments (Turekian, 1971; NAS, 1971a). Turekian (1971) noted that small amounts of the trace metals introduced in streams can be found in the water leaving the estuaries. This implies that they become complexed, associated with mineral particles, and settle close to shore. If this is so, it is necessary to learn the actual method of transport of the metals to the ocean bottom; the role played by scavenging precipitates, such as iron hydroxide; the methods for determining rates of sedimentation on a time scale related to human modification; the specific form of retention; the geographical locations of trace-metal enrichment; whether the heavy metals redissolve and, if so, how rapidly. Increased inputs from the domestic and industrial disposal of waste in local areas can be expected to result in environmental levels of contamination that are detrimental to sensitive species of the ecosystem, including man.

7.3.2 Plant Nutrients

Man's addition of plant nutrients, especially phosphate and nitrate, to the ocean is of considerable effect on the quality of the local coastal zone. These nutrients are found in domestic sewage and are also produced by runoff from agricultural areas (Weiss and Wilkins, 1969). In perspective of the world ocean, however, these additions are probably of little significance: it has been estimated that a world population of 6×10^9 injecting their metabolic wastes into the ocean would require over 10,000 years to double the fixed nitrogen compounds present.

The greatest impact of nutrient additions to the coastal zone is seen in excessive growth of non-endemic plants. The resulting reduction in species diversity tends to change the food-chain components for indigenous organisms, frequently increasing the biochemical oxygen demand, lowering the transparency of the water because of excessive phytoplankton growth, and generally lowering the aesthetic, recreational, and commercial value of the area.

Nutrients discharged into the world's drainage system may reach the marine environment or they may be incorporated into bottom sediments in freshwater streams or lakes. They may cycle through several biotic stages on their way through the freshwater system into the estuaries. By either this indirect route or by means of direct transport, a large portion of the nutrients will eventually reach the open ocean to participate in oceanic processes along with the abundant supply of nutrient material already naturally present.

Wide disparity exists in the literature on the amount of phosphorus and nitrogen contributed to the oceans through transport processes. A summary of the available literature is presented in Table 7.9.

The maximum acceptable discharge of nutrients into a confined body of water is extremely difficult to estimate, because the level depends upon the type of organisms that result from the enrichment of the water. If these resulting organisms are utilized in the food chain of higher trophic organisms, then the productivity of the water increases. Conversely, if noxious species that are not effectively utilized by higher organisms accumulate and eventually decompose, secondary pollution or eutrophication results. The concentration of nutrient elements in much of the coastal waters near urban regions of the world is now excessive. The Hudson River estuary receives five to ten times more nutrients than it is capable of assimilating (Ketchum, 1969). Some raw sewage is still discharged into the coastal waters around metropolitan centers, introducing a large biological oxygen demand on the receiving waters. In many cases, even secondary sewage treatment will not solve the problem, but removal of some of the nutrients in a tertiary phase to reduce nutrient concentration or direct discharge into the open ocean through outfalls may be essential to restore the health of the coastal waters; however, great care must be taken so that such open ocean discharges do not create other problems.

The problem of heavy nutrient loads is greatly complicated in the coastal waters by other wastes that may compete for the assimilative capacity of the receiving waters or reduce this capacity through toxic inhibition of metabolic processes and physical interference, particularly through turbidity and silting of benthic habitats. Wastes derived from modern household practices are complex, as shown in Table 7.10. Much

Table 7.9. Nutrients Added to the Oceans in River Discharge and through the Atmosphere (million tons/year)

	Nitrogen		Phosphorus	
Estimated by	River	Air	River	Air
Emery et al. (1955)	19	59	2.	0
Bowen (1966)	78.5	—	0.18	—
Holland (unpublished data)	10	10	0.8	—
Tsunogai and Ikeuchi (1968)	7	210	—	—
Total ocean reserve[a]	920,000		80,000	

Source: NAS, 1971a.
[a] From Holland (unpublished data).

Table 7.10. Average Waste-water Quality during Specific Periods of Three Different Test Homes in the United States

Characteristics of test homes	Home I		Home II		Home III	
Total people	5		5		5	
Children	2		3		1	
Home laundries	1		1		1	
Dishwashers	1		1		1	
Baths	5		3		1	
Analysis	**Without disposer**	**With disposer**	**Without disposer**	**With disposer**	**Without disposer**	**With disposer**
Total solids[a]	866	997	788	859	1249	1536
Suspended solids[a]	363	478	293	360	473	602
Total volatile solids[a]	468	571	414	485	659	942
COD[a]	705	959	540	640	882	1133
BOD_5[a]	542	518	284	356	479	598
Detergent[a]	5.3	23.2	5.2	5.7	6.9	7.4
Total nitrogen[a]	69	63	61	62	121	189
NH_3-nitrogen[a]	53	47	48	41	92	154
Total PO_4[a]	47	40	70	51	65	79
Ortho PO_4[a]	31	36	40	32	40	52
Grease[a]	95	134	33	41	66	92
pH	8.0	7.6	8.0	8.2	8.3	8.4
Flow[b]	388	374	331	258	123	148

Source: Føyn, 1971.
[a] Given in milligrams/liter.
[b] Given in gallons/day.

more complicated wastes are derived from industry. Today and certainly in the near future, these waste loads together demand a massive effort on all parts of society to improve disposal methods and recycling in order to maintain coastal water quality, especially near highly developed urban areas. There are several ways, most of which require research, in which these problems of nutrient additions and associated waste introduction to the coastal zone might be resolved. (1) Better use of the natural flushing, dispersion, and transport systems of the coastal waters is practical in some areas. In certain cases, this may be aided by use of highly developed pipeline dispersers that deliver wastes into suitable current systems to provide rapid transport to the open sea. (2) Additions must be kept within the assimilative capacity of the

The Coastal Zone Workshop recommends
Improved environmental impact statements should be prepared for each new or additional activity and structure in the coastal zone to determine the extent to which they would have social as well as environmental effects. More stringent requirements for the preparation, detail, and use of such statements in making specific decisions should be developed.

system. Better methods of predicting the effect of additions on coastal ecology must be developed by use of mathematical and physical models. (3) Treatment methods which greatly reduce important growth-stimulating nutrients can limit excessive phytoplankton growth, and its deleterious effects. (4) Closed-cycle systems could be developed in which the major blowdown concentrates are disposed of on land or in the deep sea. (5) Nutrients can be balanced by means of addition or deletion, so that desirable food species are encouraged and productivity of seafood harvests are increased. (6) Nontoxic, dilute organic wastes can be sprayed on land, where they can be utilized by plants.

7.3.3 Organic Additions

PETROLEUM. Petroleum is the major raw material for the petrochemical industry and one of the primary sources of energy in the United States. A significant fraction of petroleum is produced in and/or transported through the coastal zones of the world. More than 1000 million tons of crude oil and oil products move through the coastal region each year; a loss of even a small fraction of this material could have a major impact on coastal zone environments (SCEP, 1970; Moss, 1971). Man-made and natural injection of petroleum into the environment is well known. The ever-increasing production of oil from offshore reservoirs (over 8000 offshore wells in the Gulf of Mexico alone) is expected to reach 30 to 40 percent of total oil and gas production by the early 1980s. Some seepage from these wells occurs, but it is estimated to be less (Table 7.11) than 10 percent of the amount lost to the sea by ship operations and less than 1 percent of that lost through vaporization of petroleum products in the course of use.

The hydrocarbons, of which there are about 350 specific compounds in crude oil, are not all strangers to the sea; hydrocarbons of similar structure are synthesized by marine plants and animals to the extent of about ten million tons per year. These natural compounds are dispersed widely among the biota and differ significantly in their chemistry from petroleum in the cyclic paraffinic and aromatic fractions, the latter of which do not occur naturally. Table 7.12 gives an estimate of the weights of the predominant organic materials found in the total ocean. Similar proportions of organic materials would be expected in the coastal zone. The natural hydrocarbon production is about 0.02 percent of the total primary production, and the hydrocarbon content of the water is about 7 percent of the total dissolved organic carbon. The amount of hydrocarbon lost at sea by transport is about one-third to one-quarter of that synthesized by plants each year. The losses due to

Table 7.11. The Involvement of Petroleum with the Marine Environment (millions tons/year)

World oil production (1969)	1820.
Oil transport by tanker (1969)	1180.
Injections into marine environment through man's activities	2.6
Offshore oil production (seepage from wells)	0.1
Tanker operations	0.5
Other ship operation	0.5
Accidental spills	0.2
Deliberate dumping	0.5
Refinery operations	0.35
Industrial and automotive wastes	0.45
Torrey Canyon discharge	0.117
Santa Barbara blowout	0.003–0.011
Atmospheric input from continents through vaporization of petroleum products	90.
Natural seepage into marine environment	<0.1

Source: NAS, 1971a.

Table 7.12. Organic Matter in the Ocean

Primary productivity	5×10^{10}	tons/year
Dissolved organic matter	3×10^{12}	tons
Particulate detrital organic matter	3×10^{11}	tons
Living organic matter (carbon)	7×10^{9}	tons
Hydrocarbons: total ocean reservoir	4×10^{11}	tons
Hydrocarbons produced annually	1×10^{7}	tons/year

Source: Hood, 1972

transport are concentrated locally, however, and found largely in the surface waters. These also include toxic materials, whereas the plant materials are part of the life system of the sea.

With the extent of petroleum leakage to the oceans, two problems of concern arise: Are the petroleum hydrocarbons entering the marine food-chain? If so, what are the long-range effects of such an accumulation? There is a considerable body of data that indicates that petroleum components indeed have entered the marine food-chain.

Two major mechanisms have injected oil into the environment. Oil spills of even moderate size are well known to the public; spills have taken place in the coastal zone and in the open sea. Less well recognized but perhaps more significant are the many small injections of oil and oil products into the near-shore marine environment. Among these sources are sewage, losses incurred during production and transportation operations, storm sewers, filling station washdown operations, and other domestic and industrial losses. A recently completed sampling program in coastal areas of the northeastern United States indicates that chronic small discharges of oil products occurring during shipment of these materials in the coastal zone are probably of greater importance than the loss of crude oil (Zafiriou, 1972). Although the problems thus generated are complex, research is underway to solve them (NAS, 1971a).

Evidence for the presence of petroleum molecules in marine organisms was reviewed at a recent conference on "Baseline Studies of Pollutants in the Marine Environment," sponsored by the National Science Foundation (IDOE, 1972); Tables 7.13 and 7.14, taken from this source, support the following conclusions.

1. Petroleum components are present in seawater and in some marine organisms from the area of offshore oil production of the Texas and Louisiana coast. No effort was made to determine whether or not this material was affecting the organisms of the area.
2. Petroleum components are present in some organisms from the open ocean; the *Sargassum* community in particular is contaminated. Again, no effort has been made to evaluate biological effects.
3. Plankton from the open ocean contain petroleum components that may be associated with tar balls.
4. The community affected by the Falmouth oil spill took up substantial petroleum components; in this case, biological damage was evident.

Table 7.13. Petroleum Hydrocarbon Contamination in the Marine Environment (Coastal Waters)

Location	ppm	Reference[a]
	(wet weight)	
Biota		
West Falmouth, Mass., USA	5–90	Burns and Teal (1971)
West Falmouth, Mass., USA	1–69	Blumer et al. (1971)
Galveston Bay, Texas, USA	236	Ehrhardt (1972)
Narragansett Bay, Rhode Island, USA	3–16	Farrington (1971)
South Queensland, Australia	24–310	Connel (1971)
Chedabucto Bay, Canada	3–545	Zitko and Carson (1970)
Sediments		
West Falmouth, Mass., USA	up to 12,400 (dry wt.)	Blumer et al. (1971) Blumer and Sass (1972)
West Falmouth, Mass., USA	21–3000 (wet wt.)	Burns and Teal (1971)
Narragansett Bay, Rhode Island, USA	50–3500 (dry wt.)	Farrington (1971)
Chedabucto Bay, Canada	0–6.8 (dry wt.)	Zitko and Carson (1970)

Source: IDOE, 1972.
[a] All personal communications.

These conclusions lead one to predict with some certainty that petroleum from production, refining, or transportation has also penetrated the food chain in industrialized estuaries. It is vital that a national program be undertaken to establish concentrations of petroleum in the various components of the coastal zone ecosystems and to begin an assessment of the biological impact of these levels of petroleum on the metabolism of organisms. The absolute and relative magnitudes of the sources contributing to petroleum in the coastal zone must be determined, and the fate of petroleum contaminants in coastal ecosystems must be examined in terms of residence time. The tools and concepts are now available to make these studies.

HALOGENATED HYDROCARBONS. Halogenated organic compounds are exceptionally rare in nature. The chemical bond between chlorine and other halogens and carbon is relatively stable and resistant to attack by most microorganisms (Nash and Woolson, 1967). Biological and

Table 7.14. Petroleum Hydrocarbon Contamination in the Marine Environment

Location	Concentration	Boiling range[a]	Data source[b]
Plankton[c], Louisiana coast	100 ppm (wet wt.)	nC_{16}-nC_{36}	Parker and Winters (1972)
Seston, open ocean 2 samples North Atlantic 1 sample South Atlantic	0.3–20 ppm (wet wt.)	nC_{16}-nC_{28}	Farrington (1972) Teal (1972)
Sargassum community (plants and animals), Sargasso Sea	1–34 ppm (wet wt.)		Burns and Teal (1972)
2 fish livers, Georges Bank (New England Continental Shelf)	5, 19 ppm (wet wt.)	nC_{16}-nC_{28}	Farrington (1972)
Water, Louisiana coast (1 sample)	0.63μg/liter	nC_{16}-nC_{34}	Parker and Winters (1972)
Water, Gulf of Mexico (2 samples)	0.03, 30μg/liter	nC_{1}-nC_{3}	Sackett (1972)

Source: IDOE, 1972.
[a] Boiling range: boiling between the n-paraffins listed.
[b] All personal communications.
[c] Contained primarily in the plankton itself. However, petroleum hydrocarbons could be contributed by small tar balls in the sample.

chemical stability of this class of compounds has made them useful in agricultural and industrial applications. These compounds include the polychlorinated biphenyls (PCBs) and DDT compounds, which are among the most abundant organic pollutants in coastal ecosystems. The polychlorinated biphenyls are used as dielectric fluids in capacitors and transformers, systems from which losses to the environment are minor. In the recent past, however, these compounds were also used in many applications that resulted in significant inputs to the environment. The DDT compounds are still used extensively in agriculture and in the control of disease vectors around the world, although they are rapidly being replaced as insects acquire resistance and as more efficient insecticides are developed.

Both the DDT and PCB compounds are dispersed by atmospheric transport and fallout to distant areas of the world. In some coastal areas, however, local input appears to be a more significant source. Exceptionally high DDT concentrations have been found in marine organisms of southern California (Shaw, 1972), and reports have been made of damage to fish-eating birds. This pollution has been traced to the effluent of a DDT manufacturer in Los Angeles, however, and cannot therefore be considered typical. Runoff of DDT compounds from agri-

---

The Coastal Zone Workshop recommends
Adequate funds be provided for activities that have developed information whose results should be analyzed and published. Where useful, raw data exist that have not been subjected to adequate analysis, funds should be provided to complete the analysis and make the results available to users.

---

cultural areas is a significant local source of DDT compounds (Risebrough et al., in press).

Large amounts of PCB enter coastal waters in urban sewage wastes. It is hoped that the restrictions already placed on PCB use, tighter controls, and effluent standards will result in decreased input. Investigations are urgently needed to determine the mass balance of PCB compounds in coastal waters, which might in turn establish the rates of accumulation with time. It is possible, however, that continued atmospheric input of both DDT and PCB compounds originating elsewhere in the world might offset the effectiveness of local controls. Only detailed studies of the sources of local input will resolve this question.

Other chlorinated hydrocarbon insecticides of a relatively stable, highly toxic nature, such as dieldrin and endrin, have also been detected in coastal ecosystems. The input rates are not presently known, unfortunately, so it is not yet possible to predict future rates of accumulation.

The manufacture of such plastic materials as polyvinyl chloride results in ocean-dumped wastes containing aliphatic chlorinated hydrocarbons. These low-weight compounds have been detected in surface seawater in the North Sea and Atlantic Ocean. They have not been investigated in coastal waters of North America, but their presence is suspected.

There are other chlorinated organic compounds that can be expected to reach pollutant status in coastal waters. Materials that are manufactured in relatively large quantities, that are relatively insoluble in water, but soluble in fats and oils, that are resistant to chemical and biological breakdowns, and that are mobile when released into the environment are of particular concern. At the present time there is no mechanism to screen new organic compounds for such properties that might classify them as potential environmental pollutants. Screening procedures are rapidly becoming a necessity for new chemicals entering industry in large quantities before their environmental hazards are known.

Fluorinated and brominated compounds can likewise be expected to interact with the environment. The stability of freon, one of the fluorinated compounds produced and used in very large quantities, suggests that it will be detected both in the atmosphere and in coastal and marine environments.

The ranges of DDT and PCB found in marine plankton (Table 7.15) span three orders of magnitude. The concentrations of DDT in zooplankton range from 0.2 to 206 micrograms/kilogram, PCB from 0.7 to 1300 micrograms/kilogram; the highest values in each case were obtained in areas of known local pollution. The presence of oil or tar droplets capable of extracting organochlorines from water or air may cause a very real problem and increase variability. The results of samples analyzed from net tows indicate, however, the ubiquitous nature of these compounds (Morris, 1972).

Concentrations of DDT and PCB have been measured in several classes of fish from various areas of the world; the results are given in Table 7.16. The contamination of fish by PCB and DDT is not especially a coastal problem, since concentration ranges tend to overlap those from the open ocean and from the continental shelf and coastal environments.

The IDOE report (1972) strongly recommends that a worldwide monitoring program be undertaken to determine the distribution of these compounds and that work be carried out to study the biological effects

Table 7.15. DDT and PCB in Plankton

Area	Data source[a]	No. of samples	Totals (μg/kg wet weight)	
			DDT	PCB
Sargasso Sea	Harvey, 1972	4	0.7	7–450
S. Atlantic	Harvey, 1972	4	0.2–2.6	19–638
N.E. Atlantic	Holden, 1972	22	2–26	10–110
Clyde, Scotland	Holden, 1972	15	6–130	40–230
California, USA	Claeys, 1972	15	0.2–206	0.7–30
California, USA	Barrett, 1972	250	—	100–1300
Iceland (phytoplankton)	Harvey, 1972	1	—	1500

Source: IDOE, 1972.
[a] All personal communications.

Table 7.16. DDT and PCB Concentrations in Fish

Region	Material	Concentration[a]		Data source[b]
		PCB	DDT	
Open North Atlantic	Pelagic fish			
	muscle	1–10	0.6–3	Harvey, 1972
	liver	1000–6000	95–4800	
	Midwater fish and crustacea	8–59	3–12	
Open South Atlantic	Midwater fish and crustacea	2–14	1–8	Harvey, 1972
Denmark Strait	Groundfish			
	muscle	2–360	3–30	Harvey, 1972
	liver	300–1000	9–260	
Northwest Atlantic shelf	Groundfish			
	muscle	37–187	3–74	Harvey, 1972
	liver	1870–21,800	390–2680	
Gulf of Mexico	Whole fish or muscle	<1–530	1–150	Giam, 1972
Northeast Pacific	Euphausiids	9.2 (mean)	2.7 (mean)	Claeys, 1972
	Pink shrimp	23 (mean)	2.5 (mean)	
	Flatfish	23 (mean)	10.8 (mean)	
Scottish west coast	Fish muscle	<100–1500	<30–480	Holden, 1972
	Fish liver	200–42,600	70–5800	
Baltic Sea	Herring	150–1500	100–1500	Jensen, 1972
	Cod	16–180	9–340	

Source: IDOE, 1972.
[a] Expressed in micrograms/kilogram (ppb) wet weight; DDT values include all metabolites.
[b] All personal communications.

of these and other synthetic organic compounds upon marine organisms.

PLASTICS AND OTHER SYNTHETIC ORGANIC CHEMICALS. Industrial production of organic chemicals in the world over the past four decades has steadily increased to a present annual rate of approximately 100 million tons. Thousands of individual compounds are synthesized in quantities ranging from a few kilograms to millions of tons per year. These materials vary widely in their effect on the environment, and it probably is not necessary to devise monitoring systems to examine all of them. In the report *Marine Environmental Quality* (NAS, 1971a), emphasis is given to compounds having characteristics of relatively large production, persistence or tendency to accumulate in the environment, or toxicity to organisms. The estimated production of some of the compounds of concern is presented in Table 7.4.

Base-line data for synthetic organic chemicals in the coastal region is almost nonexistent, except for the chlorinated hydrocarbons; this is a serious gap in our knowledge of environmental chemistry. It is vital that those coastal regions that receive effluents from major chemical industries be surveyed for selected petrochemicals, and those compounds found should be investigated for their biological effects.

Plastics have been detected recently in coastal areas of the northeastern United States (Carpenter, 1972) in the form of hard, polystyrene spheres ranging between 0.1 and 2 millimeters in diameter in concentrations from 0.05 spheres/cubic meter (in Narragansett and Buzzards Bays) to 2.25 spheres/cubic meter (in the Niantic Bay area). The spheres contain 5 parts per million of PCBs, apparently absorbed from ambient seawater. These plastics have been observed in the gut contents of larval fishes from Niantic Bay and have been found also in winter flounder (*Pseudopleuronectes americanus*), grubby sculpin (*Myoxocephalis anaeus*), and young herring. Ingestion of the spheres may cause intestinal blockage, and concern as to their effect on larval fish is indicated. The source of the plastic spheres is unknown, but they appear to be "suspension beads" derived from an intermediate stage in the manufacture of polystyrene pellets. Plastics, predominantly polyethylene, have also been found floating on the Sargasso Sea in concentrations of about 300 grams/square kilometer (Carpenter and Smith, 1972).

Further investigations are needed to determine the amounts, distribution, and biological effects of plastics in the marine environment, particularly in coastal areas where the concentration appears to be the highest.

7.3.4 Solid Wastes (Domestic and Industrial)

Several general types of materials are derived from the disposal of solid wastes in the coastal zone: Solids, by their nature of high density and large volume, cover benthic habitats. In some cases the waste dump will be inhabited by organisms, but in others it may remain barren of marine life (Pearce, 1970). Toxic materials persist in the ocean dump area and include heavy metals, petrochemicals, PCBs or pesticides, and other industrial wastes (Gross, 1970). Organic wastes have high biological oxygen demand (BOD) properties and thereby deplete the dissolved oxygen. Nutrients may cause unfavorable imbalance of nutrient composition, resulting in eutrophication of the waters. Pathogens derived from sewage sludge pose a potential health problem (see also Sections 5.3, 6.2).

The pollution aspects of solid wastes depend on the ability of the waters surrounding the discharge area to dilute the noxious materials and to supply the required oxygen demand without dangerously lowering the dissolved oxygen. If the oxygen is depleted by the operation that results from dumping in excess of assimilative capacity, degradation of the bottom habitat takes place; an irreversible anoxic condition, under which organic matter is stable, may be produced. Factors that determine the recoverability of an area after dumping operations include the supply of dissolved oxygen present, wave or current energy, type of waste dumped, and sedimentation processes of the area. In general, the surface layers of the sludge will begin to oxidize at a rate dependent upon the supply of oxygenated waters. The organic material buried beneath the surface layers will stabilize and remain essentially unchanged unless it is mixed by physical forces, because of the slow processes of diffusion occurring in the sediment layers. A similar stabilizing effect will result if the organic solids are covered with inert material such as clay, sand, or plastic.

Because of these phenomena, toxic materials that are trapped in solid wastes will not escape to the water column except by occasional release from the surface layers. The major problems, then, with solid wastes are the effect brought about through increase in turbidity from suspended particulates; destruction of bottom life by physical burial; and contribution of toxic substances to the water column.

7.3.5 Radioactivity

The continued expansion of the use of nuclear energy, coupled with an occasionally planned surface explosion for excavation purposes or for weapons testing, will cause a gradual increase in coastal zone radioactivity over the next few decades. Since the first historic explosion of a nuclear device on July 16, 1945, weapons testing has been the principal source of man-made radioactivity. This source of radioactivity has been drastically reduced, however, and may eventually cease completely if and when a worldwide moratorium is adopted.

Radioactivity released by nuclear weapons testing has been widespread following its dispersion in the atmosphere and prior to its entry into the ocean, resulting in a low and rather uniform concentration geographically. The largest amounts of radioactivity enter the oceans in the surface waters of the Northern Hemisphere (Rice and Wolfe, 1971; Miyake, 1971). For a general discussion of radioactive contamination of the oceans see "Radioactivity in the Marine Environment" (NAS, 1971b).

Peaceful uses of nuclear energy by power reactors, fuel fabrication plants, and reprocessing plants will represent point sources of radionuclide release. The rates of release of radioactivity in these particular areas have varied from a few curies to 10^5 curies per year (NAS, 1971b). The release of radionuclides from a pressurized-water power reactor of 1050 megawatts electric capacity is given in Table 7.17.

Possibly the largest releases in the future will occur in the vicinity of fuel reprocessing plants. Since fission and activation products are exposed and may escape in the reprocessing of fuel elements, there will be a need for careful monitoring of iodine 151, strontium 90, and

Table 7.17. Annual Release of Radionuclides Estimated for a Pressurized-Water Power Reactor of 1050 megawatts Electric Capacity

Isotope	Half-Life	Microcuries/yr	Isotope	Half-Life	Microcuries/yr
Liquid Wastes					
^{3}H	12.26 yr	4×10^9	^{131}I	8 days	6.61×10^3
^{54}Mn	314 days	9.7×10^{-1}	^{132}Te	78 hr	6.99×10^2
^{56}Mn	2.58 hr	2.64×10^1	^{132}I	2.3 hr	2.8×10^2
^{58}Co	71 days	2.95×10^1	^{133}I	21 hr	5.13×10^3
^{60}Co	5.26 hr	3.48	^{134}I	53 min	2.16×10^1
^{82}Sr	50.4 days	9.1	^{135}I	6.7 hr	2.6×10^3
^{90}Sr	28 yr	5.76	^{134}Cs	2.1 yr	8.69×10^2
^{90}Y	64 hr	1.06	^{136}Cs	13 days	8.36×10^1
^{91}Sr	9.7 hr	2.49	^{137}Cs	30 yr	4.58×10^3
^{91}Y	59 days	2.11×10^1	^{140}Ba	12.8 days	2.28
^{92}Y	3.5 hr	5.13	^{140}La	40.2 hr	2.35
^{99}Mo	66 hr	1.25×10^4	^{144}Ce	285 days	7.82
Gaseous Wastes					
^{85}Kr	10.4 yr	5.62×10^3			
^{133}Xe	5.27 days	1.58×10^3			

Source: Preliminary Facility Description and Safety Analysis Report, Salem Nuclear Generating Station, Burlington Co., N.J. Docket No. 50–272.

cesium 137. Tritium, argon 39, and krypton 88 are other radionuclides that should be monitored in and around nuclear power and reprocessing plants.

Two problems of radioactive contamination of the environment will increase in scale over the next few decades. The first, introduction of tritium to the environment from reactor operation and fuel reprocessing, will lead to an estimated inventory in the ocean of 10^8 curies by the end of this century. Tritium is of low radiotoxicity and does not accumulate to any extent in biological materials. It is probably not a significant problem in coastal pollution. More important is the problem created by long-lived, biologically dangerous plutonium wastes in which containment for 10^5 years will be necessary. This is a worldwide problem and must be dealt with on a worldwide basis, involving all countries capable of production and use of nuclear fuels. The projected worldwide content of artificial radionuclides in the ocean is shown in Table 7.18.

The amounts of radioactivity which would be released in a nuclear explosion would be tremendously greater than the small amounts released during routine operation of nuclear power plants. Accidental releases, although they may occur infrequently, would also free much larger amounts of radioactivity than would be released during routine operation. Other significant sources of coastal zone contamination could occur from collisions of nuclear ships or submarines, from

Table 7.18. Total Content of Artificial Radionuclides in the World Ocean

	Year 1970 (curies)	Year 2000[a] (curies)
Nuclear explosions		
Fission products (exclusive of tritium)	2–6 × 10^8	? × 10^{8a}
Tritium	10^9	? × 10^{9a}
Reactors and Reprocessing of Fuel		
Fission and activation products (exclusive of tritium)	3 × 10^5	3 × 10^7
Tritium	3 × 10^5	10^8
Total	10^9	10^9

Source: NAS, 1971b.

[a] Estimates for contributions from nuclear explosions are made with the assumption that atmospheric nuclear testing will continue at about the rate of 1968–1970.

the accidental dropping of nuclear weapons from aircraft, or from the crashing of the aircraft. Although the possibility of nuclear explosion is extremely remote, the shock of a crash can cause rupture of the weapons casing and thus introduce plutonium 239 into the environment, as happened at Palomares, Spain, on January 17, 1966 (Lewis, 1967; Szula, 1967). Such occurrences are completely unpredictable, and the consequences of the resultant contamination depend upon the extent of damage in each instance.

We believe there is an urgent need for more information about the genetic effect of low levels of radiation on coastal zone organisms. The monitoring program in the vicinity of nuclear power reactors, fuel fabrication plants, and reprocessing plants is generally considered adequate at this time.

7.3.6 Pathogens

The problem of pathogen transfer through oral-anal cyclic transmission is of special concern to public health authorities in the coastal zone. Table 7.19 provides a listing of those enteric bacteria and viruses that are of most interest. The injection of human pathogens into coastal waters can be expected to continue until waste treatment practices are adopted to eliminate them. Seasonal fluxes due largely to temperature effects can also be expected.

There is an immediate need to define the problems of virus pollution

Table 7.19. Pollutants, Actual or Potential, of Microbiological Nature, Occurring in Coastal Waters as a Consequence of Contributions of Human or Animal Waste Products

Enteric viruses		Enteric bacteria, human or animal origin
Human origin	Animal origin	
Polioviruses	Simian	Salmonellae
Coxsackie-viruses A	Porcine	Shigellae
Coxsackie-viruses B	Bovine	Vibrio parahemolyticus
Echoviruses	Foot and Mouth	
Reoviruses	Vesicular Exanthema of Swine	
Adenoviruses		
Hepatitis viruses		

Source: NAS, 1971a.

in the coastal environment. Most needed are reliable methods for isolating and measuring small numbers of viruses in large volumes of seawater. Epidemiological studies are needed to determine the incidence, duration, and number of virus types found in water; the interaction between human viruses and marine organisms; and the transport of these pathogens through food organisms (NAS, 1971a; Føyn, 1971).

7.3.7 Thermal Pollution

The need for electrical power in the United States is doubling about every ten years as discussed in Chapter 3. Since most of the prime inland sites for generating plants are already in use, more plants will probably locate on estuaries and coastlines in the future. By 1980 it is estimated that 32 percent of all power stations will be in coastal areas. Effects on the local ecosystem can be minimized by siting a power plant in an area of adequate flow where entrained water would be no more than roughly 20 percent of new dilution water in the waterway adjacent to the plant, and by rapidly cooling the entrained water with high velocity mixing. Unfortunately, cooling towers and cooling lagoons, because of the attendant salt spray and water vapor problems, have serious drawbacks that limit their use in coastal areas. It is thus imperative that power plants along the coast be properly sited in areas of adequate dilution capacity. At the cooling water intake, velocities should be less than one foot per second to reduce mortality of fish from impingment onto intake screens.

Chemical changes are often induced in the water passing through the cooling system of the power plant, and some of these alterations have a direct effect on the biota. The dissolved oxygen in seawater is highly temperature dependent, decreasing markedly with increased temperature. In addition, dying or decaying material from organic overloading, due sometimes to an organic-rich estuarine condition, can be increased by an elevation in temperature, thus aggravating the low dissolved oxygen content. Especially in the summer, when the ambient temperature is high, this can result in marginal-to-poor quality water. Deep water (below the thermocline) that is used for cooling purposes in the summer in temperate regions may already be low in dissolved oxygen; if it is then passed through the cooling coils, the dissolved oxygen content is further lowered.

Besides these chemical effects, there is the well-established biological correlation of increased temperature with accelerated rate of respiration—an oxygen-demanding condition in both plants and animals. In the summer months competitive needs arise for oxygen during the night, when the plants are not photosynthetically active, and can result in catastrophic kills in thermal effluents (Nugent, 1970).

Chlorine is often used as a biocide to control and prevent fouling of the cooling system and to kill entrained organisms. High chlorination renders toxic effects to certain of the exposed benthic organisms as well as the entrained target organisms.

Trace concentrations of heavy metals can be significantly increased in the water because of corrosion of materials in the cooling system or erosion of these metals from the sediments. For instance, if there is a low oxidation-reduction potential at the mud-water interface, dissolution of iron and manganese into the water can occur, as in the case of the South Holston Reservoir (Krenkel and Parker, 1969).

Other effects of power plants on water quality may be contributed by turbulence and high current flow, which can cause erosion of sediment necessary to benthic organisms and can disrupt settling and current-orienting behavioral patterns. A second physical water condition induced by power plants is a change in light transmissivity; turbidity and silting during dredging of the cooling canal system can have severe effects on the benthic biota. Changes in water color have been noted (Bader, Roessler, and Thorhaug, 1970), with resulting decreased light penetration.

The various coastal waters of the United States (including the Great Lakes, Hawaii, and Puerto Rico) vary greatly in temperature requirements. Freshwater algae may have quite different temperature requirements than marine varieties (Patrick, 1969). A seasonal cycle must be maintained for reproductive and growth needs of the biota; in addition, the rate of temperature change in spring and fall should not exceed the natural rate of heating and cooling. It is doubtful that one standard for thermal effluent could be applied to the whole United States or even to some individual states which span several ecological regions. The entire food web, with its most temperature-sensitive members, must be considered when setting standards for power site effluent. Allowance must be made also for safety factors to maintain sublethal levels of elevated temperature (Thorhaug, 1972; Todd et al., 1972). Temperature dependencies of organisms include upper and lower temperature thresholds for survival, growth and reproduction; preferred temperature ranges in gradients; and restricted temperature limits for migration. Rate of temperature change may be particularly important for intertidal benthos (Hedgpeth and Gonor, 1969) and entrained organisms. In some of the southern ecosystems, a one-to-two degree rise above midsummer ambient temperatures can cause significant death in an ecosystem (Thorhaug, Stearns, and Pepper, 1972). In other regions, far greater temperature changes cause but little effect. Time of exposure to any given temperature is of vital concern, because mortality produced by temperatures close to the naturally occurring temperature require a relatively long time period.

7.3.8 Dredging, Filling, and Marine Mining

Pollutants contributed by disturbing the sea floor are of general concern in causing marine pollution. This subject is discussed in detail in Chapters 4, 5, and 6. All sediment-disturbing operations impose similar effects on the environment. The relatively steady state that has been established between the water-sediment interface is interrupted, allowing new chemical reactions to occur at the newly exposed sediment surface and in the water column. The suspended solids are transported to new sites, where serious siltation and habitat destruction may then develop. When reduced sediments are dredged, the release of sulfides may cause the oxygen level in nearby waters to fall low enough to endanger marine life. This process of oxidation will change the solubility relations of the heavy metals, and concentrations may reach the toxic level. In areas where redredging of industrial channels is carried out, resuspension of petroleum, petrochemicals, and other organic material may result in high levels of toxicity.

7.4 Material Balance Criteria

The management of coastal zone pollution has two distinct but related parts. First is the identification of existing problems, many of which are now acute, together with the design and implementation of remedial actions. The second is the management of a much subtler set of problems which centers around the interaction between our rapidly growing technology and population and the finite capacities of the coastal zone.

Many existing pollution problems have been recognized, and at least short-range solutions have been proposed and, in some cases, implemented. High-level radioactive wastes, for example, are not released into the marine environment. In-plant procedures have been changed to remove DDT contamination of a coastal region, and the sale of

---

The Coastal Zone Workshop recommends
Basic biological, chemical, and physical research directed toward the following types of problems in the coastal zone:

a. **Transport, dispersion, upwelling and cycling of nutrient and hazardous chemicals as they affect the functioning and stability of coastal zone ecosystems;**
b. **Surveillance of input levels of contaminants, especially chlorinated hydrocarbons, petroleum, and heavy metals;**
c. **Effects of solid waste disposal;**
d. **Effects of chronic, long-term, sublethal contaminants on organisms and ecosystems;**
e. **Assimilative capacity of the coastal zone for all kinds of wastes;**
f. **Epidemiologic and virologic studies;**
g. **Recovery processes in damaged ecosystems;**
h. **Factors affecting stability, diversity, and productivity of coastal zone ecosystems;**
i. **Techniques for increasing production of desirable species or systems.**

PCBs has been restricted by the manufacturer to nondispersive uses.

There remains the much more difficult problem of recognizing those actions that are intermediate in effect and slow in developing: for example, actions that remove 1 percent, 5 percent, or 10 percent of a resource, rather than destroying it completely; processes that reduce to perhaps 75 percent the reproductive success of a given fish population, or change the species composition of an ecosystem in a small but significant way so as to affect the food chain or competitive capacity of important organisms. At the present time there is no clear-cut way of discovering what might be an acceptable fractional loss of a resource; neither is there a way of predicting what loss will be associated with a given set of actions.

Although the problem identification procedure requires value judgments which lie outside the scope of this report, the development of predictive capabilities for the interactions stemming from society's use of its resources is an integral part of the problem.

Our use and management of coastal zone resources are now well enough developed to assume that massive, high-level, acute waste disposal problems can be satisfactorily resolved. We can recognize some cases in which free release of a substance to the environment must be completely prohibited. We do not know, however, how to limit the release of substances in a way to optimize trade-offs or to ensure that some important fraction of a resource will not be damaged. In short, we can reduce the flow completely, but we do not know what will happen, or even whether we should, reduce it only partially.

A broad framework is proposed within which the chronic or low-level contaminant-pollutant problem can be attacked. Except under exclusive-use designation, it is unnecessary and even undesirable to prevent the introduction of any and all contaminants into the coastal zone. We recognize that all coastal zone areas have a characteristic called "assimilative capacity." We view this characteristic as a dynamic property of the region—one that includes biodegradation rates, geochemical removal, and physical dispersion and transport processes that occur. This capacity will vary and will depend upon restraints made by uses of the area in time and space. Contaminants become pollutants when the assimilative capacity is exceeded. Management must provide for the introduction of controlled quantities of waste into coastal receiving waters, so that the input rate does not exceed the assimilative capacity.

The basis for management decision to alter the rate of disposal of substances into the marine environment must be the ability to predict the effects of any proposed action. This requires that it will be possible to predict the behavior of all of the material being discharged: Where does it go? What reactions does it undergo? Does it enter into biochemical reactions which are a part of man's food web, or does it become a part of the sediments and so become removed from further cycling? Is it moved by currents and mixing to another coastal area? Is it fractionated between several routes, and if so, what are the routes and what are the fractions? We must know the pertinent sources and sinks, the rates, routes, and reservoirs of the dispersed substances; we need a material balance of the wastes that are being discharged.

A part of the coastal zone may be characterized by the manner in which material enters or leaves that part, for example, by commercial transportation, by atmospheric fallout and rainout, or by transport by rivers and nontidal drift in coastal marine waters. Within an isolated region, the movement of a substance is assumed to be of primary concern. If a substance is completely immobile and inactive, it is not likely to produce a hazard to man or to parts of the environment.

A model for the routes, rates and reservoirs within an isolated coastal zone region is indicated in Figure 7.1. Reservoirs are designated, within

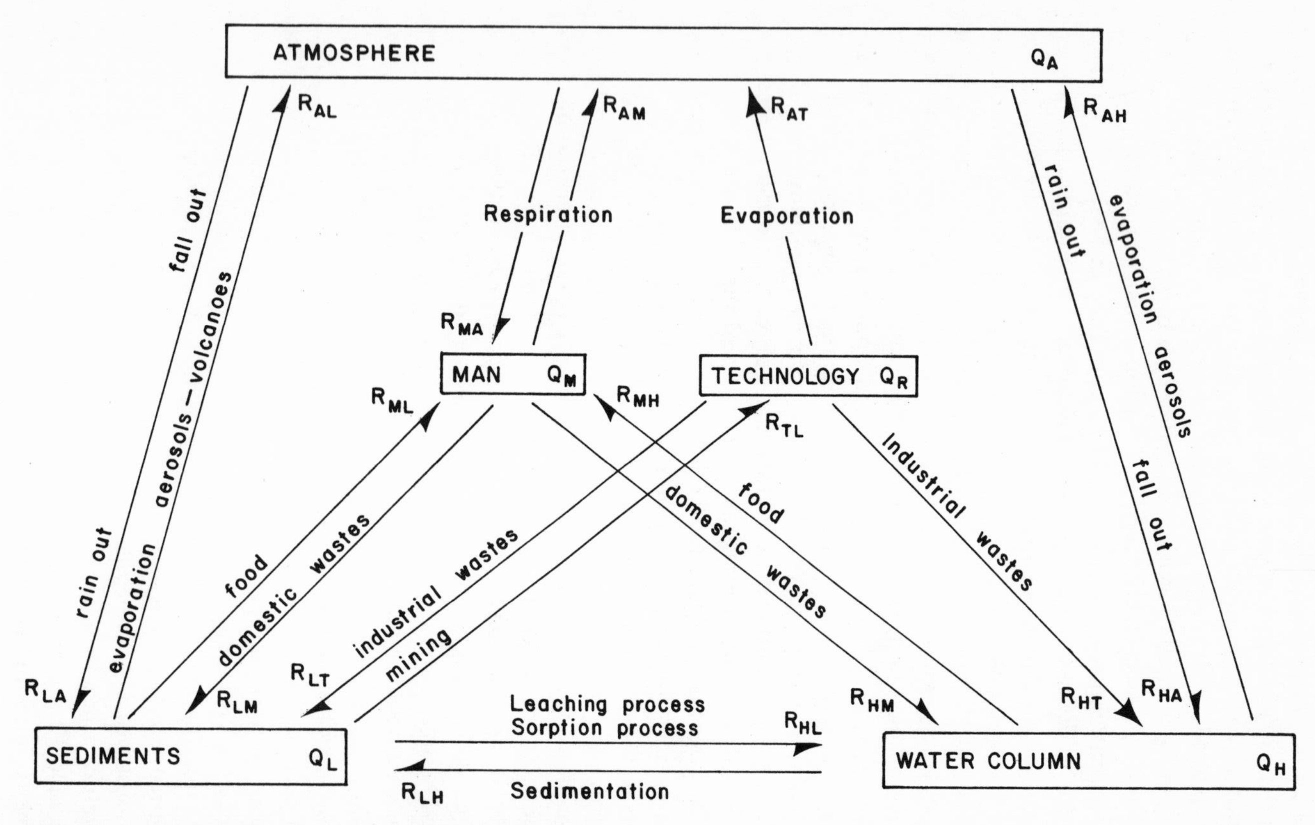

Figure 7.1 Geobiochemical reservoirs, rates, routes, and processes.

which a substance may be found, and the routes between reservoirs and the processes that tend to drive or attenuate the movement of the pollutant from one reservoir to another are noted. Man occupies a unique position in this model: He is shown as a reservoir, even though from a quantitative point of view, he is an insignificant element in the model. Nevertheless, the transfers to him via the routes shown are the features that management must control. Both the water column and the benthic organisms are included—plants and animals which are food sources for man. Within each of these reservoirs there obviously may be a complex set of reactions involving the waste materials.

For each specific waste substance it will be necessary to construct a detailed system such as that shown schematically in Figure 7.1. Only through manipulation of these features can control of the system be achieved by management.

The proposed criteria for waste management will demand extensive insight into the input levels of materials, transport, and dispersion, degradation rates of substances, and chronic effects of low levels of waste on ecosystem relations.

The following actions will be necessary in order to accomplish the goals of this plan for the management of the coastal zone:

1. Detailed mass balance and dynamic interaction studies of potentially significant pollutants in selected near-shore environments should be conducted. The current state of scientific knowledge is such that it is impossible to predict reliably the fate of most pollutants that enter a selected region of the coastal zone. A pollutant may react in different ways in different environments and have a different effect on the ecosystem. A contaminant may be either diluted below or accumulated above critical concentrations for aquatic life and man. It is vital therefore to estimate inputs and outputs of potential pollutants and their interactions within the system in order to develop predictive capability.

2. Surveillance programs should be established to ascertain the levels of such contaminants as halogenated hydrocarbons (PCBs, DDT, and derivatives), petroleum, and heavy metals (lead, cadmium, mercury) in all coastal zones receiving domestic and industrial discharges through outfalls or rivers. Monitoring our water courses and coastal regions for elements of pollution already known to be hazardous and found in the environment in various forms is essential in order to follow the buildup, degradation, and ultimate fate of these threats to marine ecosystems. The data obtained are essential in order to establish performance-based

standards in water quality and in management of any coastal resource.

3. Efforts should be accelerated to develop criteria for critical concentrations of pollutants in the coastal zone. There is an urgent need for research laboratories devoted to establishing the critical concentrations of various contaminants for aquatic life in estuarine and marine waters. This function is currently the responsibility of the Environmental Protection Agency, with its Duluth Water Quality Laboratory for fresh waters and a laboratory at Narragansett, Rhode Island, for marine waters. Greater effort is needed in the marine area, however, and the resources of research organizations and agencies should be directed to the problem.

4. Management capabilities (legal, economic, social, and technical) for screening new chemical and degradation products are necessary. It has been estimated that approximately 500 new compounds are produced each year in the United States that could become significant environmental contaminants. These new compounds and their degradation products must be screened for potential environmental impact on natural waters. A series of chemical crises has occurred in the aquatic environment caused by the presence of such compounds as DDT, mercury, and PCBs. At the present time no agency is reviewing the newly produced compounds with respect to their environmental impact, although several bills exist in Congress which are designed for this purpose. The current administration bill, as developed by the Council on Environmental Quality and presented by the President in February 1971 as part of his environmental message, represents a significant step toward developing the procedures by which this type of screening can be accomplished. A concentrated effort should be made to ensure passage of legislation of this type in the near future with adequate funding to implement the necessary programs.

5. Any manipulation of the uses of the coastal zone must be accompanied by an evaluation of its effect on coastal zone pollution and utilization. In some parts of the coastal zone large amounts of funds are being expended in an effort to manage for optimum utilization for certain of man's activities. Often the decisions are made without an adequate technical basis for prediction of the effect of a certain type of management on the behavior of pollutants and for best utilization of the resource.

6. Closed-cycle, minimum-discharge waste water disposal systems should be adopted where possible for municipal and industrial wastes.

It is becoming recognized that domestic and industrial waste waters contain chemical compounds such as heavy metals, PCB, and DDT which may be found through further research to have a significant deleterious effect on the aquatic ecosystem in the region of discharge. It is recommended that closed-cycle systems be attempted where possible for a particular municipal or industrial waste. Often it is not possible to achieve completely closed-cycle operations; in these instances maximum technically feasible treatment is the recommended practice, that is, minimum discharge in terms of concentration of various contaminants and total load.

7. Laboratories should be established in the United States to look for unrecognized pollutants in natural waters and aquatic organisms. The chemical crises that have occurred in the aquatic environment over the past few years have in part been discovered by accident. It is recommended that research laboratories be designated with the primary goal of examining natural waters and selected aquatic organisms for potential pollutants. The laboratories should be staffed with specialists in various branches of analytical chemistry, water chemistry, aquatic toxicology, aquatic ecology, and related disciplines. Their activities should be restricted to looking for widespread contaminants which are not now recognized as being significant or even present in aquatic ecosystems. These laboratories should be immune from any enforcement activity that could divert their attentions to a particular localized problem.

8. Epidemiologic and virologic studies should be undertaken to determine the incidence, duration, and number of virus types found in water. Domestic and, in some instances, industrial waste waters contain large numbers of organisms that are potential pathogens for man. These organisms include viruses causing infectious hepatitis and typhoid and other bacteria and protozoans that cause dysentery in man. Traditional criteria that have been used to judge the potential health hazard for using waters contaminated with these organisms are not reliable. Detailed studies are needed on the significance of pathogenic organisms in waste water discharges with respect to their effect on the utilization of coastal waters for recreational purposes, possible feedback through food fish, and pathogenic effects on marine life.

9. All dischargers of municipal and industrial waste waters directly to the coastal zone should be required to report typical chemical composition of the discharge to the coastal zone managing agency. Large amounts of municipal and industrial wastes are discharged to coastal

zone waters, and the composition of these waters is largely unknown at this time in many parts of the country. It should be the established responsibility of each discharger to take representative samples of this waste, to determine in detail the chemical composition, and to keep an accurate record of the total quantity of waste water discharged each day. The frequency of sampling and the particular chemical analyses required for a certain type of discharge would be selected after consultation with appropriate water pollution control authorities.

10. Any disposer of solid wastes in the coastal zone should be required to demonstrate with a reasonable degree of reliability the effects of such disposal practice on the utilization of the coastal zone for other purposes. Solid municipal and industrial wastes, including concentrated liquid wastes from some types of industries, can be disposed of in the oceans without significant ecological damage to beneficial uses of the coastal zone. The limits on this activity are determined by the assimilative capacity of the area and other uses that might be restrained by this activity. Each disposer who wishes to use a coastal zone for solid waste dumping should be required to conduct studies to determine the overall effects of such practices on the aquatic ecosystem receiving the waste. The discharger should be required to maintain a monitoring program to support adequate evaluation of the effects of the dumping practice based on the initial study.

7.5 Conclusions

1. *Current cultural patterns and expected trends coupled with current technology for waste management are leading in the destruction of many coastal zone resources.* Current technology for removal of pollutants from domestic and industrial waste waters usually specifies reduction of the level to some arbitrary maximum. For example, the present practice of the United States is to require that domestic waste waters receive at least secondary treatment, which removes 90 percent of the biological oxygen demand. Secondary treatment represents a significant reduction in the amount of waste discharge to many natural waters, especially near urban centers. Even with complete secondary treatment, however, the waste assimilative capacity in the region receiving the waste will eventually be exceeded as population numbers and individual discharge needs increase.

Similarly, the projections for the use of coastal zone waters for once-through cooling at electric generating stations point to a significant

---

The Coastal Zone Workshop recommends
The establishment of regional *Coastal Zone Centers* to develop and coordinate natural science, social science, and legal research and to provide relevant information about the coastal zone to government agencies and the public. These centers should cooperate with existing research organizations to resolve basic questions of the environment in that region, help appraise management techniques, and provide inventories of coastal resources. Coastal Zone Centers should be established in regions corresponding to the major types of coastal environments and may be international in character.

---

deterioration of water quality due to heating and associated effects. General degradation of the environment through release of toxic substances has occurred in heavily populated areas and is still on the increase, although notable remedial efforts are being asserted in certain localities.

2. *The aquatic ecosystem of the coastal zone has a finite assimilative capacity for a particular contaminant without significant deleterious effects.* There are some who advocate that all waste discharge to natural water systems should be eliminated. This approach is neither technically feasible nor in the best interests of the public. The assimilative capacity of any particular part of the coastal zone is determined by physical processes such as currents and mixing, geomorphology, types of sediments, types of water chemistry, and biology. The "no waste discharge" approach ignores the fact that nature does provide for waste treatment without significant harm. The problem develops when man attempts to apply a certain waste load without giving consideration to the characteristics of the particular receiving area. This is the problem associated with using universal effluent standards, as is commonly advocated today. Each region of the coastal zone should be considered on its own merit, its own uses and characteristics, and the contaminant load should be determined on a performance basis—an approach that requires a much better understanding of the aquatic ecosystem than exists today.

3. *The scientific information available today does not enable a complete assessment of the assimilative capacity of the coastal zone for certain contaminants.* In some cases the relationship between the effect of a contaminant on the aquatic ecosystem and the concentrations of the contaminant in a particular discharge can be predicted. For ex-

ample, reasonable estimates can be made of the effects of the discharge of oxygen demand materials from domestic waste waters on oxygen depletion in the coastal zone. Additional research is required on the aqueous environmental chemistry of many other contaminants in order to predict the contaminant assimilative capacity. Included in the group requiring much additional study are the chlorinated hydrocarbons, such as DDT and the PCBs, and the toxic trace metals.

References

Bader, R. G., Roessler, M. A., and Thorhaug, A., 1970. Thermal pollution of a tropical marine estuary, *FAO World Conference on Pollution in the Ocean,* Dec. 9–13, 1970 (Rome: FAO fisheries report no. 99), FIR:MP/70E-4:1–6.

Bertini, K. K., and Goldberg, E. D., 1971. Fossil fuel combustion and the major sedimentary cycle, *Science, 173:* 233–235.

Blumer, M., Sanders, H. S., Grassle, J. F., and Hampson, G. R., 1971. A small oil spill, *Environment, 13:* 2–12.

Blumer, M., and Sass, J., 1972. The West Falmouth oil spill II. Chemistry (Woods Hole, Mass.: Woods Hole Oceanographic Inst.), technical report no. 72-19.

Bowen, H. J. M., 1966. *Trace Elements in Biochemistry* (New York: Academic Press), 241 pp.

Brooks, R. R., and Rumsby, M. G., 1965. The biogeochemistry of trace element uptake by New Zealand bivalves, *Limnology and Oceanography, 10:* 521–527.

Burns, K. A., and Teal, J. M., 1971. Hydrocarbon incorporation into the salt marsh ecosystem from the West Falmouth oil spill (Woods Hole, Mass.: Woods Hole Oceanographic Inst.) technical report no. 71-69.

Carpenter, E. J., 1972. Personal communication to Workshop on Critical Problems in Coastal Zone.

Carpenter, E. J., and Smith, K. L., 1972. Plastics on the Sargasso Sea surface, *Science 175:* 1240–1241.

Connel, D. W., 1971. Is the Mediterranean dying? *New York Times Magazine,* Feb. 21, 1971.

Egorov, V. V., Zhigalovskaya, T. N., and Malokhov, S. G., 1970. Microelement content of surface air above the continent and the ocean, *Journal of Geophysical Research, 75:* 3650–3656.

Ehrhardt, M., 1972. Petroleum hydrocarbons in oysters from Galveston Bay, *Environmental Pollution,* in press.

Emery, K. O., Orr, W. L., and Rittenberg, S. C., 1955. Nutrient budget in the ocean, *Natural Sciences in Honor of Capt. Allan Hancock* (Los Angeles: University of California Press), pp. 229–309.

Farrington, J. 1971. Ph.D. thesis (Providence: University of Rhode Island), unpublished.

Føyn, E., 1971. Municipal wastes, *Impingement of Man on the Oceans,* edited by D. W. Hood (New York: Wiley-Interscience), pp. 445–459.

Gross, M. G., 1970. Analysis of dredged wastes, flyash, and waste chemicals—New York Metropolitan Region (Stony Brook: Marine Science Research Center, SUNY), technical report no. 7.

Hedgpeth, J. W., and Gonor, J. J., 1969. Aspects of the potential effect on thermal alteration on marine and estuarine benthos, *Biological Aspects of Thermal Pollution,* edited by P. A. Krenkel and F. L. Parker (Nashville: Vanderbilt Univ. Press), pp. 80–118.

Hoffman, G. I., 1971. Trace metals in the Hawaiian marine atmosphere (Honolulu: University of Hawaii), Ph.D. thesis.

Hood, D. W., 1972. Pollution of the world's oceans, *Topics in Ocean Engineering,* vol. 4, edited by C. L. Bretschneider (Houston, Texas: Gulf Publ. Co., in press).

Hood, D. W., Stevensen, B., and Jeffrey, L. M., 1958. The deep sea disposal of industrial wastes, *Industrial and Engineering Chemistry, 50:* 885–888.

Hunt, J., 1972. Personal communication to Workshop on Critical Problems of the Coastal Zone.

IDOE, 1972. Baseline studies of pollutants in the marine environment and research recommendations, *The IDOE Baseline Conference, May 24–26, 1972* (New York: IDOE Baseline Conference), 54 pp.

Ketchum, B. H., 1969. Eutrophication of estuaries, *Eutrophication: Causes, Consequences, Correctives* (Washington, D.C.: National Academy of Sciences), pp. 197–209.

Krenkel, P. A., and Parker, F. L., 1969. Engineering aspects, sources and magnitude of thermal pollution, *Biological Aspects of Thermal Pollution,* edited by P. A. Krenkel and F. L. Parker (Nashville: Vanderbilt Univ. Press), pp. 10–61.

Lewis, F., 1967. *One of Our H-Bombs Is Missing* (New York: McGraw-Hill), 270 pp.

Merlini, M., 1971. Heavy-metal contamination, *Impingement of Man on the Oceans,* edited by D. W. Hood (New York: Wiley-Interscience), pp. 461–486.

Miyake, Y., 1971. Radioactive models, *Impingement of Man on the Oceans,* edited by D. W. Hood (New York: Wiley-Interscience), pp. 565–588.

Morris, B. F., 1971. Petroleum: tar quantities floating in the northwestern Atlantic taken with a new quantitative neuston net, *Science, 173:* 430–432.

Moss, J. E., 1971. Petroleum—the problem, *Impingement of Man on the Oceans,* edited by D. W. Hood (New York: Wiley-Interscience), pp. 381–419.

Nash, R. G., and Woolson, C. A., 1967. Persistence of chlorinated hydrocarbons in soils, *Science, 157:* 924–927.

National Academy of Sciences (NAS), 1971a. Marine environmental quality (Washington, D.C.: Ocean Sci. Comm., National Academy of Sciences/National Resource Council), 107 pp.

National Academy of Sciences (NAS), 1971b. Radioactivity in the marine environment (Washington, D.C.: National Academy of Sciences).

Nugent, R., 1971. Elevated temperatures and electric power plants, *Transactions of the American Fisheries Society, 99:* 848–849.

Okuba, A., 1971. Horizontal and vertical mixing in the sea, *Impingement of Man on the Oceans,* edited by D. W. Hood (New York: Wiley-Interscience), pp. 89–168.

Patrick, R., 1969. Some effects of temperature on freshwater algae, *Biological Aspects of Thermal Pollution,* edited by P. A. Krenkel and F. L. Parker (Nashville: Vanderbilt Univ. Press), pp. 161–190.

Patterson, C., 1971. Artifacts of man: lead, *Impingement of Man on the Oceans,* edited by D. W. Hood (New York: Wiley-Interscience), pp. 245–258.

Pearce, J. B., 1969. The effects of waste disposal in the New York Bight—interim report for 1 Jan. 1970.

Rice, T. R., and Wolfe, D., 1971. Radioactivity—chemical and biological aspects, *Impingement of Man on the Oceans,* edited by D. W. Hood (New York: Wiley-Interscience), pp. 325–379.

Risebrough, R. W., Menzel, D. W., Martin, J. D., and Olcott, H. S., in press. DDT residues in Pacific marine fish, *Pesticides Monitoring Journal.*

SCEP, 1970. *Man's Impact on the Global Environment* (Cambridge, Mass.: MIT Press), 320 pp.

Shaw, S. P., 1972. DDT residues in light California marine fishes, *California Fish and Game 58*(1): 22–26.

Stanford Research Institute, 1969. *Economics Handbook* (Menlo Park, Calif.).

Szula, T., 1967. *The Bombs of Palomare* (New York: Viking Press).

Thorhaug, A., 1972. Macroalgae and grasses, *An Ecological Study of South Biscayne Bay and Cord Sound, Florida,* edited by R. G. Bader and M. A. Roessler (Miami: University of Miami, School of Marine and Atmospheric Sci.), rep. to U.S. Atomic Energy Comm., pp. 1–84.

Thorhaug, A., Stearns, R. A., and Pepper, S., 1972. The effect of heat on *Thalassia testudinum* in Biscayne Bay, Florida, *Proc. Florida Academy of Sciences* (in press).

Todd, J. H., Engstrom, D., McClaney, W., and Jacobson, S., 1972. An introduction to environmental ethology: preliminary comparison of sublethal thermal and oil stresses on the social behavior of lobsters and fishes from freshwater and marine ecosystems (Woods Hole, Mass.: Woods Hole Oceanographic Inst.), rep. to U.S. Atomic Energy Commission.

Tsunogai, S., and Ikeuchi, K., 1968. Ammonia in the atmosphere, *Geochemical Journal, 2:* 151–166.

Turekian, K. K., 1971. Rivers, tributaries, and estuaries, *Impingement of Man on the Oceans,* edited by D. W. Hood (New York: Wiley-Interscience), pp. 9–73.

United Nations, 1968. *Statistical Yearbook,* 1967 (New York: Statistical Office of the United Nations).

U.S. Department of Interior, 1970a. *Mineral facts and problems* (Washington, D.C.: U.S. Govt. Printing Office), Bureau of Mines bull. 650, 1291 pp.

U.S. Department of Interior, 1970b. *Minerals yearbook, 1968,* vols. 1–4 (Washington, D.C.: U.S. Govt. Printing Office), 3029 pp.

U.S. Tariff Commission, 1968. *Synthetic Organic Chemicals, U.S. Production and Sales* (Washington, D.C.: U.S. Govt. Printing Office).

Weiss, C. M., and Wilkes, F. G., 1969. Estuarine ecosystems that receive sewage waste, *Coastal Ecological Systems of the United States: Source Book for Estuarine Planning,* edited by H. T. Odum, B. J. Copeland, and E. A. McMahan (Morehead City, N.C.: Institute of Marine Science, University of North Carolina).

Zafiriou, O., 1972. Personal communication.

Zitko, V., and Carson, W. V., 1970. The characterization of petroleum oils and their determination in the aquatic environment, Fisheries Research Board, Canada, tech. rep. no. 217.

Part III Management of Coastal Resources

Increased human utilization of coastal resources has now reached the stage where conflicts between opposing uses are commonplace, and the effects of these uses on natural processes are in many cases disastrous. If negative environmental impacts are to be reduced and if user demands are to be satisfied in an equitable way, then a comprehensive system of management is necessary.

The coastal zone contains aesthetically valuable resources such as beaches and open space as well as the nonrenewable resources such as gravel and oil. Many of the important resources are self-renewing, and if the natural systems that support them are not destroyed, their benefits can be realized over an indefinite period. These resources are finite, however, and if overutilized or misused, future generations will be deprived of their benefits.

The development of refined legal institutions and procedures and the establishment of new management agencies can be expected to make the implementation of rational decisions, based on relevant data, a reality. We must design new public institutions that incorporate existing social processes and are cognizant of existing natural constraints if we are to utilize our coastal resources in a sustainable way that will keep them available for future generations.

Chapter 8
Strategies and Research Needs

PREPARED BY
J. L. McHugh, Gerard A. Bertrand, Robert A. Ragotzkie

WITH CONTRIBUTIONS FROM
Robert L. Bish, James Bolan, William D. Marks, Mitchell L. Moss, John S. Steinhart, Linda Weimer

Chapter 9
Allocating Coastal Resources: Trade-off and Rationing Processes

PREPARED BY
Robert Warren, Robert Bish, Lyle E. Craine, Mitchell L. Moss

WITH CONTRIBUTIONS FROM
Clare Gunn, Claes Rooth, Russel Stogsdill, Durbin Tabb

Chapter 10
Coastal Management and Planning

PREPARED BY
John Armstrong, Russell Davenport, Joel Goodman, William V. McGuinness, Jr., Jens Sorensen

WITH CONTRIBUTIONS FROM
William Aron, Harold Bissell, Brad Butman, William D. Marks, H. Dixon Sturr, Thomas Suddath, Anitra Thorhaug

Chapter 11
Legal Aspects of the Coastal Zone

PREPARED BY
Dennis M. O'Connor, Walter J. McNichols

WITH CONTRIBUTIONS FROM
John Clark, Lyle Crain, Joel Hedgpeth, Marc Hershman, H. Crane Miller, Garret Power, Jens Sorensen, Robert Strauss

Chapter 12
A Systems View of Coastal Zone Management

PREPARED BY
Rezneat M. Darnell, Demitri B. Shimkin

WITH CONTRIBUTIONS FROM
Bostwick H. Ketchum, William Niering, Ruth Patrick, Claes Rooth, H. Dixon Sturr, Malcolm Shick

8 Strategies and Research Needs for Coastal Zone Management

8.1 Introduction

An adequate fund of information is essential for effective planning and management of the coastal zone. The issues are so complex and span such a wide variety of disciplines and activities that the supply of information must be diverse and voluminous. Moreover, the information management system must be comprehensive, systematic, and flexible, for it must be equally capable of dealing with qualitative data, expressed verbally, in the social sciences, economics, and law, as it is in the natural sciences of physics, chemistry, geology and geophysics, and biology, and in applied sciences such as engineering and oceanography.

The existing data base for coastal zone planning and management is inadequate. This is not because the problems of the coastal zone are new or entirely neglected. Critical problems of water quality, navigation, and fisheries, among others, have been recognized and attacked since colonial days, and a vast store of information relating to these types of problems already exists. Detailed records of the fisheries of the nation date back to 1880, and for some fisheries much earlier than that. Records of tidal rise and fall, river runoff, and even water quality have been collected in many areas around our coasts for a long time. References to many other problems and phenomena of the coast also are of long standing. It is the term "coastal zone" that is new, not the problems nor the attempts to document and resolve them.

Nevertheless, the growth of modern technology since the Second World War has focused the eyes of the nation on coastal zone problems. The critical problems of the coastal zone are national, not local. For the first time the nation has recognized the dangers and our lack of adequate information and management capability to deal with them.

To ensure the long-term conservation of the coastal zone and wise use of its resources, there must be a rational management plan which must also take into account immediate human needs and desires. The uses man makes of the coastal zone, their impacts, and constraints imposed on them by the environment and by other uses must be incorporated. There are several ways in which these complex interactions might be approached.

8.2 Uses of the Coastal Zone

Man's demands on the coastal zone include some uses that are compatible with each other and some that are not compatible and require exclusive use of an area. Some uses now made of the coastal zone may be displaceable to other areas, but care must be taken to avoid moving these uses to another area if the environmental impact there would be increased. Decisions on each usage will require better information than is now available in almost all cases. We must learn to predict the effects of these decisions in both scientific and sociological ways.

Management for the wise and rational use of the coastal zone and its resources must seek to maintain the integrity of the environment while attempting to meet man's needs and desires. Only when man's current uses of the coastal zone are assessed to see which are compatible, which are exclusive, what conflicts arise, and what technological solutions to these conflicts exist, can this goal be met.

8.2.1 Compatible Uses

Compatible uses are those that can coexist without conflict on the same section of coastline. The compatible- or multiple-use strategy, frequently applied to public lands, is too often adopted as the easiest and most acceptable management course. Though the doctrine of multiple use is politically attractive, it should be applied only when the uses under consideration can be demonstrated to be genuinely compatible in the long run.

Some widely diverse uses are compatible with one another. For example, waste disposal, shipping, recreational boating, and urban housing can usually coexist along the same section of the coast. Other uses, such as high-rise housing and wildlife preserves, obviously cannot.

Multiple use is also conditioned by economic and political factors. For example, industrial uses, urban housing, and shipping tend to be concentrated in limited areas because of their intimate interrelationships and they cannot be moved or dispersed by management decisions. At the same time, when new centers of industrial activity on a coastline or estuary are created, the existing uses or the potential alternate uses of the site must be taken into account.

Though the criterion of a multiple-use planning strategy is compatibility, increased use of the coastal zone inevitably leads to conflicts because of its unique and linear nature. Conflicts arise when previously compatible uses increase to the extent that they begin to interfere with each other. For example, organic waste disposal at a low or moderate level may not interfere with commercial or sport fishing or even sports such as swimming, but as the level of waste disposal increases with increased population and as the assimilative capacity of the receiving water is exceeded, fish populations are adversely affected and, for sanitary and aesthetic reasons, swimming is no longer possible.

Conflicts can also arise from changes in use practice. For example, DDT added to agricultural runoff may concentrate in fish at high levels and render them useless as a food source, whereas previously the runoff nutrients may have been harmless or even beneficial to fish production in the area.

New uses imposed on a previously compatible set of uses can also create serious conflicts. Industrial development of a recreational area or near a prime fishing ground can cause extreme polarization of interest groups and lead to violent conflicts. Many controversies and litigations over coastal zone use stem from such unplanned growth situations and represent the most difficult managerial problems of all.

There may be technological solutions to conflicts, potential conflicts, or environmental constraints and these should not be overlooked. Well-known examples of successful technical solutions are usually the result of solving single-use problems such as treatment of wastes, protection from waves by breakwaters and seawalls, dredging of channels and harbors, or automatic control systems on offshore oil rigs.

Looking beyond the single-problem solution approach, one might seek to convert several problems to an opportunity. For example, the valuable nutrients contained in sewage and agricultural runoff should not be wasted—these are valuable resources and ways to recycle them back into the system should be sought. Domestic sewage and possibly also waste heat from power plants might be managed to increase biological production either in existing natural ecosystems or through aquaculture.

The idea of converting problems to opportunities, especially with respect to waste treatment, is an old one. The closed man-agriculture system as practiced in Japan is a familiar example. Yet large-scale attempts to employ this strategy in the United States are rare. Environmental management agencies should encourage and support pilot projects to test the feasibility of these kinds of approaches to multiple-use problems.

It is important that the benefits from reduced damage to the environment and the savings realized by eliminating other direct technical

---

The Coastal Zone Workshop recommends
The creation of a national system of *Coastal Area Preserves* for the permanent protection of the basic genetic stocks of plants and animals and the essential components of their environments, which together constitute ecosystems. These Coastal Area Preserves should be severely restricted in use. Some other coastal areas should be developed for recreational usages that are compatible with the natural life of the area.

---

solutions be weighed along with the value of the product or direct benefits of the system when considering the economic feasibility of these solutions.

8.2.2 Exclusive Uses
A few uses of the coastal zone are exclusive. They are, by their nature, totally incompatible with any other use. Wildlife preserves, for example, can tolerate almost no other use in the same area, with the possible exception of some limited recreational activities, such as wildlife observation or photography. Aquaculture, intensively pursued, is also an exclusive use, since it requires a highly controlled environment to maximize the production of one or more species of food fish. Some less intensive aquacultural practices, such as oyster and clam beds, may allow sport fishing or boating in the same area.

8.2.3 Displaceable Uses
With increased use pressure on the coastal zone and the resulting conflicts, man must begin to be more selective in which activities he chooses to allow there. Some human activities can be moved elsewhere if the economic, social, and political pressures are sufficiently great. But to do this, innovative engineering and in some cases actual changes in human behavioral patterns are necessary.

Electric power generating plants, for example, are sited on coastlines or estuaries to take advantage of a large and convenient supply of cooling water. With appropriate engineering and sufficient capital investment, they could be sited far from the coastline, either inland or offshore. Water can be conveyed inland by aquaducts for once-through cooling, or plants can be designed for closed cooling systems using lagoons or cooling towers. Offshore sites would greatly simplify the waste heat disposal problem.

Marinas can be constructed inland and connected to the sea by canals, thus eliminating both the commercial sprawl that often accompanies a recreational boating center and the destruction of natural habitat in the bays or inlets where harbors are developed. Moving ports offshore is a displaceable use which, out of necessity, is already being seriously considered with respect to the commercial shipping industry (see Section 6.2).

The economic costs of displacing present or planned activities must be balanced against the benefits of lowering the pressure on highly valued sections of the coast and leaving them for nondisplaceable uses such as commercial fishing, aquaculture, mining, recreation, and wildlife and habitat preserves. Engineering and managerial ingenuity and the persuasion of the public and political bodies will determine the success of this management strategy.

8.3 Information Needs

Information needs for effective management of the coastal zone fall into two broad categories. The first is a pressing need to make full use of information already available. The past is history, and past events cannot be monitored now. Yet an extensive and relatively disorganized body of information on past events and environmental changes exists in various statistical publications, scientific papers, reports, unpublished files, and in the accumulated experience of many individuals. Often this information has not been adequately analyzed and reported, because planning and funding of information collecting activities have not made adequate provision for analysis and dissemination. Funds must be provided to recover this invaluable store of information, to analyze it as far as is practicable, and to make the results available to users. This would produce, at modest cost, a window into the past that can be obtained in no other way.

Much information on coastal zone problems has been, and will continue to be, presented in qualitative form. This is especially true of the social, economic, legal, and political aspects of coastal zone management. This is one of the reasons why historical data remain largely undigested and obscure. Modern electronic data processing methods are designed primarily to handle quantitative data. They are not so well adapted to storage, analysis, and retrieval of qualitative information expressed verbally. Thus, legal, economic, social, and many kinds of biological information are not readily available to decision makers.

The Coastal Zone Workshop recommends
Adequate funds be provided for activities that have developed information whose results should be analyzed and published. Where useful, raw data exist that have not been subjected to adequate analysis, funds should be provided to complete the analysis and make the results available to users.

This deficiency now is clearly recognized, and encouraging steps have been taken to resolve the problem. O'Connor (1972) shows that it is possible and feasible to take information expressed in words, sentences, formulas, citations, and so on, and reduce it to a form compatible with electronic data processing.

Most coastal states are making vigorous efforts to describe the resources of their coastal zones, the threats to these resources, the problems of multiple use, and the requirements for rational management. In the forefront have been the states of California and Florida, which are producing comprehensive surveys and plans, various parts of which have already been published. One of these studies described a permanent coastal zone data inventory and information system (Nash, 1970). The essential broad conclusions and recommendations are that a permanent data and information system is necessary to support decisions on coastal zone management, and that continued lack of sufficient data forestalls any possibility of effective planning and management. This conclusion is applicable nationwide, and the California study should be examined carefully for its implications for national policy, as should the studies completed or underway in other states. One of the impressive features of the California inventory is that it has relied upon compilation and assessment of existing information rather than on new field work.

Thus, as far as the availability of pertinent information for solving critical problems of the coastal zone is concerned, we may be closer to some solutions than we realize. Great masses of data already exist, in places already cited. To some extent this accumulation of raw data and experience has been tapped and subjected to analysis. But much remains unused, or only partially used, and no systematic nationwide effort is being made to correct this deficiency.

Data gathering continues at an accelerated pace at local, state, and national levels and in various agencies, and these activities are still relatively uncoordinated. For example, many coastal states have made, or are making, assessments of the various activities and processes affecting the coastal zone and the present quality of the environment and are developing plans for coastal zone management, but these efforts are often based on widely different concepts of what the coastal zone includes. Some definitions are very narrow, including only a strip of land and wetland which adjoins the shoreline. California, for example, has defined the coastal zone as "the area extending inland approximately $\frac{1}{2}$ mile from the mean low tide and seaward to the outermost limit of the state boundaries" (State of California, Resources Agency, 1972). Others may include wider areas of land and sea, but few, if any, have adopted a broad definition such as has been accepted by the Coastal Zone Workshop. These incompatibilities may seriously limit the usefulness of local sources of information, and solutions derived locally from such information may fall short of solving the existing problems.

The existence of large amounts of information relevant to coastal zone problems should not be allowed to obscure the fact that this information is far from adequate. A single example will illustrate. The effects of sudden catastrophic events, such as a violent storm or a large accidental oil spill in the coastal zone, are there for all to see in the form of land erosion, damaged buildings, or dead fish and shellfish. It does not need much analysis to determine cause and effect and to assess the damages accurately. These are the kinds of things that catch public attention, but they are not the most dangerous aspects of environmental change. Much more disturbing are the chronic, relatively low-level changes created by man which are not obvious to the eye but which are steadily degrading the quality of the environment. These things have been of concern to scientists and other technically qualified people for a long time, but they have come to public attention only recently, as wide and sometimes exaggerated publicity has been given to the dangers of persistent toxins such as DDT and mercury. Public awareness of these dangers, now awakened, has been heightened by the very circumstances that hide these insidious effects from direct view, and the national sentiment now is shaped by the fear of the unknown. This can lead to dangerous overreactions and false alarms. Our knowledge of the concentrations of harmful substances in the coastal zone environment, of their fate as they spread out from the source, and of their effects on marine life and on man, is fragmentary, and this lack of knowledge cannot be allowed to continue. The present status of knowledge on such substances as trace metals, chlorinated hydrocarbons, and petroleum, and their effects and the kinds of research needed, have

The Coastal Zone Workshop recommends
The acceleration and expansion of comparable surveys and complementary inventories of coastal resources, including demographic patterns, ownership and land-use patterns, and other socioeconomic data in addition to new base-line ecological studies of natural and modified coastal systems.

been summarized in a recent report (IDOE, 1972), and the needs for further information clearly identified (see Chapter 7).

Existing information and information yet to be gathered is of no use if institutional barriers prevent action. Such barriers do exist as ignorance, tradition, and vested interests and are often expressed in the form of local, state, or even federal laws, and these barriers can be formidable. The general failure of the coastal states to manage the fisheries of the coastal zone is an example. Such barriers can probably never be removed entirely, especially when self-interest is the motivation, but several things can be done to minimize their effects. One such solution is to gain as complete an understanding as possible of the effects of natural and man-made changes upon the coastal zone environment, but the most important needs lie in the field of public education. Responsibility for environmental education needs to be placed in the hands of trained people. It must be pursued vigorously at all levels of government and in all the communications media. The emphasis being placed by the National Sea Grant Program on information and extension services is an important and encouraging step in the right direction.

8.3.1 Inventories and Base Lines

Two of the most critical information needs for land-use planning and effective coastal zone management are environmental inventories and ecological base-line studies. Inventories include mapping of those resources and amenities which make up man's physical, biological, and cultural environments. The present lack of compiled information on these features could prevent effective land-use planning in many states.

Regional zoning, protection of wetlands, and regulation of uses of the coastal zone are all contingent upon a thorough understanding of the location, quantity, and qualities of our coastal resources and the processes critical to their maintenance and renewal. These inventories, prepared in the form of large-scale maps including coastal resources, cultural and demographic features such as transportation routes, population distribution, and land ownership patterns, would be of tremendous help in formulating projections and assisting planned and orderly growth in the coastal zone. The initial gross impacts on coastal resources of, for example, a planned road could be estimated from the inventory maps. The general compilation could serve as a basic tool for regional and national planning, although more detailed inventories of ownership and jurisdictional lines would be needed on the local level to enforce land-use policies such as dredge and fill operations in wetlands. Each mapped feature could be identified in supporting material by coordinates, acreage, any unusual features, and an indication of state, regional, and national significance.

Numerous environmental resource inventories are in progress, and many have been completed. In addition to the example cited in the introduction, Virginia has mapped and is presently evaluating its wetlands (Wass and Wright, 1969). New Jersey has completed a wetland mapping program (Anderson and Wobber, 1972), and Delaware and Massachusetts have programs in progress. Inventories of natural areas, including coastal areas, have been completed in Maryland (Metzgar, 1968) and in Maine (Natural Resources Council of Maine, 1972). Federal programs include The National Estuary Study (USDI, 1970), a national natural resource information system in the Department of the Interior (USDI, 1972) and the Environmental Reconnaissance Inventory program being conducted by the Department of the Army, Office of the Chief of Engineers (1972). We encourage such inventories and emphasize the need for continued development of coastal zone inventories.

Ecological base-line studies provide equally valuable sources of information for planning. Present knowledge of the composition and basic biological, physical, and chemical processes is not adequate for accurate prediction of change caused by modification of the coastal environment. Information is needed on the functional processes in modified and unmodified coastal systems. Assimilative capacities, rates of repopulation of disturbed areas, responses of currents and sediments to physical modification of the shape of an estuary, uptake and incorporation of heavy metals, and the general health of coastal marine life, are all examples of basic information necessary for accurate understanding and predicting environmental change. This understanding is critical for establishing criteria for coastal use. It is evident that many areas of the coastal zone are under stress, but more basic ecological informa-

The Coastal Zone Workshop recommends
The further development of predictive models to aid in understanding the effect of activities and structures upon the coastal zone environment and to improve the management of the coastal zone by evaluating the impact of alternative actions.

tion is needed before we have the understanding to regulate and identify compatible uses of the coastal zone and preserve its essential characteristics.

It will also be important to standardize and quantify methods and techniques for measuring the characteristics of the coastal zone. For example, it is extremely difficult to make a comparative analysis of benthic studies made by different people because such a variety of sampling devices and techniques is used, and the mesh sizes of sorting screens are not standardized. Species diversity indices are such useful tools for measuring the effects of water quality on aquatic plant and animal communities that some standardization is mandatory.

As stated previously, cognizance should be taken of existing information before embarking on new inventories. An inventory of existing inventories should be made first, and judgment should be made of the values of each, their accuracy, scope, applicability to present problems, and how the data can fit into the overall scheme. Appropriate state and federal agencies should be consulted, as well as universities and private institutions. Only then should new comprehensive inventories be started.

8.3.2 Predictive Models

One of the consequences of our lack of information about the coastal zone environment is our inability to predict accurately the effects of particular actions and structures.

It is obvious, for example, that a dam will produce changes in the salinity regime downstream, in transport and deposition of heavy metals and other contaminants, and perhaps in the supply of fresh water to the coastal zone (as described in Section 6.3). Yet we cannot say with any assurance what the ultimate effects will be in determining abundance and edibility of recreational and commercial fish and shellfish, in producing ecological changes that will reduce or enhance the value of the adjacent coastal zone for swimming or water skiing, or in affecting public health. It has been asserted so often that the living resources of the coastal zone are dependent upon the marshlands and estuaries for their survival that the proposition has been accepted as a truism. Yet for many important species it is difficult or impossible to find clear scientific evidence that this dependence is critical. For example, we cannot say how much of a marsh is essential to preserve any given species.

Public works and residential or industrial development in areas that will affect the coastal zone are usually presented in the context of space and time constraints that make it extremely difficult to demonstrate potential damage to other users. A ten-acre industrial development here or a five-acre housing development there present relatively small threats in themselves. But the transformation of small, independent changes such as these into a large, adverse change in the environment comes about through the steady, unplanned proliferation of such small changes. It is standard practice for many who advocate industrial or other developments in the coastal zone to place such stringent limits of time and space on the decision makers that legitimate opposition forces are severely handicapped. For example, a project to deepen a navigation channel may be presented in such a way that only the immediate effects can be considered by the decision makers. Yet the really serious effects may be the long-term consequences of the action, such as pollution from new industries that may be able to come in for the first time, increased ship traffic with all its hazards, and domestic wastes from the increased human population that is attracted to the area.

Even when the information needed to predict the effects of such changes seems complete, it is seldom used effectively to assist in making the best possible judgments. One reason is that empirical consideration of complex data is not adequate for maximum use. Through modern techniques, predictive models can be constructed and computers used to assess existing information thoroughly, to test the sensitivity of certain variables, and to judge what new information is needed most urgently (see Chapter 12). In doing so, however, care must be taken to recognize the limitations of machine methods. Even the most sophisticated computer lacks the capacity to think, but some operators forget this in their preoccupation with complicated models and intricate electronic instrumentation. The thinking must be done by humans, and this is why such methods must be directed by people with great competence and understanding.

8.3.3 Environmental Impact Statements

The fragility and interdependence of natural processes and characteristics of the coastal zone have been described in the early chapters. It has already been decided by Congress that action must be taken to preserve and conserve the essential features of the coastal zone for continued use by man (NEPA, 1970). The coastal zone management proposals made by the Coastal Zone Workshop indicate the necessity for accurate methods of predicting significant impacts associated with human activities which could alter these characteristics and processes to the detriment of man. However, accurate methods have yet to be fully developed and utilized.

The environmental impact statements required by section 102(c) of the National Environmental Policy Act of 1970 (NEPA) are concerned with the desired planning processes in which all reasonable alternatives are defined and evaluated. Agency implementation has devoted far more effort to collecting and compiling data into extensive reports than to the thoughtful assessment of impacts. A major analysis, supported by the Corps of Engineers, of more than 300 impact statements submitted up to mid-1971 (Ortolano and Hill, 1972) demonstrated that the statements lacked substance and were inadequate for making intelligent decisions. Attempts are now being made to change existing guidelines for preparing environmental impact statements and to develop new procedures and methods.

Much of the confusion of the past two years was probably necessary to test the system and to develop procedures and methodologies of assessment. Complete assessment should include information on political, social, legal, and economic issues as well as on the scientific and engineering aspects. Final assessment and evaluation of this information is essential before submission to the decision maker. The impact assessment process is difficult, and, as yet, no system of procedures has been developed to address adequately all impacts of a major coastal activity or structure. The procedures developed should eliminate the wasted effort that has been spent on insignificant items.

An example of the complexity of impact assessment is provided by one aspect of the impacts of a dam on an estuary downstream (see Figure 6.2). It is evident that many impacts are not accounted for in present procedures. Impacts may be of first, second, third, or lower orders, and these may be interrelated at any level with other impacts, some detrimental, some possibly beneficial to some users. Because impacts may be social as well as environmental, the real situation is far beyond the provisions of NEPA and of our present methods of analysis. The principles and standards proposed by the Water Resources Council (1971) for evaluation of water resource projects are broader than NEPA because they include social and economic analyses. Even these guidelines, however, fall far short of evaluating the total short- and long-term *costs* and *benefits* of man's effects on the environment and on society.

The major impediment to thorough analysis lies in the identification and evaluation of impacts. Shortsighted planners have perhaps been reticent to pursue identification of secondary and lower-order impacts because the difficulties of evaluation are so well recognized. Quantification, or at least qualification, of many types of impacts has as yet been largely ignored. We all know that wetlands have significant value to the coastal ecosystem, but these values have yet to be expressed in terms that fit the present decision-making system. If decision making for the coastal zone is itself faulty, it must be modified to measure the value of wetlands if the goal of a proper balance between use, conservation, and preservation is to be achieved. Complex problems will not yield to a realistic understanding of total costs and impacts unless all aspects of physical, biological, and cultural environments are considered and assessed. It also is evident that if we are to succeed in the goal of rational use of coastal zone resources, these improved environmental impact statements must be extended to all state and local activities and structures which in themselves or through their cumulative effects have significant impacts upon the coastal zone. Improving impact assessment procedures and analyses at all levels of federal, state, and local government will require both a major commitment of resources to the planning process and redirection of thinking on what plan formulation means.

8.3.4 Biological, Chemical, and Physical Research

Answers to many of our most pressing coastal problems cannot be found without basic investigations into ecological processes. Particular topics are in urgent need of study because they represent potential hazards to human health and to the health of the ecosystem.

One subject of increasing urgency is the need to evaluate the sources, effects, and fate of the numerous chemical compounds being added to the aquatic environment. This problem has been discussed in detail in Chapter 7. Application of the 1899 Refuse Act (Executive Order No. 11575, 1970) has demonstrated the basic lack of information by industry about the quantities and composition of chemical wastes. Major effort is needed to identify these unknowns and to understand the effects and

---

The Coastal Zone Workshop recommends
Basic biological, chemical, and physical research directed toward the following types of problems in the coastal zone:

a. **Transport, dispersion, upwelling, and cycling of nutrient and hazardous chemicals as they affect the functioning and stability of coastal zone ecosystems;**
b. **Surveillance of input levels of contaminants, especially chlorinated hydrocarbons, petroleum, and heavy metals;**
c. **Effects of solid waste disposal;**
d. **Effects of chronic, long-term, sublethal contaminants on organisms and ecosystems;**
e. **Assimilative capacity of the coastal zone for all kinds of wastes;**
f. **Epidemiologic and virologic studies;**
g. **Recovery processes in damaged ecosystems;**
h. **Factors affecting stability, diversity, and productivity of coastal zone ecosystems;**
i. **Techniques for increasing production of desirable species or systems.**

---

fates of pollutants of known concern, such as DDT, polychlorinated biphenyls, and heavy metals (IDOE, 1972).

It appears that the frequency and severity of oil spills and oily discharges into the marine environment will continue to increase (see Sections 3.2, 5.3, and 7.3). The gross effects of a few oil spills have been investigated in some detail, but not enough has been done on the sublethal effects of long-term low-level exposure of marine organisms to oil and its breakdown products. Behavioral and metabolic responses also need to be explored. Because these topics are still virtually unexplored, there is little or no information on the assimilative capacity of the marine ecosystem, and particularly the coastal zone, for oil, solid wastes, or any other contaminant. Such information is necessary to establish criteria for regulation of such wastes in the coastal zone.

Although the problem of sewage disposal has been with us since man began to live in cities, and methods have been developed and are in use to reduce the hazards to man and to the coastal environment, serious problems of sewage disposal remain unsolved. Research needs to be done on the cycling and recycling of nutrients from sewage and on the effects of domestic wastes upon metabolism, diseases and parasites, and reproductive capabilities of marine life. Another problem of concern is the fate of viruses that pass with treated sewage into the coastal zone, and that may be transmitted by fish and shellfish to man (Berg, 1971). More attention should be given to the possibility of bold new methods of domestic waste disposal, including low-cost individual units that would place the onus for sanitary disposal upon the individual.

Some wastes might be used to improve the value of the coastal zone if their effects are understood, and if they can be distributed to improve rather than degrade the environment. Nutrients in proper concentrations can increase biological productivity of natural waters but are detrimental if added in excess of the receiving capacity. Solid wastes can be used to build artificial reefs which attract sport fishes. Dredge spoil can be used to create islands and other artificial structures which can improve the amenities of the coastal zone. When possible, such positive approaches to waste disposal should be followed, but procedures for beneficial uses need study.

8.3.5 Legal, Political, Economic, and Social Research

Complete scientific information will not automatically provide solutions to the problems of the coastal zone. These problems also have important legal, economic, social, and political implications. They arise from such questions as jurisdiction over the land, water areas, and seabed; interpretation of statutory authorities and administrative directives; and conflicts between interests. Disagreements and disputes between parties with different interests tend to become strongly polarized, particularly with reference to conservation and development in the coastal zone. In part, the degree of conflict expresses the inadequacy of the information base, and it follows that some of the problems might be resolved if the available information were complete and accurate. But information on such complex problems is never complete, and even if it were, many direct conflicts between users could never be resolved by scientific and technical information alone, for the dispute may be based upon incompatible demands. Examples would be a proposal to discharge domestic or industrial wastes, or to dredge a navigation channel through a piece of oyster ground. Such disputes are usually resolved in one of several different ways which have little to do with the assessment of scientific data. Formal action on such issues, if all interested parties receive a fair hearing, can uncover extremely complicated interactions, and reasonably equitable solutions are likely to be dependent upon qualitative information in law, economics, sociology, or politics. For example, the failure of fishery management in the coastal zone has been caused largely by dominance of such qualitative issues over scientific and technical facts. There must be a mechanism that correlates natural science data with socio-legal information if more rational decisions are to be expected.

Problem solving is made more difficult in the coastal zone because information on many of the important environmental factors is fragmentary or lacking. When the principal need is to develop new knowledge, it is difficult or virtually impossible to evaluate systematically what potential future benefits may be derived. At best, bits of knowledge, even of the simplest kind, may be interpreted differently by different observers. Diversity of interpretation becomes more likely as the available knowledge moves from simple facts to questions of relationships among elements, or as the professional or educational backgrounds of interpreters diverge. At worst, there may be a conscious effort on the part of those affected by certain decisions to change their behavior to distort the plan or to alter the outcome of predicted results. In such circumstances the new information may not be compatible with what was gathered before.

The quality of information and the efficiency of its use may be affected greatly by the size and structure of the organization that makes the decisions. In general, individuals or very small organizations appear to be able to use time and place information for such purposes more effectively than large organizations. Also, in large organizations, the loss of information as it passes through organizational levels may be substantial.

8.3.6 Siting of Power Plants and Deep Ports

Siting, construction, and operation of power plants, development of deepwater terminals, and dredging and disposal of spoil are among the most important pending problems of the coastal zone, and they have been given widespread publicity. Problems of energy supply are discussed in Chapter 3, and the impacts of heat disposal in Chapters 5 and 7, but the need for information is a critical part of these questions.

The principal problem with power plants is the requirement for large quantities of water for cooling. Entrainment in the cooling water system can destroy eggs and larvae of fish and shellfish and also the plankton on which animals of the coastal zone feed. Currents created by the

The Coastal Zone Workshop recommends
The cooperation of industry, public utilities, state agencies, and the federal government in the development of regional planning and utilization of energy, including fossil, nuclear, or other fuels in the coastal zone so that costs and benefits of alternative sites of development within and outside the coastal zone can be compared. Public authorities should be guided by both the urgency of protecting the environment and the demand for energy in the United States.

intake and discharge of large volumes of water affect the behavior of marine life. Of equal concern is the effect on the coastal zone of the discharge of hot water. Much more information than now exists is needed on the magnitude of the mortality of aquatic life, the effects upon abundance of the marine community as a whole, and the possibilities for reducing or eliminating adverse effects. With these problems, as with other problems caused by man's uses of the coastal zone, the possibility of using waste heat for increasing biological productivity and producing commercial and recreational benefits must be considered. These questions are not being pursued with sufficient vigor or coordination at present.

Deep ports may not present the same kinds of difficult problems as power plants, but the public is justly concerned about the possible effects. Deep ports are deemed necessary as a means of providing increased supplies of oil for the production of power at reasonable cost, and the indirect impacts may be of great local concern. Direct effects will be caused by leakage or accidental spills during transport and unloading of cargoes. A preliminary analysis of possible impacts on the environment (Cronin, 1971) shows that supership operation and deep port siting and construction will have great potential for ecological disruption unless the necessary information is available for planning. Direct impacts would include shoreline disruption during construction; indirect effects would include changes in drainage patterns and increases in domestic and industrial wastes associated with the increased population attracted by the development. Specific information needs will include base-line studies of the dynamics and stability of aquatic biological communities, the fate of leakages from anticipated cargoes and their biological and social effects, and improved techniques for identification and evaluation of ecological change.

The Coastal Zone Workshop recommends
The monitoring of activities in the coastal zone not only for their effect upon the near-shore waters, but upon the seas and oceans. Chemicals, airborne and waterborne, from the coastal zone, as well as certain drilling, dredging, and dumping, may cause serious harm to the marine environment and should be regulated to avoid serious damage to oceanic ecosystems.

8.4 Monitoring Systems

Man-made changes in the coastal zone environment must be viewed against a background of widely variable natural changes. Some natural changes are seasonal and regular; some are longer term, related to external influences, such as climatic variations; some are sudden and sharp, such as the effects of violent storms. Changes created by man may also vary in their time scales and their severity.

The steady discharge of wastes from an industrial plant or a sewage treatment outfall may produce no effects noticeable to the casual observer, but the hidden, sublethal effects upon the plants and animals of the coastal zone and the quality of the water may be much more harmful than the clearly obvious effects of a sudden catastrophic accidental spill. Only continuous monitoring can evaluate these long-term effects.

Studies of the environment after an effect is noticed may be helpful in deciding what may have caused the problem, but they seldom can identify the cause with certainty. Changes that do not produce obvious effects may be very difficult to detect by casual, infrequent surveys. Even more important, if certain chronic sublethal concentrations of pollutants offer unusually severe environmental threats, it may be necessary to take extraordinary precautions to ensure that such substances do not get into the coastal zone. Research to determine immediate and long-term effects requires an adequate system of regional and national monitoring, automatic insofar as possible, which is capable of making observations at appropriate intervals of time and space. Adequate monitoring can give advance warning of impending damage. It will be important to determine carefully what should be monitored, and where, since it will be impractical to monitor all possible pollutants at all places.

Some monitoring systems have been operated for some time by federal

---

The Coastal Zone Workshop recommends
A sustained national commitment to education and training of the necessary talent for the management of the resources of the coastal zone. The goal should be a widespread awakening in the public to the importance of maintaining a sound coastal zone environment as well as the preparation of a future generation of natural and social scientists to manage wisely their environmental heritage.

---

and state agencies, but there is no coordinated national plan of sufficient geographic coverage or capability to measure the variables necessary to provide an adequate record of the status of environmental quality and its changes in the coastal zone environment.

8.5 Education and Training

Education and training for management of the coastal zone are needed to provide the necessary talent for resource management. People with broad skills in the natural and social sciences are hard to find. Few universities train such people, and indeed, the level of skill required for management decisions is gained not so much by training as by practical experience. The growing public awareness of environmental problems and governmental reactions at all levels in response to public pressures have so increased the need for people with broad understanding of science and human affairs that the universities must respond positively to the demand.

Federal agencies with an interest in the quality of the coastal zone, like the National Sea Grant Program of the Department of Commerce and Research Applied to National Needs (RANN) of the National Science Foundation, must encourage institutions of higher learning to develop responsive teaching programs and see that those already underway receive adequate support.

Another important need is to produce an informed and responsive constituency. A public atmosphere of emotion and crisis is as unhealthy as an attitude of lethargy and complacency. Extreme public attitudes based upon ignorance, fear, emotion, or self-interest are apt to be counterproductive and destroy the organizations charged with management responsibility without providing acceptable substitutes. All segments of society must have an opportunity to participate in making decisions on important public issues, but if rational management of the coastal zone is to succeed, we must be sure that the public is well informed as well as willing to speak out. Wise use, not protection for its own sake, was an important theme of the Coastal Zone Workshop. For example, the arguments that led to the present U.S. position on killing of marine mammals were heavily based on emotion. Perhaps the question should be reconsidered and a decision made about whether a philosophy of complete protection for groups of species is realistic or warranted.

Many ways are available to create an informed public, and all should be used to full advantage by appropriate public and private institutions. These ways include elementary, high school, and university education; authoritative articles and news items in the press and other public media; radio and television presentations; and organized information and extension systems which can translate technical knowledge into language that everyone can understand. The emphasis on extension services by the National Sea Grant Program is a model that might be expanded. All agencies with educational responsibilities, federal, regional, state, local, and private, should give priority attention to the needs of information dissemination on problems of the coastal zone and their rational solutions.

8.6 Management of Coastal Fisheries

The United States has maintained a domestic fishery production at an annual level of about 5 billion pounds despite the increasing needs of a growing population. Per capita consumption of edible and industrial fishery products has increased steadily in the last 25 years or so, and this nation is now one of the most important consumers of fishery products in the world. These growing needs have not been met by our own fishermen, however, but by imports from other fishing nations. In the last few years the United States has imported between two-thirds and three-quarters of the fish and shellfish products that it consumes.

This state of affairs has often been cited as a problem that needs to be corrected, but this is not necessarily true. In fact, the most prosperous segments of the American fishing industry are those processing and distributing firms that do not rely entirely upon domestic fishery production. The popular picture of the American fishing industry as a sick industry is only partially true.

Because the primary producing segments of our fishing industry, with some notable exceptions such as the tuna fishing operation, are primarily land based, they rely largely upon the resources of the coastal zone. Coastal fishermen have many problems, among them competition from foreign fishing fleets which operate outside our 12-mile fishery jurisdiction on the rich fishing grounds of the North American continental shelf. This competition is especially severe off Alaska and the Pacific Northwest and on the fishing banks off New England, but in other areas the effects of foreign fishing have been much exaggerated. Where the United States can present scientific evidence that demonstrates beyond reasonable doubt that foreign fishing is damaging the coastal fisheries, satisfactory arrangements usually have been possible with other nations to mitigate the conflict. For adequate resolution of these conflicts, and especially to protect the right of American fishermen to have access to the living resources, a much better understanding of the effects of fishing and of natural environmental variables upon the health of coastal fishery resources is needed.

Difficult as it is to reach satisfactory agreements with other nations that fish off our shores, international agreements for fishery management have been far more successful than domestic arrangements, even with respect to resources that are confined entirely to coastal waters within national jurisdiction. In the state of New York, for example, most fishermen will identify foreign fishing as the cause of their problems, but in reality the problems are of domestic origin, caused by a mass of restrictive laws and policies which place severe economic restrictions on fishing operations (McHugh, 1972).

Total landings of fish and shellfish in New York have been declining for the last 20 years or more. Even as it has declined, this yield has been maintained only by a constant shifting from one resource to another as, one by one, traditional resources have declined in abundance. This loss of resources has been due to many causes, some of which are known, some unknown. In New York, the two most important fisheries of earlier days, for oysters and for menhaden, are now of minor importance. With proper management and a general understanding of the scientific requirements for effective management these declines need not have occurred. The same lack of understanding, which has led to tragic mismanagement of most fisheries of the coastal zone, is responsible for the poor economic status of the fishermen of most coastal states. With few exceptions, the states have been successful in managing coastal fisheries only when they have been forced to do so under solemn international agreements. An important contributing factor to the economic difficulties of American coastal fishermen has been overinvestment of capital and labor, which has eliminated the earning of adequate profits by the fishermen.

References

Anderson, R., and Wobber, F. J., 1972. Wetland maps in New Jersey, *Proc. 38th Annual Meeting of the American Society of Photogammetry* (Washington, D.C.: American Society of Photogammetry), pp. 530–536.

Berg, G. J., 1971. Integrated approach to problem of viruses in water, *Journal of the Sanitary Engineering Division, Rec. Amer. Soc. Civil Engineers, 97* (SA6): 867–882.

Council on Environmental Quality, 1970. Ocean dumping, a national policy (Washington, D.C.: U.S. Govt. Printing Office), report to the President.

Cronin, L. E., 1971. Preliminary analysis of the ecological aspects of deep port creation and supership operation (Washington, D.C.: U.S. Army Engineer Institute for Water Resources), IWR Report 71-10, 31 pp.

Dept. of the Army, Corps of Engineers, 1970. Waterborne commerce of the United States, calendar year 1970, part 5, National summaries (Washington, D.C.: Dept. of the Army, Corps of Engineers).

Dept. of the Army, Office of the Chief of Engineers, 1972. Environmental reconnaissance inventory (Washington, D.C.: Dept. of the Army, Office of the Chief of Engineers), circ. EC-1165-2-113.

Executive Order No. 11575, Dec. 23, 1970. Instructs the U.S. Army Corps of Engineers to use 33-U.S.C.-407 Mar. 3, 1889 to control dumping.

IDOE, 1972. Baseline studies of pollutants in the marine environment and research recommendations (New York: IDOE Baseline Conference, May 24–26, 1972).

Livingstone, R., 1965. A preliminary bibliography with KWIC index on the ecology of estuaries and coastal areas of the eastern United States (Washington, D.C.: U.S. Dept. Interior, Fish and Wildl. Serv.), spec. sci. rept., Fisheries 507, iii + 352 pp.

McHugh, J. L., 1972. Marine fisheries of New York State, *U.S. Dept. Commerce, Natl. Marine Fish. Serv. Fishery Bull.,* vol. *70*(3): 570–604.

Metzgar, R. G. (ed.), 1968. Catalog of natural areas in Maryland (Annapolis: Md. State Planning Dept.), 148, 108 pp.

Nash, R. A., 1970. Permanent coastal zone data inventory and inventory system (Sacramento: State of Calif. Resources Agency, Dept. of Navig. and Ocean Devel.), prepared by Space Division, North American Rockwell, SD 70-636, v + 133 pp.

National Environment Protection Act (NEPA), 1970.

Natural Resources Council of Maine, 1972. Natural areas inventory of Maine (Augusta: Natural Resources Council of Maine), state tech. rept.

O'Connor, D. M., 1972. Information management for environmental decisions. A final report on development of improved methodologies and advances in the state of the art for law/economics in environmental data management (Washington, D.C.: National Tech. Info. Serv., U.S. Dept. Commerce), GURC/NASA (MTF) contract no. NAS8-26897, GURC/NASA(MTF)-OL1, ii + 24 pp.

Ortolano, L., and Hill, W. W., 1972. An analysis of environmental statements for corps of engineers water projects (Palo Alto: Civil Engineer Dept., Stanford University), Nov. 1971, revised May 1972.

State of California, Resources Agency, 1972. California comprehensive ocean area plan. Background, history and administrative framework, organization, responsibilities and policies, definition of coastal zone (Sacramento: State of Calif., Resources Agency, Dept. of Navig. and Ocean Devel., COAP Development Program), 12 pp.

U.S. Department of Interior (USDI), 1970. *National Estuary Study* (Washington, D.C.: U.S. Govt. Printing Office), 7 volumes.

U.S. Department of the Interior National Natural Resource Information System (USDI), 1972. Final draft (Washington, D.C.: Director of Regional Planning, U.S.D.I.), 100 pp.

Wass, M. L., and Wright, T. D., 1969. Coastal wetlands of Virginia, interim rep. to Governor and General Assembly of Virginia (Gloucester Point: Va. Inst. Marine Sci.), spec. rep. 10, 154 pp.

Water Resources Council, 1971. Proposed principles and standards for planning water and related land resources, *Federal Register, 36*(245), part 2: 24144–24194, Tuesday, Dec. 21, 1971, Washington, D.C.

9 Allocating Coastal Resources: Trade-off and Rationing Processes

9.1 Introduction

The coastal zone is inhabited by a variety of forms of life, only one of which is human. This area, where water and land meet, contains unique physical characteristics. These physical features of marine life and human behavior and activities may interact with both negative and positive effects. Important and necessary attributes of the natural ecosystem are being threatened by the type and magnitude of effects generated by the human activities. Deliberate changes in the way humans utilize coastal resources must be made if negative environmental impacts are to be controlled or reduced.

Public action to protect the natural coastal processes will not be initiated simply by making information about the problem known. The introduction of new information into a decision-making system does not automatically result in new policies. Measures which require a large and heterogeneous body of people to forgo or reduce the level of utilization of valued resources will not be implemented easily, but this is the task which must be accomplished in allocating the resources of the coastal zone.

Marine biologists, oceanographers and ecologists have been instrumental in raising the coastal zone issue to a public forum. Their ability to make the complexity and fragility of marine ecosystems understandable has been of major importance in creating a national awareness of the problems. Social scientists have not performed the comparable function of making it clear that the social, economic, and political processes which exercise important influences over the formulation and administration of governmental policy are also highly complex. The passage of new laws or the establishment of another regulatory agency does not guarantee the achievement of the legally defined purposes of either.

Any serious effort to create rules and public institutions that will reflect desired environmental and social values in allocating coastal resources must consider several factors which have received little attention in the past. One is an understanding of the way in which social, economic, and political processes mediate between the conflicting demands which individuals and groups believe require the use of governmental authority and public action. The content and effectiveness of any coastal zone policy that is adopted will be influenced by these processes.

A characteristic of the complex American system of government is the large number of entry points into the system; points at which influence on policy can be exercised. This allows well-organized groups to become especially adept at circumventing regulations which interfere with their goals. Hence, the placement of all authority under the aegis of a single coastal zone authority has little chance of success, as existing interaction of the many resource allocation mechanisms will be in competition with any new agency that is devised. A second factor is the restrictions these processes place upon the capacity of any governmental or private agency to modify existing patterns of human behavior in relation to coastal resources. There is a high value placed on access to the coast and no adequate public policy can be formulated without first understanding the immense social and economic pressure for increasing human activities of all types in the coastal zone. Consequently, agencies with authority to manage the coastal zone cannot unduly reduce ways of human access. There must also be a responsibility for increasing the carrying capacity of the coast for human activities in ways that produce maximum benefit with minimum environmental impact.

The success of efforts to manage coastal areas will depend to a large extent upon the ability of the overall political process to provide general guidelines, and the capacity of governmental agencies to administer them in a dynamic environment. The physical, biological, social, psychological, and economic values and the interplay of these values concerning uses of the coastal zone are not static. The number and strength of groups associated with particular values will not remain the same, but will change with time as will their strategies for influencing policy.

The formulation of a comprehensive federal- or even state-level coastal zone policy will not be a matter of simply deciding on rules to preserve marine ecosystems, since there must be choices made among environmental, social, and economic values. There will also be a need to ration or choose among the possible alternative uses of the resources and determine who will be able to use them. These decisions must be made within a framework that allows fair access to the policy making process, a socially equitable distribution of coastal resources among groups in the nation, and adequate information for determining the social, economic and environmental effects of the policies and rules that are considered and adopted.

The following general assumptions are considered fundamental to the development of methods for the allocation of coastal resources:

1. Social, economic, and political systems relevant to the coastal zone

are complex and varied, cannot easily be modified, and major changes in their structure may produce negative as well as positive effects.

2. There is no consensus in the nation or in various regions on the extent to which social and economic values should be foregone to maintain physical and biological systems in the coastal environment.

3. The base-line and process data available on social, economic and psychological, as well as physical and biological, variables relevant to the coastal zone are limited in many cases and do not exist in some.

4. No single organizational arrangement for coastal zone management is uniformly applicable to all parts of the nation.

5. Governmental management agencies constitute only one of several means by which decisions affecting the coastal zone are made and function concurrently with market transactions, bargaining, voting and adjudication in the allocation of resources.

9.2 The Governance and Management of the Coastal Zone

The large and rapidly growing concentration of people and economic activity in the coastal zone is producing a range of pressures upon its resources. Demands for public and private goods and services will become greater because of increases in the absolute number of humans in proximity to the coast. Higher personal income levels, additional leisure time, and new recreational technology will also mean that more people in the country will seek to engage in coastal-related activities. Current demographic and economic projections indicate that these growth patterns will continue in the foreseeable future. Scale-economies, employment, cultural opportunities, and climate, as well as transportation advantages, will continue to produce a higher growth rate for people and industry in metropolitan areas on the coast than for the nation in general.

These increases in the magnitude and variety of demands have not occurred without costs. Their cumulative effect has produced a deep concern on the part of many about how a balance can be struck between maintaining the natural coastal ecosystems and satisfying strong social preferences for goods and services which rely upon coastal resources. If, as is commonly proposed, public agencies are to be created to establish and administer policy concerning the coastal zone, the way in which they are designated becomes a matter of some importance. Far more attention has been given to identifying and publicizing the environmental issues in the coastal zone than to seriously working through the implications of forming special public bodies for dealing with coastal problems.

9.2.1 The Existing System

Decisions concerning the use of the coastal zone are now made by a variety of governmental units of all sizes and types; many are at the local and substate level. Coastal cities and counties exercise considerable control over what happens in the land-water interface within their boundaries. Specialized agencies, such as port authorities, largely determine policies about specific functional uses. State and federal agencies produce some goods and services, such as parks, dredging, marine rescue services, harbor improvements, and defense. Their major influence in the coastal zone, however, is through regulating the activities of others in both the public and private sectors and providing incentives for positive actions by governmental units through grants and other types of funding. Most commonly, a number of administrative agencies within each state and the federal government are involved in these functions. These may include state agencies dealing with health, fisheries, water quality, recreation and game departments; and the Corps of Engineers, Coast Guard, Federal Power Commission, Atomic Energy Commission, and Environmental Protection Agency, among others, at the federal level.

Many direct allocative decisions are made through market transactions in the private sector. The pricing system plays a major part in determining what use will be made of particular parts of the coastal zone and who will have access to the market. A final mechanism that is used in allocative decisions is adjudication. State and federal courts, for example, often play a major role in creating policies concerning the utilization of the coast, which are binding upon private citizens as well as public agencies.

The present problem is not the absence of public agencies for making policy about the coast. Rather, the difficulty in many people's minds is that the use patterns for coastal resources produced by the existing system are not desirable in environmental terms. This failure to give adequate weight to the ecological effects of human activity is usually attributed to the existence of many decision points which can influence how the coast is used. Some, such as local units of government, are seen as too small in scale. At the state and federal level, the belief is that there are too many agencies involved and that their responsibilities overlap and even conflict in some cases. This type of analysis is the basis of the frequently made recommendation that there should be new,

---

The Coastal Zone Workshop recommends
The development of a vigorous and comprehensive *National Coastal Zone Policy* by the federal government in cooperation with the states that will provide for the wise use of the marine, estuarine, wetland, and upland areas bordering the American shores. All future uses of the coastal zone must be designed to maintain the natural ecosystems and to provide for the use of contiguous resources by the people of the United States. Cooperative action by cognizant federal, regional, state, and local governments will be required. The integral element of the National Coastal Zone Policy should be the focus of management responsibility at the state level, with the active participation of local governments, under federal policies that provide grants and set guidelines for creative and effective programs.

---

fewer, and larger public agencies and that they should have special and clear-cut authority to manage the coastal zone.

As a consequence, much of the attention that has been given to governmental arrangements has centered on the scale of the governmental unit to be used and the formal authority it will exercise. The following dominant issues have involved—

1. the relative roles of the federal and state governments;
2. the need for some kind of multistate agency in some cases;
3. the question of how comprehensive the powers of new agencies should be; and
4. what internal organization or administrative structure should be adopted for these agencies.

9.2.2 Social, Economic, and Political Processes

The discussion about coastal zone management agencies tends to leap from the recognition of the problem to detailed consideration of the legislation believed necessary to solve it. This type of jump, however, usually means that no consideration has been given to the social, economic, and political processes that will influence whatever public action that is taken. Failure to give attention to these mediating forces is not unusual in efforts to design public agencies. The result is often no action at all or a new unit that is either ineffective or that reflects extremely narrow values.

One reason for this failure to anticipate such problems in efforts to implement "obvious solutions" is that the proponents, in this case marine-oriented scientists, planners, and administrators, mistake agreement among themselves for a national consensus on the issue. There may be general support in the United States for the idea that some limitation is necessary upon human exploitation of coastal resources. However, there is no consensus as to which social and economic values should be foregone and in what amounts to protect various physical and biological systems in the coastal environment. Consequently, advocates who attribute their own values and understandings to the public at large expect the new governmental agency they call for will administer widely accepted guidelines for allocating resources. If widespread public agreement does not exist, the agency turns out to be poorly designed to deal with basic value conflicts that must be resolved in their decisions. The danger of this occurring in relation to the coastal zone arises from a confusion of management and governance functions among proponents of new public action, as well as from the absence of a national consensus.

Dealing with the coast is most frequently referred to as a "management" function. However, if there is not a consensus as to how coastal resources should be allocated and a number of value conflicts remain to be resolved, more is involved than the management of resources for predetermined uses. The term *management* refers to a specialized part of the public sector and should be distinguished from governance. It is the governance process that provides the framework and sets the parameters for management actions. It is through the components of the governance process, voting and legislative, executive and judicial decisions, that public policies concerning the coastal zone will be formulated.

Public management functions are performed through hierarchically organized agencies, established to carry out policies determined by the governance process. In some cases, such agencies may exercise substantial delegated authority. It is also true that a management agency that has an interest in an electoral, legislative, or executive decision will usually find ways of having its position represented. However, the management function is not the governmental process itself and is normally viewed as means for carrying out rather than making public decisions.

9.2.3 The Management Syndrome

The coastal zone, at least at present, is more than a management problem in the above sense. If substantial authority over the coastal zone is given to governmental agencies, they must have mechanisms to reconcile conflicting values, allow individual and group preferences to be articulated, and provide for citizen representation in the decision process. Agencies designed to engage primarily in "management" functions almost never have the structure or incentives to deal with questions of access, conflict, resolution, or equity.

Management-oriented responses to complex value problems also have built-in assumptions that require careful evaluation. These concern scale, standardization, and what is known about the phenomena to be managed. It is commonly assumed in the management field that the larger the scale of the agency to be created, the more effective it will be. This is reflected in the primary emphasis upon the states and national government for coastal zone management noted previously. However, this must be reconciled with the fact that the production of most of the public and private goods and services that utilize coastal resources occurs at the local and substate level, as do their environmental and social effects. Consequently, many of the day-to-day activities involving public regulation of coastal areas, as well as aspects of decision making, will be organized locally.

The scale of a governmental unit affects the access people have to it, its ability to generate and use information, and the types of value biases that exist in its decision processes. Standardization of procedures and uniformity in organizational arrangements where branch offices exist are common elements in the design of large governmental agencies. This approach assumes that there is a standard treatment of the phenomema of concern to the agency and that it should be uniformly applied by all units of the agency. The variations in the physical, social, and economic characteristics in the coastal zone of the United States do not easily support a standardized treatment. For example, the fact that over 50 percent of the nation's population lives within 50 miles of the seacoast or Great Lakes (Woodward, 1972) is often used to justify the creation of coastal zone agencies, but it can also be misleading. It gives the impression that all portions of the coast are being uniformly affected by population and economic pressures. The opposite is true.

Approximately 55 percent of the total coastal population of the country lives in the 25 largest coastal counties in 11 states (Center for Urban Affairs, 1972). There are also important differences within and among states. A third of the state of Washington's population resides in one coastal county while twelve others have less than 5 percent each. In North Carolina, 17 of 20 coastal counties have less than 1 percent each

The Coastal Zone Workshop recommends
Research on the environmental social, economic, and legal effects of:
a. Siting, construction, and operation of coastal and offshore power plants and deep ports
b. Dredging and deposition of spoil

of the state's population. In California and New York the counties around Los Angeles and New York City have over half of the population of the nation's two largest states.

Clearly the problems of coastal development in a megalopolis and a cluster of counties with a small and declining population and economy are not the same. Between these two poles there is a range of variations. The public sector should not only be able to respond differently in these cases but also be able to prevent the values and preferences of developed portions of the coast from being uncritically imposed upon less developed areas.

The establishment of a governmental agency to deal with a problem area can also result in misleading assumptions concerning the adequacy of the information available about the phenomena that are to be regulated. In the case of the coastal zone we have enough data to know that important and fragile natural ecosystems are being threatened by human activity. This is adequate to justify public intervention. At the same time, however, there are serious limits on the amount and quality of physical, biological, demographic, and economic information that exists or is in usable form for making judgments about which specific policies should be adopted. The base-line and process data necessary for identifying the full costs and benefits of alternative choices for coastal use are frequently inadequate or lacking. This is particularly true of the information about the social, economic, and psychological transactions of humans that relate to the coast. A major function of coastal zone management agencies will not be to administer an existing body of knowledge but to facilitate and provide incentives for the generation of substantially more information than now is available (see Chapter 8).

A final point to be mentioned is the tendency for management-oriented agencies to be designed as if they would be able to operate in a political and societal vacuum. As noted earlier, the market, voting, bargaining, and adjudication, in addition to administrative agencies with overlapping responsibilities, will all affect and be affected by any new unit with authority over the coastal zone. There are a variety of processes by which individual and group preferences concerning the coast can be articulated and satisfied. With these issues defined, it will be useful to consider some general value questions, the concept of a rationing policy, and the importance of devising methods of increasing coastal carrying capacity before moving to a more detailed discussion of the trade-off process and organizational requirements for coastal zone agencies.

9.3 General Values in Coastal Zone Decisions

Whatever their scale and formal characteristics, there are certain similar functions which agencies with a major responsibility for coastal areas will carry out. These include—

1. regulating or prohibiting phenomena that are determined to have negative effects of an unacceptable degree;
2. allocating or rationing coastal resources among potential users when the demand exceeds the supply; and
3. determining ways of increasing the carrying capacity of the coastal zone to increase the supply of socially preferred or necessary resources.

The processes by which these functions will be carried out must be based on some form of calculation by which the costs and benefits of one use or another by one set of users or another can be balanced and judged. The ability to make these trade-offs accurately and equitably is a central requirement of any agency that makes or administers policies concerning the coast. A public decision to encourage one use rather than another should be based on procedures that allow the identification of what is being foregone and gained, by whom, and in what amounts. A variety of trade-off combinations is usually possible in relation to any particular policy question. In general terms there must be a set of criteria for coastal zone decision processes to respond to and to allow a balance to be struck among various uses. The criteria should include economic, physical, biological, social, and psychological values. The means of measuring each of these and their "standing" in public decisions about the coastal zone vary considerably at present.

Cost-benefit measures in economic terms are by far the best developed and most accepted form of calculating values. Measurement techniques for physical and biological variables are also well refined. Two problems exist, however, in their use. In specific cases, such as a dredging project or offshore oil drilling, the information necessary to evaluate ecological effects may be possible to collect, but only at a high cost. When the data are available, they may not be given adequate weight with other variables in a decision. There are formal federal and state requirements for environmental impact studies which are intended to "force" consideration of these values. Yet the general agreement on the limited success of impact studies in meeting this goal reflects the resistance of decision makers to take ecological values into account.

Social values present further problems. There is no well-established agreement on methods for measuring and reflecting social values, such as equity, in policies concerning the coastal zone. Apart from measurement difficulties, which are formidable, there has been little serious inclination to identify the potential social costs of particular allocative policies. In some instances the character of the problem has been noted. Perhaps the most frequently mentioned aspect is the problem the public has in obtaining access to beach areas in a number of places (Report to the Governor's Task Force on Oceanography, 1969; Finley and Vantel, 1970). Access to the coastal zone is a serious problem for the poor and minority groups who have limited resources for competing in either the political process or market, but this receives little attention (Nehman and Whitman, 1971). Little, in fact, can be found which attempts to trace the increasingly complex social impact of present allocative patterns as, for example, how nondominant groups in a community might be affected.

If a number of luxury high-rise apartments, for example, are located close to one another on the coast there are several kinds of costs that can fall disproportionately on certain groups (Moss, 1972). The construction of the apartments may require the replacement of housing utilized by low-income persons, the aged, and members of fringe subcultures. The value of adjacent land with moderate-cost housing normally increases and results in additional sites close to the coast being put to higher use. Where the development fronts on public beaches, it can reduce the street parking space available and may change the general social character of the area so that use of the beach by low-income or minority groups is discouraged. Other similar cases are not difficult to find. The replacement of a beach by a marina would normally reduce the number of individuals who could utilize the space and change the socioeconomic characteristics of those who do.

Means should be devised to protect the interest of individuals who cannot enter into the political process either because of a lack of re-

---

The Coastal Zone Workshop recommends
Recognition of the interest of people dwelling outside the coastal zone, but who are directly affected by its environmental conditions or its productivity. The needs of individuals and groups who have limited resources for competing in the political bargaining process in reaching coastal zone policy decisions must be considered.

---

sources or because they are not aware of the particular action and its potential effects. Emerging land-use trends in coastal counties and cities will make it important that attention is given to means by which social costs and benefits are more systematically considered in policies concerning the coastal region. If this does not occur, market pressures for higher use of the coast combined with conservation activists seeking to protect the environment can very well have the effect of further limiting access for people generally and producing a rationing system primarily based on pricing.

Failure to deal with these issues now may have important social and political consequences in the future. Many of the large urban coastal centers will have substantial increases in the percentages of nonwhite population in the next decade (Taeuber, 1971; Farley, 1971). Given the present policy patterns, deep social stresses could arise if black and chicano citizens become a major political influence but are largely excluded from access to coastal resources because of pricing or a set of public facilities designed to meet the values of middle- and upper-class whites.

Psychological factors involved in the association of humans with marine environments are the least understood of this set of values and have seldom been subject to serious study or played a direct part in decisions concerning the coastal zone. The intensity of the involvement of people with the water-land interface of the coast is reflected in literature, music, and paintings, and in the symbolism of psychoanalytic theory (Huxley, 1962; Sauer, 1967). Having the opportunity to live in proximity or to view periodically bodies of water or engage in water sports may have psychologically therapeutic effects as well as provide aesthetic benefits (Gunn, 1972).

During the urban riots of the middle 1960s, portable swimming pools became a popular service provided by city governments in ghetto areas to "cool" things. The importance of mass public use of coastal zone related facilities, however, has received almost no attention. The importance of the question can be underscored by trying to imagine the psychological effects and, in turn, the social consequences of completely closing some or all of the beaches in the New York City area during a hot summer.

The creation of a decision-making system to utilize coastal zone resources and support the articulation of the values discussed here will not guarantee a particular outcome unless there is a public mandate to give some more weight than others. The absence of a firm consensus in most cases will increase the opportunity to participate in a trade-off process for individuals and groups who wish to initiate public action and others who will be affected by the results of a particular allocative choice.

9.3.1 Rationing Policies

If it is decided that the total demand for a coastal resource cannot or should not be satisfied, there are other choices to be made about how much of the particular resources will be used, by whom, and at what cost. It will be through these rationing decisions that some of the most basic policies concerning the coast are determined. Rationing assumes that the preferences of some will not be accommodated. Unless the rules for these allocative choices are well defined and treated as important policy questions, the probability is high that the most influential individuals and groups interested in the resource will gain disproportionate benefits.

Rationing may take several forms. One option is to ration among uses, for example, between direct human use and ecosystem maintenance. California has enacted laws to forbid the removal of sea shells and certain types of marine life from beaches and tidal pools except for educational purposes (*Santa Monica Evening Outlook,* 1972). In Delaware, certain industries have been restricted from locating in any part of the state's coastal zone (Coastal Zone Act, 1971). When two or more activities occur or can potentially occur on the same site, rationing may involve the elimination or proportionate reduction of one in favor of another. In a small boat marina, for example, swimming or water skiing are often excluded, restricted to a small and segregated portion of the total usable area, or limited to specific hours during the day (Jaakson, 1971).

A rationing policy can also involve the determination of a set of rules which discriminate among potential users of the same resource. Pricing is a traditional means of achieving this end. The imposition of fees for access to public beaches is used in some places and is increasingly advocated (Goldin, 1971). Quotas can also be set administratively and different techniques utilized to fill them. A state can set a limit on the number of offshore oil wells for a given surface area and then allocate leases through bidding. Overnight public camping grounds are normally restricted to a specified number of occupants. A first-come first-serve policy can be used either through direct queuing or a reservation requirement. Use of camp sites can also be rationed by imposing a limit on the number of nights an occupant can stay. The use of lotteries to allocate space in high demand areas may not be too far in the future.

The restriction of certain activities to specified areas is a final type of rationing that will be noted here. High-rise apartments, power plants, or offshore airports could be feasibly located at a number of sites along many urbanized coastlines. However, there are regulations, usually in the form of zoning, that restrict the range of sites that can be utilized.

These examples of rationing among uses, users, and sites do not exhaust the range of variations that is possible nor fully explore the implications of rationing policies. The discussion does indicate the nature of the choices that are made and that a large number of public agencies—cities, counties, districts, and state and federal agencies—are involved in determining such policies. It is also clear that the use of rationing as a policy instrument means that each choice results in some value being foregone in relation to another. Consequently, the adequacy of the information upon which such decisions are made and the articulation of the extent by which values will be beneficially or adversely affected are two crucial questions which agencies formulating general policy must consider.

9.3.2 Alternatives to Rationing

Scarcity of a resource can be responded to in several ways, in addition to rationing per se. In some cases, substitute inland sites can be used for activities that otherwise would be located in the coastal zone. In others, technological and management solutions can be used to increase the carrying capacity of the area.

There can be deliberate public decisions in large population centers to expand inland swimming and boating facilities, waterfront residences, and industrial sites as part of an overall policy to utilize resources of the total region to relieve pressure upon the coast. Such an orientation would, however, require an ability to think in spatial terms beyond the conventional definition of the coastal zone and the public authority to take such actions.

Site substitution, in some cases, can occur in the private rather than the public sector. Population growth does not necessarily mean a straight line increase in the demand for a particular coastal resource. In response to strong pressure for housing adjacent to water, private developers in Southern California are building an increasing number of single-family dwellings, town houses, and condominiums around artificial lakes. Individual pools or neighborhood association swimming clubs are relatively common in many places. It is also true that the inconvenience of queuing for access to coastal beaches and boating sites in congested population centers will produce pressure for public investment as well as make private investment attractive for the development of inland recreation sites. People may also be encouraged to substitute one activity for another, such as golf for boating. Tax benefits and other incentives can be means of encouraging private responses to redistribute certain kinds of activities inland from the coast.

9.3.3 Coastal Carrying Capacity

In spite of rationing and site substitution strategies, future population increases still promise to involve far greater demands for access to coastal areas than are now made. The social, economic, and political processes through which societal decisions are made are not likely to stop the growth of the number of people living in and using the coastal area. Consequently, whatever one's view of the desirability of introducing more people into the coastal zone, the question of how its carrying capacity can be enlarged to satisfy social preferences and needs is one that must be explored. Technology provides one option and management strategies another.

On the technology side, the construction of offshore facilities will represent a major factor in future coastal development. The first attempts to use offshore locations grew out of offshore mineral exploration on the continental shelf. Heightened demand for energy resources made offshore exploration financially feasible, and building technology has responded to the requirements of offshore commercial activity by developing materials to resist the special elements of wear, indigenous to the marine environment. Concrete has the durability and strength desirable for underwater construction. The Brookhaven National Laboratory has recently developed a process for injecting liquid plastic into the voids of hardened concrete. Once the plastic hardens, the concrete is impermeable to water and more durable and stronger for marine applications (Wallen, 1971).

This technology has made it feasible to explore the concept of floating cities or satellite communities offshore as a means to accommodate the population pressures in high-density coastal areas. In England, there are plans for a self-sufficient floating community of 30,000 inhabitants to be built fifteen miles offshore at Haisborough Tail (Nettleton, 1971). In Hawaii, a scale model of a floating platform is being tested as the initial step in the construction of a large residential community over water. The entire project is scheduled for completion by 1977 (Lear, 1971a,b). A floating airport has been suggested as a means to serve Chicago area travelers.

To accommodate supertankers, offshore deepwater ports are under consideration for both the Texas (Architecture Research Center, 1971; Bragg and Bradley, 1971) and Delaware coasts. In New Jersey, the Public Service Electric and Gas Company has proposed construction of two nuclear power plants on floating platforms eleven miles northeast of Atlantic City. The firm has been unable to find land sites acceptable to the public and now considers the water sites to be more environmentally satisfactory (Lapp, 1972). Other technological developments have been made in providing transportation access to on-water and underwater facilities, through hovercraft and structures such as bridges and submerged tubes or tunnels.

On-water and underwater sites may also be used to enhance the carrying capacity of the coast for residential space and recreational activities. The New York City Planning Commission recently approved plans for a $1.2 billion community to be built on a series of platforms extending into the water, along one mile of shoreline (Siegel, 1972). In Santa Monica, California, an offshore island has been proposed for new residential complexes. Artificial reefs can be built to improve surfing conditions or sport fishing, and underwater parks may permit tourists more intensive experiences with natural phenomena. Underwater parks already exist in Hawaii, the Virgin Islands, and in California.

Management strategies can also serve to enhance or expand the carrying capacity of the coast. This can involve altering the rules that regulate coastal activities, reducing negative externalities from existing coastal uses, and redistributing transportation access to the coast. More stringent requirements for sewage treatment plants could, in some cases, reduce negative aesthetic and biological spillover effects so that the shore adjacent to such plants could be usable and attractive to humans for recreational activities. Pigeonhole storage of smaller pleasure boats and the encouragement of multiple ownership of boats through fee incentives could allow both more craft to operate from a marina and for their more efficient utilization.

Public subsidies for bus service to coastal recreation area sites and the provision of more adequate parking facilities for beach goers can increase the use of beaches that are now underutilized. The installation of fluorescent lighting in selected recreation areas in highly urbanized regions could allow citizens more latitude in allocating their own time budgets for beach use, as well as encouraging usage after working hours. The pricing of beach parking and entry could also serve to increase the use of shoreline recreation facilities during nonpeak periods and thus relieve pressure on peak-use days.

Industrial facilities can also be integrated with other uses and activities. One proposal for an agro-industrial nuclear complex would produce fertilizers and desalinated water on site and then irrigate and fertilize the adjacent coastal desert for farming. According to preliminary plans, such a "nuplex" could ultimately provide settlement and employment opportunities for 250,000 to 400,000 persons (Meier, 1969).

Methods to increase the carrying capacity of the coastal zone will receive additional attention in the immediate future and, in turn, will become the focus for major policy conflicts. It is likely, for example, that proposals to construct offshore ports, energy plants, or cities will generate some of the most intense controversies over the use of the coast during the next decade. The same will be true of questions about the rationing of access as the percentage of the population adversely affected increases. How efficiently and equitably these issues will be articulated and resolved again leads us back to the matter of how adequate our public mechanisms will be for allocating coastal resources.

9.4 Organizational Requirements for Coastal Agencies

Any new set of public agencies with responsibilities for the coastal zone will be in an anomalous position. They will be created to correct inadequacies in the existing system by which coastal policy is made and administered. Most of the local, state, and federal agencies found to be inadequate, however, will remain in existence and still exercise some

---

The Coastal Zone Workshop recommends
Public authorities at all levels should consider methods of increasing the carrying capacity of the coastal zone through technical and managerial means, utilizing airspace over land and water as well as submerged areas in order to achieve community goals.

---

degree of authority in relation to the coast. Similarly, the other major allocative mechanisms will continue to function. The new coastal zone authorities will become part of a preexisting, albeit modified, system rather than replace it.

The success of new units will depend, in part, on their capacity to do some tasks more adequately than other public agencies and, in part, on their ability to interact effectively with other institutions that are making or influencing allocative decisions concerning the coastal zone. Some of the issues that must be considered in creating effective public agencies can be identified by reviewing selected characteristics of existing small-scale policy making units and then considering how larger-scale coastal agencies would differ.

9.4.1 Small- and Large-Scale Effects in Organizations

Discussions of what coastal zone agencies are expected to do often imply that they would be assuming all of the policy-making functions of integrated political units. If this logic was carried out, it would be necessary to establish a governance system exclusively concerned with the coast. Officials would be chosen by an electorate. The former would be responsible for policy formulation and management functions to enhance the preferences of the citizens. Officials would moreover undertake intergovernmental negotiations with other public agencies at the local, regional, state, and national levels.

The proposals that are made, however, do not explicitly cross the line from a management to a governance system. As a result, a number of serious questions concerning how policy will be formulated tend to be left out. For example, the existing governmental units located on the coast which do constitute integrated political units, cities and counties, are rejected without real attention to what they do well and how their desirable attributes could be maintained in agencies that would assume some of the decision-making authority now possessed locally. To consider this matter further, the way in which cities and counties, particularly small municipalities, deal with questions of access, information, preference articulation, and equity will be reviewed and then compared with how large-scale organizations might function.

With the exception of small coastal cities, it can be argued that no governmental organization now exists which has coastal resources as a primary concern. The observation that only small coastal cities have an integrated governance system is a revealing one, because coastal zones are also contained in large cities, counties, and states. Yet none of these larger units has the focus on coastal zone problems that may be necessary if valuable coastal zone resources are to be efficiently, effectively, and equitably used in the future.

The fact that larger units have not become polities for dealing with coastal zone problems, however, does not mean that decisions on these issues are not made. Instead it means that decisions regarding the coastal zone within the larger-scale organizations have often become dominated by well-organized interests and there may be little direct awareness by citizens and unorganized groups. Thus, many individuals or groups that are affected by coastal zone trade-off decisions are unable to make their preferences known or to participate actively in the trade-off process. Consequently, it is possible, and even likely, that when special interests dominate coastal zone trade-off decisions they will foster developmental patterns that are neither efficient nor equitable.

If the trade-off process in larger governmental units may not be equitable and efficient, what comparative observations can be made relating to the coastal zone decisions made by small coastal cities? Small coastal cities are more able to focus on coastal zone problems because the coastal area is the major resource of the community and its use is critical to most of the city's citizens. Further, because the political unit is small, the citizens are likely to feel that their views can be taken into account in the political process and thus have greater incentive to be politically active than the average citizen in a larger more diverse political system.

Knowledge of the internal governance system of the small city is not, however, sufficient to understand its behavior vis-à-vis coastal zone resources. The small city is part of the entire economy and polity of the state and of the nation, and it must respond accordingly in many kinds of intergovernmental relations. It is usually influenced in its actions by superior political jurisdictions in order to achieve supralocal, state, regional, or national purposes.

Federal and state governments impose water quality standards with which local authorities and industrial dischargers must comply; federal and state grants-in-aid for construction of waste disposal facilities may influence design and specification of treatment plants and location of outfalls; state and federal governments may contribute to costs of local parks in return for removing restrictions on nonresident visitors, and in some cases agreements may be negotiated for city parkland to be transferred to county or state operations because so many users come from beyond the city's boundaries. Nevertheless, most of the problems perceived to exist in the coastal zone are commonly attributed to the relatively small scale of coastal cities and the lack of fit between their boundaries and the problems of a region.

9.4.2 A Coastal Zone Management Authority

In order to deal with the problems attributed to this decentralized structure, proposals have been made to create new governmental units —special coastal zone agencies large enough to take into account most potential trade-offs for coastal zone uses (CMSE&R, 1969). Such an agency, it is argued, would overcome the small-scale problem of small coastal cities and the neglect of the coastal zone by existing larger general-purpose governmental agencies. Most recommendations are for an administrative agency with appointed officials rather than for a political entity through which a political process could operate to deal with the societal trade-offs associated with coastal zone and related problems throughout the region. Proposals for an administrative agency of this kind raise several critical questions:

1. Will the recommended organization provide knowledge and access for all relevant individuals and groups to make their preferences for the use of coastal resources known and considered?
2. Will the recommended organization provide opportunities to consider the wide range of alternative courses of action in response to these preferences, and to compare and choose among the trade-offs inherent in each?
3. Will the entire process be regarded as fair and one that will promote a high level of agreement on the decisions that result?

Two related but conceptually separate problems continually arise in the decision-making processes of administrative organizations. These are related to gathering, transmitting, and using information per se, and questions as to which societal groups or individuals are best able to articulate the preferences and to ensure consideration in the trade-off decisions that are made. Both of these are treated in the following two sections.

9.4.3 Information (Hayek, 1945; Lindblom, 1965)

The relationship between the scale or size of an organization and the kind of knowledge it can effectively develop or use for decision making is generally neglected in both social science and management literature. For simplification of this analysis we shall consider knowledge as running along a continuum from that defined as "time and place" to that defined as "of general scientific value." Time and place information is that knowledge of particular circumstances or occurrences, such as the rate of eutrophication in a specific estuary. Information of general scientific value includes knowledge of scientific laws or functional relationships that appear to exist in a large number of places, such as the relationship between wind velocity and wave height.

Individuals and very small scale organizations appear to be able to use and produce time and place information in their decision-making calculations more effectively than large-scale organizations, in which time and place information is often distorted or ignored. New scientific knowledge of general utility may most efficiently be produced by a large-scale organization which can support the costly research efforts necessary. Individuals and small-scale organizations may be able to use general scientific information in interpreting time and place observations, but the individual or small-scale organization is unlikely to support and conduct the kind of basic scientific research which provides the basis for improved information necessary for coastal zone decisions. At the same time large-scale organizations may not be able to utilize time and place information because of the simple nature of the hierarchical or bureaucratic process used for internal organization.

9.4.4 Preference Articulation

Individuals and groups articulate preferences to political units primarily through bargaining procedures such as lobbying, direct contact with elected or appointed officials, and letters and voting for representatives and on specific issues. The size and scope of a particular political unit does affect the cost and efficacy of preference revelation. The larger the unit's size and scope and the more heterogeneous the preferences of individuals within it, the less influence an average citizen will likely have. For those functions that affect him most strongly, he will find it efficient to form functionally oriented interest groups to enter into direct bargaining with elected and appointed political officials.

Because of the size problem of hierarchies, these clientele groups will find it most efficient to deal with the level of bureaucracy that can satisfy their demands, thus bypassing formal voting or demand-articulation structures. This implies that the larger the size and scale of a political organization, the more issues will be decided within the organization and between the organization and its most interested clients outside the view of the general public and press. This process may be efficient where there is high agreement among citizens, but severely biases the political process against the average citizen and unorganized groups where interests of citizens differ considerably.

On the other hand, if political units are generally of limited scope or small scale, individuals will be able to make their preferences known to the unit more effectively, and many crucial issues will be resolved in intergovernmental bargaining among political units. The bargaining is more likely to be visible and public than are the agreements struck between interest groups and administrators in larger units, and the smaller political units will make it less likely that the average citizen or unorganized groups will not be taken into account when they feel strongly affected by some aspect of the political process. Thus, political systems composed of units of relatively small size and scope will provide more opportunities for effective preference articulation and open resolution of conflicts than will larger units which are organized hierarchically.

The delegation of specialized authority to an agency can compound this type of problem. While it can be expected that well-organized and knowledgeable groups will generally have better access and exercise more influence in decisions, this tendency has often been exaggerated in cases where governmental units are created for limited regulatory purposes. A history of regulatory commissions provides the more classic examples (Noll, 1971). Commissions are usually established by a political authority in response to widely shared demands for "public" regulation to better assure that specific goods and services meet consumer preferences, as well as to prevent monopolistic nonresponsive behavior. Shortly after the regulatory agency is created, however, the concern of citizens tends to lag, or turn to other issues, and those groups with the greatest interest in what the regulatory agency controls invest more and more resources to influence its behavior. The ultimate result is that the regulatory agency is virtually taken over by those with the strongest interests, almost always the groups that it is supposed to regulate. The agency then provides a strong legal basis for monopolistic and nonresponsive behavior under the guise of "public regulation." If the regulatory agency performs too poorly, the possibility exists that consumer interest will be rekindled, new regulators or a new regulatory agency will be set up, and the process will begin all over.

Characteristics of regulatory agencies may also be observed in large public administrative bureaucracies which substitute producer preferences (that is, their own preferences or those of producers who are well organized to support the agency) for consumer preferences and consequently do not respond to demands of citizens in the polity (Bish and Warren, 1972). The more specialized and politically detached an administrative agency is, the less likely it will respond to individual demands. A real question then is whether incentives will exist for the agency to continue to pursue its original goals. In the light of the history of regulatory agency and other bureaucratic behavior in the United States, this question cannot be neglected in relation to the coastal zone.

9.5 Alternative Allocative Mechanisms (Dahl and Lindblom, 1953)

In any complex system, a variety of trade-off processes are involved in the allocation of resources and the rationing of output. The most common of these are the market or price system, voting in the political system, administrative decision making, bargaining, and adjudication. Some characteristics of the administrative process, particularly with regard to information and preference articulation for coastal zone resources, have been identified. To point out weaknesses in one process does not, however, imply that process should not be used. In reality, all five processes interact in a system of public decision making. Accordingly, the objective should be to establish institutional relationships that will provide the mix of processes most appropriate for the conditions and requirements of coastal zone management. The significant advantages and disadvantages of each process for allocating coastal resources and for rationing their use will therefore be considered.

9.5.1 Markets

The market system is the major process for making trade-offs associated with resources allocation and rationing decisions in the United States.

Three market prerequisites are of special relevance to coastal zone management. If they are not present in economic transactions, a mutually satisfactory trade-off process is not likely to occur.

1. Both resources and benefit outputs must be divisible. Unless units of the resource and the product outputs are separable so that ownership can be transferred, they cannot enter uninhibited into market transactions.

2. All costs associated with production must be absorbed and all benefits from production must be captured by the producer.

3. The specific production enterprise must be open to entry, and competitive conditions are maintained in all steps of producing and marketing.

Although these requisite market conditions impose important limitations on utilizing them in coastal zone management, three special features of the market should be recognized: the efficacy with which preferences are revealed; the minimum information requirements for achieving coordination of resource uses; and the potential for using prices for evaluative purposes. These will be discussed in turn.

If the prerequisite market conditions are satisfied the market will express demand and reveal individual preferences with a high degree of fidelity. Each individual will allocate his resources to acquire the goods available by making direct purchases up to the quantities where his marginal satisfaction per dollar spent is equal among the goods purchased. In the process of making transactions the individual expresses his preferences and reveals his valuation by giving up resources (by paying the price of the goods) in order to obtain what he wishes.

The market reconciles values with little organized information and analysis. Each producer needs only know the price he can receive for an output and the price he must pay for an input to make his decisions for efficiently combining inputs to produce goods. If each consumer knows the prices of the goods and services for which he has desires, he is able to allocate his income among the available consumer goods to achieve his maximum satisfaction. If all conditions necessary for market operations prevail, the market can achieve an allocation of resources and a rationing of products where no one could be better off without making someone else worse off in the process. This is the same as saying that all trade-offs resulting in net gains have been taken.

The market system, however, cannot be fully understood apart from the governmental system. It operates within a set of parameters determined by the decisions in the public sector. The assignment and enforcement of property rights and agreements by government is an essential prerequisite to the use of markets for resource allocation. A well-understood legal structure, within which actions of other individuals are reasonably predictable, is necessary. Moreover, government decisions and market decisions are made interdependent in many other ways. The setting of interest rates, regulating business practices, determining levels of taxation and government investment, and committing government purchases, represent a few.

Of particular importance to coastal resources allocation are the instances in which government intervenes in the production of specific goods and services where necessary market conditions do not exist. The types of "market failures" that are most likely to result in public action in relation to the coastal zone include the following situations. (1) Resource use or the production of goods and services from resources results in external effects, that is, goods whose production or consumption affects others than those directly participating in the voluntary transaction, in which case prices will not reflect valuations of all effects of the good. Such spillover effects would occur when effluents discharged upstream wash downstream and reduce opportunities for swimming and fishing. (2) There exist public or collective goods, that is, goods that when available can be consumed by anyone, whether or not they contribute payment for the goods provided. Such a "free rider" situation occurs, for example, when a public beach, supported by the tax revenues of the governmental unit that owns and operates it, is used by persons who live outside the taxing unit's boundaries. (3) There are common pool resources, that is, resources that are owned only upon capture. In such a case there tends to be overutilization of the resource, because if any single user reduces his consumption, other users may simply increase theirs. A fishery is a classic example of a common pool resource and is generally regulated by a public agency or agreements among governments within the United States and internationally.

9.5.2 Voting (Bish, 1971; Downs, 1957)

A second trade-off process is voting, either for a delegate to undertake actions on behalf of his constituents, or directly on specific issues. A primary advantage of voting, especially when used to select a delegate, is to reduce decision-making costs for the individual citizen. The citizen can vote for someone with whom he agrees and then let that person obtain specialized knowledge and make decisions on particular public policy issues. Voting for representatives also has the potential for raising the general level of knowledge on public issues. This is because competition for political positions provides delegates with an incentive to seek out new solutions to issues and to identify potential public goods for citizens and unorganized groups.

There are several ways in which voting can be used by citizens to determine policy for coastal resources. Candidates for public office may be chosen on the basis of their position on environmental issues and on the degree of their committment to ecological values. Local bond issues which require the support of the electorate also affect decisions on coastal resources. Funds for beach acquisition, parks, recreational facilities, and sewage treatment plants often depend on funds obtained through the issuance of state or local bonds which require voter approval. Furthermore, the resolutions which authorize issuance of the bonds may impose substantial conditions of use for a particular coastal facility.

Popular referenda can also provide citizens with a way to affect allocation of coastal resources for particular uses. For the past several years, there has been intense discussion in the town of Falmouth, Massachusetts, concerning sewage treatment disposal facilities. A 1971 Town Meeting authorized the construction of a treatment plant which would result in sea outfall into nearby Vineyard Sound, a decision opposed for a variety of reasons by several citizen groups. One highly organized group opposed to effluent discharge into the Sound subsequently circulated a petition to have a referendum on the sewage treatment facility placed on the general election ballot. After obtaining the required number of signatures, the referendum was placed on the ballot, and in November 1971 the citizens overruled the Town Meeting decision. The town is now considering the use of a "living filter" which would return the effluent to the water table by spraying nutrient-rich secondary treated sewage onto grassland and forest land (Ahern, 1971).

In states where voting rules permit it to be used, the initiative has been a particularly important mechanism for allowing citizens to participate directly in the coastal decision-making process. In the state of Washington the threat of citizen initiative on shoreline protection sponsored by the Washington Environmental Council, WEC (Washington Laws, 1971), was a catalyst for passage of the Shorelines Management Act by the State Legislature in 1971. In 1970, WEC conducted a campaign to have an environmentally oriented proposal, the Shoreline

Protection Act or Initiative 43, presented to the Legislature and also placed on the November 1972 ballot. The threat of voter approval of Initiative 43 forced the Legislature to respond with its own Shorelines Management Act, passed in the 1971 extraordinary session (Spencer, 1971). Initiative strongly influenced the specific content of the state act, since the voters would still have the option, if they did not consider the new law adequate, to choose between the initiative measure and the state legislation in the November 1972 election.

An almost parallel situation is occurring in California, where coastal management legislation has been considered but not approved by the State Legislature during the past few sessions. In 1972, when it appeared that Senate Bill 200, which was supported by environmental groups, would not be passed by the legislature, a campaign was begun to have the measure placed on the November 1972 ballot as the "Coastal Initiative," so that citizens could directly consider the issue of establishing a coastal management system (Endicott, 1972).

In addition to its useful features, the voting process does present some problems in allowing accurate articulation of citizen preferences. In many cases, for example, one vote must be used to indicate preferences on a variety of issues. This is particularly true with the selection of elected officials. No single candidate is likely to be completely in accord with a voter on all issues, which makes it difficult for the citizen to reveal specific preferences. Another type of problem is that voting normally requires an all or none choice, instead of permitting marginal adjustments. A person who favors a marina but not the amount of the bond issue or only part of an initiative coastal management act must face the dilemma of saying yes or no to the total package. Similarly, while there may be great variations in the intensity of voter feelings about a ballot issue, there is no way this can be reflected in the election itself.

Apart from these problems, voting remains a major way in which public decisions are determined, either directly or through the election of representatives. The nature of the coastal zone agencies that are created, the policies that guide them, and the general political climate in which they operate will be affected by this process.

9.5.3 Bargaining (Lindblom, 1965)

Bargaining is interaction among individuals who have opportunities for increasing their benefits by exchanging resources over which they have control. The market system, voting, and administrative decision making are all supplemented by bargaining as a process through which trade-offs among interested parties are made. Bargaining, for example, is frequent between political leaders and administrative officials, among administrative officials, between executive agencies and their clientele, and between any number of interested participants in decision-making processes.

This process plays a pervasive part in the allocation of coastal resources and involves both public and private groups. A recent study of federal decision-making identified six bargaining areas in decision-making: (1) interagency bargaining, (2) intra-agency bargaining, (3) agency-clientele bargaining, (4) state-federal bargaining, (5) state-state bargaining, and (6) market bargaining (Schmid, 1971).

Bargaining provides government with a wider range of responses for utilization of coastal resources, rather than just approval or rejection. In Kitsap County, Washington, the County Planning Commission imposed twenty-two restrictive conditions on a proposed recreational community and marina to be built on Hood Canal, while still approving the zoning and subdivision applications for the development (Bish et al., 1972). In order to proceed with the development, sewage treatment facilities had to be built and special precautions had to be taken to maintain fish spawning areas. Bargaining among governmental units often plays an important but overlooked role in allocating coastal resources. In California, the State Department of Parks and Recreation is trying to increase public access to beaches by providing funds for beach facilities and improvements to municipal and county governments responsible for day-to-day operation of the beach. In order to obtain state funds, the local government units are being asked to provide additional opportunities for public access to and utilization of the local beach.

The American public economy, both the private and especially the public sector, relies heavily upon bargaining (Ostrom, 1969). The federal system with its separation of powers at the national level and 80,000 state and local organizations necessarily depends upon a great deal of voluntary cooperation to function. Unfortunately, adequate research on the outputs from the complex bargaining aspects of this system is lacking. What evidence has been compiled indicates that the decentralized structures possess many advantages to respond to preferences of diverse groups of individuals and to adapt to a changing environment when compared with organizations more centrally organized economically and politically. Given the fact that allocation of resources in the coastal zone must occur within the federal system, it is almost guaranteed that many of the crucial trade-offs and rationing decisions will be made through bargaining rather than through markets, voting, or hierarchical management decision making.

9.5.4 Adjudication

Perhaps no single decision-making process has had as much importance in coastal resource allocation as adjudication. Legal action through the courts often amounts to the final step in a particular decision-making cycle. A variety of recent court cases have substantially influenced federal administrative procedures, state legislation, and individual activities that pertain to coastal resource use. Despite the importance of the courts in affecting coastal policy making, it must be noted that adjudication requires enormous financial resources which relatively few citizens possess. Therefore, court action is most frequently used by highly organized groups that can afford the time and money to institute law suits. Two recent judicial decisions will be briefly examined here to indicate how the courts can affect particular activities on the coast and the behavior of other decision-making institutions.

On July 23, 1971, the U.S. Circuit Court of Appeals for the District of Columbia reached a verdict in the case of *Calvert Cliffs Coordinating Committee et al.* v. *U.S. Atomic Energy Commission et al.* The court ruling was the final step in an eighteen-month conflict between environmental groups and the Atomic Energy Commission concerning AEC regulations on nuclear power plants in general and specific issues pertaining to a nuclear plant under construction at Calvert Cliffs, Maryland (Barfield, 1971). The court ruled that the AEC had incorrectly and inadequately interpreted the requirements for environmental protection procedures set forth in the National Environmental Policy Act of 1970 (National Environmental Policy Act, 1970). According to the court, the AEC regulation did not comply with the NEPA requirements to consider environmental factors in the administrative agency's decisional process. More importantly, the commission was ordered to adopt a series of environmental review procedures: (1) an environmental cost-benefit analysis was required for each facility; (2) all environmental factors had to be independently reviewed in all licensing actions; (3) the effects of nonradiological pollution had to be assessed even if a facility met other state and federal standards. Consequently, on September 3, 1971, the AEC revised rules in accordance with the Calvert Cliffs decision (Barfield, 1971).

In California, the State Supreme Court ruling in *Gion* v. *City of Santa Cruz* and *Dietz* v. *King* has dramatically affected opportunities for public access to private beaches (*Gion* v. *City of Santa Cruz,* 1970). The court ruled that when the public has used a private beach and the access routes to it for a period exceeding five years, with the landowner's knowledge, but without asking or obtaining his permission, it constituted an "implied dedication" of the beaches to the public (O'Flaherty, 1971). The intent of this ruling, to increase public access to beaches, has been in sharp contrast with its actual impact. Property owners have begun constructing fences and posting "No Trespassing" signs along their beachfront property in order to discourage public use of their property and to show they have no intent to dedicate their property. Furthermore, the Legislature in 1971 attempted to reduce the impact of the decision on property rights by setting forth new requirements for implied dedication to occur, through modification of the Civil Code (Krueger, 1972).

Adjudication does provide an alternative for individuals who are dissatisfied with the consequences of market, voting, hierarchical, or bargaining decisions. The availability of this option may in and of itself encourage both governmental agencies and private groups to make more of an effort to take divergent interests into account or bargain for agreements in order to avoid the costly court processes into which they may be drawn by parties who cannot prevail in the other allocative systems. There are, at the same time, limits as to who can utilize the courts and on what issues. Beyond the problem of cost, there is a question of standing. While new precedents are emerging, the court still may require that an individual or group demonstrate that it will be immediately, directly, and negatively affected by the action it wishes to halt or modify. Further, in many instances only procedural issues rather than substantive effects can be considered when appeals are made from the decisions of governmental units.

9.6 The Design Task

It is clear from the preceding discussion that an agency established to manage coastal resources must operate within a complex political, social and economic environment that is dynamic and not static. If such units are created, market forces will not vanish, the courts will not close, and other agencies will continue to exercise powers that affect coastal development. Consequently, the "location" of a new agency within this ongoing structure must be considered in its design.

The market, voting, bargaining, and adjudication, as well as other administrative units serve important social and political functions. They provide citizens with a wide range of opportunities for preference articulation and access to decision making and increase the amount of information available in policy decisions. To the extent that authority for allocating coastal resources is shifted to specialized coastal agencies it must be done with a deliberate effort to maintain the opportunity for individuals and groups to gain access to policy making and ensure the production and use of adequate information upon which to base decisions. This can be done successfully, however, only if the effects of scale on organizational arrangements are known and taken into account. A set of boundaries that is optimal for ecological purposes may not be optimal for representation and decision making.

9.7 Conclusions

The problem of how to design institutions so humans can collectively make decisions on an equitable and informed basis about the allocation of resources is a pervasive one in society. The rich set of resources within the coastal zone heightens rather than reduces the magnitude of the task. The strength of people's preferences for continuing and increasing their involvement in water-related activities is a major factor in the difficulty of establishing a clear-cut mandate for any public agency about the allocation of coastal resources. The complexity of social, cultural, psychological, and economic interaction with marine environment which generates threats to ecosystems also imposes constraints on the way in which human behavior can be modified.

One source of concern about the coastal environment is that actions of people can endanger ecosystems by simplifying their structure and making them more fragile and unstable. To a degree, it is this complexity that coastal zone policies are asked to protect. A similar observation can be made about human social, economic, and political organizations. Productivity, cultural achievement, the variety of experiences available, and the generation and utilization of knowledge all tend to be related to organizational complexity. The analogy suggests that recommendations that would have the effect of reducing the complexity of the public sector should be made only with an understanding of how the present allocative system functions and with a caution aimed at preserving the beneficial aspects of complexity.

A biological scientist would seldom be willing to support an action for substantially changing the habitat of a major species of marine life without some understanding of how the ecosystem would be affected. Those involved in efforts to change substantially and simplify the formal mechanisms for making and administering policies concerning the coastal zone must follow a similar rule. There are many instances in which new governmental agencies have been created without any real change occurring in the behavior they were supposed to regulate (Schmid, 1971). Political, as well as natural, systems have adaptive capacities. The need for caution in approaching the question of coastal zone governance and management, then, is not to delay action, but to increase the possibility of its success.

A number of suggestions are made which should receive priority in efforts to design organizations that will be involved in coastal zone policy.

1. Public mechanisms or processes must exist for individuals and groups to indicate their preferences for utilizing coastal resources to the extent that such decisions are made or regulated through the exercise of governmental authority.

2. The responsible management agency, itself or in combination with other institutional components, must be able to identify socioeconomic and psychological, as well as physical and biological, effects of existing and proposed uses of the coastal zone and be able to take the information into account in resource allocation decisions.

3. Procedures by which allocative choices are determined should make it possible for all relevant trade-offs (what will be foregone and gained, in what amounts, and by whom) to be taken into account.

a. When a decision involves a choice between the coastal zone and inland sites, as with airports or nuclear power plants, there should be mechanisms to evaluate the comparative costs and benefits of alternate sites.

b. When a decision is made by a coastal zone agency that reduces or eliminates access to the coast for activities that are socially desirable, there should be positive responsibility to identify inland sites where the same or substitutable activity can occur and support their development.

4. Any public agency concerned with the coastal zone should have a built-in mechanism for—

a. Representing the interests of individuals and groups with limited capacity to compete in the political process or market for access to the coastal zone and utilization of its resources;

b. Providing for the representation of the interests of developed and nondeveloped coastal areas; and

c. Providing for the representation of the interests of groups and areas that will be disproportionately affected by a policy or action.

5. Major attention should be given to devising methods of increasing the carrying capacity of the coastal zone for socially preferred or necessary activities.

References

Ahern, W., 1971. The Falmouth sewage disposal problem. A paper produced for the public policy program, Harvard University (unpublished).

Architecture Research Center, 1971. Port and harbor development system (College Station, Texas: College of Architecture and Environmental Design, Texas A & M University).

Barfield, C. E., 1971. Energy report/Calvert Cliffs decision requires agencies to get tough with environmental laws, *National Journal, 3*(38): 1925–1933.

Bish, R. L., 1971. *The Public Economy of Metropolitan Areas* (Chicago: Markham), chapters 3 and 4.

Bish, R. L., and Warren, R., 1972. Scale and monopoly problems in urban problems in urban government services, *Urban Affairs Quarterly,* to appear.

Bish, R. L., Crutchfield, J. A., Harrison, P., Moss, M. L., Warren, R., and Meschler, L., 1972. *Coastal Zone Resource Use: Decision-Making in the Puget Sound Region* (Seattle: Univ. of Washington Press), chapters 4, 5, and 8 (in press).

Bradley, E. H., Jr., and Armstrong, J. M., 1972. A description and analysis of coastal zone and shorelands management programs in the United States (Ann Arbor: Univ. of Michigan Sea Grant Program, Michigan University), section 72-284, pp. 267–298.

Bragg, D. M., and Bradley, J. R., 1971. Work plan for a study of the feasibility of an offshore terminal in the Texas Gulf region (College Station, Texas: Sea Grant Program, Texas A & M University).

The Center for Urban Affairs, 1972. 1970 coastal data file (Los Angeles: Univ. of Southern California).

The Coastal Zone Act (Act 175), 1971. Chapter 70, title 7 of the Delaware Code, section 1 (paragraph 7003).

Commission on Marine Science, Engineering and Resources (CMSE&R), 1969. *Our Nation and the Sea* (Washington, D.C.: U.S. Govt. Printing Office), chapter 3.

Craine, L. E., 1971. Institutions for managing lakes and bays, *Natural Resources Journal, 11:* 519–545.

Dahl, R. A., and Lindblom, C. E., 1953. *Politics, Economics and Welfare* (New York: Harper and Row).

Downs, A., 1957. *An Economic Theory of Democracy* (New York: Harper and Row).

Endicott, W., 1972. Coastline issue going on ballot, *Los Angeles Times,* June 28, p. 1.

Farley, R., 1971. Indications of recent demographic change among blacks, *Social Biology, 18*(4): 341–358.

Finley, S. P., and Vantel, D. J., 1970. Californians need beaches—maybe yours, *San Diego Law Review, 7:* 605–626.

Gion v. *City of Santa Cruz, Dietz* v. *King,* 1970. 2 Cal. 3d 29, 84 Cal. Rptr. 162, 465 P. 2nd 50 (1970).

Goldin, K. D., 1971. Vacationscope: designing tourist regions (Austin, Texas: Bureau of Business Research, University of Texas).

Gunn, C. A., 1972. Concentrated dispersal, dispersed concentration—a pattern for saving scarce coastlines, *Landscape Architecture, 62*(2): 133–134.

Hayek, F. A. von, 1945. The use of knowledge in society, *American Economic Review, 35:* 519–530.

Huxley, A., 1962. *Oceans and Islands* (New York: G. P. Putnam's Sons), p. 11.

Jaakson, R., 1971. Zoning to regulate on-water recreation, *Land Economics, 47:* 382–388.

Krueger, R. B., 1972. Coastal zone management, the California experience (Woods Hole: Coastal Zone Workshop presentation), Civil Code section 1008 cited.

Lapp, R. E., 1972. One answer to the atomic-energy puzzle—put the atomic power plants in the ocean, *The New York Times Magazine,* June 4, p. 20.

Lear, J., 1971a. Floating people get it all together, *Sea Grant Newsletter, 2*(3): 3.

Lear, J., 1971b. Cities on the sea, *Saturday Review, 54*(49): 80–90.

Lindblom, C. E., 1965. *Intelligence of Democracy* (New York: Macmillan).

Meier, R., 1969. The social impact of a nuplex, *Bulletin of the Atomic Scientists, 25*(3): 16–21.

Moss, M. L., 1972. The urban coastal zone: a case study of Marine Del Rey (Los Angeles: Center for Urban Affairs, Univ. of Southern California).

National Environmental Policy Act, 1969. Public Law 91-190, 91st Congress, 1st Session, January 1, 1969, 83 Stat. 852.

Nehman, G., and Whitman, I., 1971. Development of Chesapeake Bay estuaries, *The Economic and Social Importance of Estuaries* (Washington, D.C.: Environmental Protection Agency), Water Quality Office, technical support division, project director D. Sweet, Estuarine Pollution Study Series-2, April 1971, Appendix H, h-18.

Nettleton, A., 1972. Cities in the sea, *Oceans, 5*(2): 71–75.

Noll, R., 1971. Reforming regulation (Washington, D.C.: Brookings Institution).

O'Flaherty, M. A., 1971. This land is my land, the doctrine of implied dedication and its application to California beaches, *University of Southern California Law Review, 44*(4): 1092–1134.

Ostrom, V., 1969. Operational federalism: organization for the provision of public services in the American federal system, *Public Choice, 7:* 1–17.

Report to the Governor's Task Force on Oceanography, 1969. Hawaii and the sea (Honolulu: State of Hawaii, Department of Planning and Economic Development), p. 58.

Santa Monica Evening Outlook, 1972. Taking sealife from coastal tide pools is a no-no, *Santa Monica Evening Outlook,* July 8, p. 17a.

Sauer, C. O., 1967. Seashore—primitive home of man, *Land and Life,* edited by J. Leighly (Berkeley: Univ. of California Press), pp. 310–311.

Schmid, A. A., 1971. Impact of alternative decision-making structures for water resource development, unpublished report to the National Water Commission.

Siegel, M. H., 1972. Plan unit backs South St. seaport, *The New York Times,* June 1, p. 38.

Spencer, W. H., 1971. Environmental management for Puget Sound: certain problems of political organization and alternative approaches, (Seattle: Division of Marine Resources, Univ. of Washington), Washington Sea Grant Program, WSG-MP71-2.

Taeuber, C., 1972. Population trends of the 1960s, *Science, 176* (4036): 773–777.

Wallen, I. E., 1971. Underwater houses and production plants (Washington, D.C.: Smithsonian Institution).

Warren, R., 1964. A municipal services market model of metropolitan organization, *Journal of the American Institute of Planners, 30:* 193–204.

Wash. Laws, 1971. First exec. sess., ch. 286.

Woodward, J. M., 1972. Population shifts toward sea, *Daily Breeze,* January 9, p. 36.

10 Structure of Management and Planning

10.1 Introduction

Regardless of the strategy chosen for use of the coastal zone, the concept of management must eventually arise. What assembly of procedures, information and organization is necessary for management of such a resource? Some basic approaches and management foundations are discussed in an effort to develop an understanding of some of the basic premises of coastal zone management and their implications as measured in the realities of our social and political system.

The process of defining the inland limits of the coastal zone and the difficulty in doing so (see Chapter 1) indicates that the coastal zone is a natural entity, but with flexible boundaries. Concentration on solutions to coastal zone problems, no matter how the zone is defined, must be tempered with an awareness of adjacent inland and ocean perspectives, if parochial decisions are to be avoided. For management purposes the coastal zone has been considered as the geographic area bounded on its seaward side by the outer limits of state jurisdiction, and on its landward side by the inland limits of significant marine influences as defined by the individual states.

This chapter deals directly with many specific principles in the text; the following are more general in character or deserve special emphasis.

1. No coastal zone management authority should exist without sufficient authority and financing to achieve its objectives.

2. Management personnel must have a high level of understanding and competence in the scientific, social, and economic aspects of the coastal zone.

3. All coastal zone users must understand that the multiple-use concept as a management principle for coastal zone resources cannot mean all things to all people. Popular misuse of this concept has led many to misunderstand the coastal zone's capacity. Regardless of the management structure, the coastal zone may not be able to support multiple uses without unacceptable conflict or irreversible damage. Multiple use is a viable principle only when the geographic or functional extent of the particular resource is large. It becomes increasingly less viable as one focuses on specific sites within the coastal zone system.

4. A single specified federal agency should have responsibility and primary authority for the development and guidance of an integrated national program for coastal zone management and use.

5. The appropriate federal coastal zone management authority should

establish a standard uniform format for mapping, resource inventory, and other data collection.

6. Effective coastal zone management must begin in the uplands and continue into the sea. It must be flexible enough to cope with the multilevel and multidimensional problems. The coastal zone management authority must possess the jurisdiction and diversity of competence to handle the coastal zone problems as a single unit.

7. State administered coastal zone programs should have the authority to specify use classifications within their coastal boundaries. The state's coastal zone management authorities should administer a permit system based on their comprehensive coastal zone plans to regulate selected activities within use classifications.

8. The coastal zone management authority should classify proposed coastal zone uses according to their dependence on coastal zone resources, and it should give preference to those uses that are exclusively dependent on the coastal zone. The coastal zone management authority should assure that all coastal zone users adhere to the principle of minimum use impact, regardless of what use policy the authority selects.

9. Each state coastal zone program should develop and maintain a resource classification and inventory system that describes the coastal zone spatially and temporally. Coastal zone management authorities will update the inventory system periodically. The inventory program should be sufficiently detailed to permit the authorities to evaluate impacts and changes due to man's activities in the coastal zone.

10. Each state's coastal zone management authority should develop a comprehensive long-range plan to guide the allocation of coastal resources. Plan implementation should include incentive policies as well as regulatory programs.

11. State authorities should place high priority and emphasis on the local governments' and the general public's involvement and participation in evolving coastal zone plans and programs. States should use incentive grant programs to promote and to encourage such involvement.

12. Each state coastal zone authority should function administratively independent of other agencies. It should utilize the information and expertise of other appropriate agencies to obtain relevant information and data to assist in the management process, to establish criteria for management, and to enforce the standards and regulations the authority establishes.

13. The coastal authority should determine standards and regulations that state and local agencies should use. The regulatory process should

be at the lowest possible administrative level within the management structure.

14. The state coastal zone management authority should require that all proposed projects, publicly or privately sponsored, file an impact statement if the authority anticipates that the project could have a significant impact on social, economic, or environmental coastal zone conditions.

15. Each state should create a separate appeal board to provide public review and hearings for individuals, firms, groups, and government agencies that the state's coastal zone management authority's programs affect.

10.2 The Need for Management and Planning

The rationale for a coastal zone management program is well stated in the Committee Report of Senator Hollings which accompanies S3507, the "Magnuson Coastal Zone Management Act of 1972." These findings may be summarized as follows:

1. There is a national interest in the effective management, beneficial use, protection, and development of the coastal zone.

2. The coastal zone is rich in a variety of natural, commercial, recreational, industrial, and aesthetic resources of immediate and potential value to the present and future well-being of the nation.

3. The increasing and competing demands upon the lands and waters of our coastal zone occasioned by population growth and economic development, recreation, extraction of mineral resources and fossil fuels, transportation and navigation, waste disposal, and harvesting of fish, shellfish, and other living marine resources, have resulted in the loss of living marine resources, wildlife, nutrient-rich areas, permanent and adverse changes to ecological systems, shoreline erosion, and decreasing open space for public use.

4. The coastal zone and the fish, shellfish, other living marine resources, and wildlife therein are ecologically fragile and consequently extremely vulnerable to destruction by man's alterations.

5. Important ecological, cultural, historical, and aesthetic values in the coastal zone which are essential to the well-being of all citizens are being irretrievably damaged or lost.

6. Special natural and scenic characteristics are being damaged by ill-planned development that threatens these values.

7. In light of competing demands and the urgent need to protect and

to give high priority to natural systems in our coastal zone, present coastal state and local institutional arrangements for planning and regulating land and water uses in such areas are inadequate.

8. The key to more effective use of the land and water resources of the coastal zone is to encourage the coastal states to exercise their full authority over the lands and waters in the coastal zone by encouraging cooperation with federal and local governments and other vitally affected interests, in developing land- and water-use programs for the coastal zone, including unified policies, criteria, standards, methods, and processes for dealing with land- and water-use decisions of more than local significance.

Coastal zone management must be considered in terms of the two distinct but related regimes of land and water. The key to more effective use of the coastal zone in the future is the introduction of management systems permitting informed choices among the various alternatives. Coastal zone management problems can be conveniently grouped into four classes, as follows.

1. *Localized major impacts.* These involve the development of individual units, possibly a nuclear power plant or some other industrial complex. In this class of problem, the issues must be resolved primarily by action at local level. The Hilton Head chemical complex provides a good example of the issues raised in this category.

2. *Localized major impacts with national involvement.* The current energy "crisis" provides an excellent example to describe this class. The increasing demand for energy and the energy projections for national use suggest that new deep ports or offshore facilities for supertankers must be developed and that a national policy must be implemented in the very near future. While the impact of a new facility would be felt in a fairly local area, the need for fuel is nationwide and the beneficiaries include a broad cross section of the country.

3. *Degradation of the coastal zone from historical usage.* The New York Bight and its role as septic system for the metropolitan area provides a classic example for this category.

4. *The rehabilitation of the coastal zone.* A major question of coastal zone management must involve a willingness to break with past traditions and to reexamine past uses of coastal regions to determine possible new thrusts for action. In this case, the possible rehabilitation and reconstruction of the Port of New York will be examined.

The following examples illustrate each of the above problem classes.

1. *The Hilton Head Case.* In the fall of 1968, Governor McNair of South Carolina and members of the State Ports Authority, and the Industry and Trade Commission, made a trip to Western Europe for the express purpose of attracting more heavy industry into South Carolina. The tax and investment incentives offered by the state of South Carolina and Beaufort County were extremely attractive and one of the companies which was enticed to locate in South Carolina was the Badishe Anilin und Soda Fabrik (BASF). The site offered to BASF was located in Beaufort County about midway between Beaufort, S.C. and Hilton Head Island, S.C., the location of Sea Pines Plantation and other multimillion-dollar residential developments.

The considerations offered to the company prompted the BASF Board of Directors to approve the South Carolina project. The announcement of the project to the press was accompanied by glowing economic forecasts by state and county officials and the BASF management. In reaction, a coalition of property owners from Hilton Head and Beaufort County called for public hearings in order to air their views. These property owners consisted of large plantation owners and officials of such enterprises as Sea Pines Plantation whose goal had been to make the Hilton Head and Beaufort County areas compatible for man and nature by preserving the long untouched beauty of lowland South Carolina plantation country and sea island area. The hearings centered around the environmental issue. The company produced what to their minds was irrefutable technical evidence that their proposed project was clean and pure and would do no harm to the air, water, and surrounding land. As the hearing proceeded, the German company became increasingly intransigent, with pride and company "face" or image at stake.

The local hearing also brought forth economic and environmental witnesses from local, state, and federal agencies. After violent and acrimonious debate for a period of six to eight weeks, the issue ended up on Governor McNair's desk after an intensive coastal management study done by the South Carolina Water Resources Commission. Because of political pressure and some observations in the study, the governor abrogated the agreement which the state had made in good faith with the German chemical company. The following observations are offered:

—There was no management structure then existing at the state level dealing effectively with the issues.

—The issue was tainted by emotion rather than an examination of the management alternatives.
—If proper alternatives had been presented, a possible compromise decision might have been rendered.
—The issue became a "yes" or "no" decision, ending up with advantage to neither side.

2. *Supertanker Ports and Offshore Mooring Facilities.* Current energy projections indicate that if future demands are to be met, oil importation must expand. In part, this increase could be met by building more small tankers capable of being handled at existing facilities. From a purely economic point of view, however, this is a poor approach. Environmentally, in view of the increased possibilities for accident through collisions, it is not clear that the operation of a large number of small tankers is better than using a smaller number of large ships. The comparative risks are poorly understood. It is possible that a single spill at a point source from a single supertanker would be environmentally more damaging than multiple spills of the same amount of oil at several separate points. The probability of accidents of small versus large ships is not known, but the supertankers do require long distances to stop or turn (see Section 6.2.1).

If one makes the assumption that supertankers will become operational, only a few U.S. port areas are possible for consideration. At the current time, a multi-agency task force (involving the Council for Environmental Quality, the departments of Commerce, Transportation, and Defense, and the Environmental Protection Agency) is examining the possible environmental impacts of supertanker facilities at seven sites along the East and Gulf coasts of the United States. If supertanker facilities in the United States become a reality, it is likely that only a few of these seven sites will be developed. The selection would affect a relatively small proportion of our population, although the benefits of maintaining an adequate supply of fuel oil would benefit many people. Major questions of coastal zone management are raised both in terms of making the site selection and then accommodating the local populations that would be negatively affected by the new industry.

3. *Degradations of the Coastal Zone from Historical Usage.* The megalopolis which borders the New York Bight evolved in a basically unplanned fashion. By virtue of good port facilities, adequate power,

and desirable geographic location, the area grew at an astonishing rate. The rate surpassed the ability of governmental processes to cope with the problem of rational planning or to assure adequate treatment of sewage, dredge spoil, and the other noxious by-products of large industrialized communities. As a result, the New York Bight area now receives about two billion gallons of sewage effluent each day (40 percent of which receives no more than primary treatment, 18 percent is untreated); about seventeen million cubic yards of dredge spoil each year; and an indeterminate but very large amount of airborne wastes. The result of this assault is that beaches are closed in much of the area because of public health concerns; Raritan and Jamaica bays, once highly productive nursery areas for commercial and recreational fishes are badly damaged; shellfish cannot be used for food; and in an aesthetic sense, the region could be described as a disaster area. The cause of the damage was not intentional—it just happened. The rehabilitation of the area or the prevention of future damage, however, will not just happen. A conscious, careful, and detailed approach must be evolved. The approach must consider the total system, involving not only the megalopolis, but the upstream areas as well. It must be coordinated to assure the positive changes forced by one jurisdiction are not cancelled by negative modifications in adjoining jurisdictions. It must also assure that alternative solutions to the problems of environmental degradation for the New York Bight area are developed, that these solutions involve more than just transfer of the damage from one site to an undisturbed area, or to a region of lesser political clout. The costs of mounting a major program to rehabilitate the New York Bight area will be enormous. These costs must be placed in the perspective of the population size which will be affected, the basic economic value of the area, and perhaps most critically, the end product which may result if current practices are allowed to continue.

4. *Rehabilitation of the Coastal Zone.* Possibly the most valuable properties in the United States are our waterfronts; these were the earliest parts of our country to be settled and developed, and, in many instances, they include our most dilapidated, worn and obsolete complexes. Highly valuable property has become a prisoner of the past. Some breaks with tradition must be considered if we are to maximize the long-term value of the land-sea interface.

For example, the New York waterfront includes a considerable array of obsolete port facilities, which are not equipped to handle efficiently either our newest ships or those which are planned. Inadequacy of port facilities does, in fact, somewhat inhibit ship design, and results in operations that may be less environmentally sound than the state of the art would otherwise permit. The construction of offshore port facilities is possible, and the development of a major offshore facility in the New York Bight area to replace a substantial portion of the Port of New York should be considered. This could include not only the facilities for unloading and loading ships, it could also include power plants, an airport, liquefied natural gas facilities, and by a proper marriage of the thermal wastes of power plants with improved sewage treatment facilities, an aquaculture complex. By starting afresh and examining the total system, it should be possible to design an industrial island that is both economically sound and not detrimental to the surrounding environment. While the costs of the above complex would be high, so would the benefits. A reconstruction of the whole New York waterfront to provide new housing, office space, and readily accessible parkland could have a significant role in the rejuvenation of the city. Obviously such planning does require a number of new approaches, both technical and institutional, and might indeed require a total revision of the Port of New York Authority.

10.2.1 How to Develop a Coordinated Approach to Management

The four examples described emphasize the need for a coordinated approach to problem solving at all levels of government. What is the best approach to initiate the development of a process to answer these and other relevant questions about the coastal zone?

The Coastal Zone Workshop has recommended the creation of a multidisciplinary task force to formulate a national management program. Some of the problems the task force might look at should include—

—What should specific national policy be?

—How should such a policy be ascertained?

—What criteria should be used to determine areas of national interest or concern?

—What is the best approach to take in fostering national interest?

—What is the best organizational structure to work with the problem?

—How should the regional coastal zone centers be organized?

—What should the areal limit of the zones and for the centers be?

—What can be done to help the states develop their plans, especially in terms of technical expertise?

—What can be done to improve the education process for the general public with respect to use and understanding of the coastal zone?

—How can interaction among federal agencies be guaranteed?

The Coastal Zone Workshop recommends
The President of the United States request the National Academies of Science and Engineering to create a multidisciplinary *Coastal Zone Task Force* to formulate a management program. The Task Force should assist the federal government in designing the national program and evolving model guidelines for state coastal zone management authorities. The Task Force should work with the coastal states, regional agencies, and the federal government in preparing specific plans for coastal regions of the United States.

In order for any complex program to be both cost-effective and viable, it is necessary for such a task force to make an overview of the entire situation and try to categorize the interactions and conflicts between the various components of the program. After this process has been completed, it will be possible to develop a coordinated and standardized approach to deal with both general and specific problems. The usefulness of a task force would be greatest in the development of such an approach. The credibility of any recommendations made will be in direct proportion to the broadness of the information base, which emphasizes the need for personal interaction.

10.2.2 What Is Coastal Zone Management?

Concern for the viability of the natural processes in the coastal zone ultimately is a reflection of man's ability to protect himself from the impact of his own use of the resources. After reviewing the ecological character of the coastal zone and man's historical impact, one must consider the policies, programs, and institutions by which man might exercise his powers of rational decision making.

A management process should be designed to achieve a set of stated objectives. The "stated objective" should be to maintain and improve the coastal zone's usefulness for man by encouraging the quality and extent of the natural system on which he depends. This should be done both for the present and for the future in a way that would be most acceptable to economic, social, and environmental goals. With such a best-use objective in mind, we can establish a basic definition of coastal zone management through the introduction of its major functions. Coastal zone management is the process of (1) developing an understanding of the coastal zone as a system; (2) using this knowledge to create a dynamic plan for its best use; and (3) implementing and enforcing that plan.

The Coastal Zone Workshop recommends
The federal government establish a national coastal zone management program, which should be vested in one of the existing federal agencies and should coordinate all agencies involved in coastal zone activities. The federal agency should administer grants to state coastal zone programs and set appropriate guidelines for such programs as well as for the management of federal coastal lands.

The sequential steps in the total management process are:
—To determine man's desires in using the coastal zone, for example, values and priorities;
—To determine the capacity of the coastal zone in relationship to these desires;
—To determine what uses are compatible with these capacities and how the various uses impact the natural coastal zone system;
—To determine what the trade-offs must be if capacities and uses are not matched;
—To determine how the need for capacity match can be communicated to the public;
—To determine what mechanisms are necessary and possible to regulate and promote the compatible uses.

10.2.3 Current Status of Coastal Zone Management

The types of shoreland management programs that have been instituted in coastal states have varied in accordance with the geographic characteristics of the different regions of the United States. A detailed survey of these programs can be found in Bradley and Morse (1972). The Atlantic Coast states have focused much of their attention on protecting their wetlands because of their great value to coastal fisheries and other estuarine life (USDI, 1970). Florida has taken action to manage one of its coastal resources through establishment of a system of aquatic reserves (Florida CCC, 1971). Delaware has taken a major step through its Coastal Zone Act which bans new heavy industry and port facilities from the state's coastal areas (Delaware, 1971).

The Gulf Coast states have concentrated on methods of allocation of exploitation rights for their estuarine resources, but they recently have begun to consider protective measures as the effects of overdevelopment have begun to threaten their coastal resources. One early instance of this concern was Texas's action in 1959 to protect public access to the state's dry beach areas.

In the Great Lakes area, the eight Great Lakes states have joined together to undertake a comprehensive planning study of the region's water and coastal resources through the auspices of the Great Lakes Basin Commission's (GLBC) work on its Comprehensive Framework Study. In addition, Wisconsin and Minnesota have enacted shoreland management programs which apply to their coastal shorelines, although the programs are primarily aimed at reducing pressure on their inland lakes (Wisconsin, 1965; Minnesota, 1969). Michigan's shoreland management program, on the other hand, is concerned with protecting high-risk erosion areas and "environmental areas" from misuse (Michigan, 1970).

On the West Coast, California and Oregon have instituted programs to protect the public's long-standing rights to use those states' extensive beach areas (California, 1972). Both states have also embarked on coastal zone planning programs as first steps to formulating coastal zone management programs. In the San Francisco Bay area, the San Francisco Bay Conservation and Development Commission (BCDC, 1965; Ralezel and Warren, 1971) has instituted a management program to protect the Bay area's resources. The state of Washington has instituted one of the first comprehensive coastal zone management programs in the United States (Washington, 1971).

Alaska is in a unique situation in that nearly all of its land area has been owned by the federal government and that its coastal resources have not been subject to as much pressure for development. This situation is likely to change somewhat with the settlement of the Alaskan pipeline issue and the lifting of the freeze on land development as a result of the Alaskan Native Claim Issue, which will result in the completion of the land selection process by the Alaska state government. When it became a state, Alaska did enact constitutional provisions relating to the protection of its coastal resources and legislation to provide basic guidelines for the management of its lands in accordance with those provisions. Whether these provisions will be sufficient to protect the state's coastal resources as they become subject to more development pressure remains to be seen.

Hawaii initiated a state-wide land use program in 1961 (Hawaii, 1961). In recent years, Hawaii has found that this program does not provide sufficient protection for its coastal resources, and as a result, it is investigating measures that can be taken to provide the needed protection.

Table 10.1 shows the various programs that each coastal state has undertaken. Table 10.2 shows the legislative acts, executive orders and court cases that the states have utilized in implementing their coastal management programs.

10.3 Allocation of Responsibilities

Who should do coastal zone management? The fundamental answer is everyone, but that is not very helpful. To pinpoint responsibilities in a more useful way, the scope of the problems and the common resources involved require that the job be done by the government in general and should include the stimulation of, placement of restrictions on, or noninterference with nongovernmental processes such as "the market-place" and "public opinion." The job should be distributed along the ladder of government:

1. Federal;
2. State (including interstate regional groupings); and
3. "Local," a collective term that we shall use hereafter to include all levels of government below the state level—counties, municipalities, towns, villages, and various groupings thereof.

As stated in the Introduction (Chapter 1), a central issue to coastal zone management is the degree of centralization necessary, a fundamental management question not unique to the coastal zone. Generally, higher levels of government lead to greater centralization, but also to more general perspective, more objectivity, more expert talent available and more funds and political impact. Moving toward decentralization in the lower levels of government, there is more intimate knowledge of the problems, a more myopic outlook, and a greater likelihood of living with the effects, whether good or bad, of the decisions. Furthermore, if higher government does not limit its own decision-making appetite, it can become hopelessly bogged down in detail at the expense of the perspective it claimed in the first place. It is appealing to seek out a middle ground, ideally one that preserves the recognized unique attributes of the extremes. Coastal decision-making can be delegated to the lowest level of government consistent with the scope of the problems, but decisions must conform to the goals and constraints articulated by the next higher level. The corollary is "management by exception."

Restrictions are generally formulated to ensure that the external effects of the local decisions are kept within tolerable bounds. A workable system incorporating the principles of delegation of authority and

Table 10.1. Coastal Zone and Shoreland Management Programs Undertaken in the United States

State	Wetlands[a]		Beach access	Power plant siting	Shorelands zoning	Site location regulation	Compre-hensive planning	Coastal zone manage-ment	Other[b]	
	Prot.	Maj. Acqu.							Dir. rel.	Indir. rel.
Alabama										
Alaska									x	
California	SF Bay		x	x			x	SF Bay	x	
Connecticut	x	x								
Delaware		x				x	x			
Florida							x			
Georgia	x									
Hawaii			x				x		x	
Illinois							GLBC study			
Indiana							GLBC study			
Louisiana	(study)						x			
Maine	x					x	x			x
Maryland	x			x						x

Massachusetts	x	x				x			x
Michigan					x	GLBC study			
Minnesota					x	GLBC study			
Mississippi						x			
New Hampshire	x								
New Jersey	x	x							
New York	x			x		GLBC study			
North Carolina	x	x				x		x	
Ohio						GLBC study			
Oregon			x	x		x		x	
Pennsylvania						GLBC study			
Rhode Island	x					x	x		
South Carolina									
Texas			x	x		x		x	
Virginia	(study)								
Washington				x		x	x	x	x
Wisconsin					x	GLBC study			x

Source: Bradley and Armstrong, 1972.
[a] Prot.: protection; Maj. Acqu.: major acquisition.
[b] Dir. rel.: directly related; Indir. rel.: indirectly related.

its corollary, management by exception, would place decision making at the lowest level commensurate with the anticipated impact of the decision. We suggest the following allocation of responsibilities to various levels of government.

10.3.1 Federal Responsibilities

The federal government should provide the leadership by developing a national coastal zone policy. This policy should articulate what is "in the national interest," and develop means to foster that national interest; designate a lead agency to coordinate the many relevant coastal functions of federal government; delegate specific decision-making authority to lower levels insofar as the decisions remain in harmony with the broad national policy and interest; provide assistance to the lower levels in the form of grants and technical guidance, and maintain overview to improve the effectiveness of the overall system in serving the basic objective of coastal zone management; and provide instruction from Congress as to overall objectives for resolution of basic issues such as the relative priority of public or private ownership or control together with a designation of the means by which equitable compensation is provided to those whose investment is lost or compromised.

If a national program of coastal zone management is to succeed, a single federal lead agency must be explicitly designated by Congress. The lead agency must coordinate the relevant coastal functions of government occurring in the coastal zone under one umbrella agency. Part of the responsibility of the lead agency would be to see that coastal goals and constraints are reflected in decisions on federal programs, impact statements, and permits. For example, the agency could be represented on the U.S. Water Resources Council and see that the numerous comprehensive plans being developed for the nation's major river basins adequately reflect coastal zone considerations. National interests must be defined in harmony with the states, as the states operate as the focal points of coastal zone management. Attempts to develop broad "cookbook" types of coastal zone management plans usually reiterate generalities and rehash history. The complexity and diversity of the interests and problems involved tend to preclude the effective, explicit definitions necessary to create specific recipes applicable to all situations. Anyone can suggest what he thinks should be in the national interest; they actually become "national interests" only when they gain sufficient support to be enacted into law. The laws, which collectively define the national interests, tell us what the federal government does, prohibits and encourages. The guidelines are found more in the body of the law than in the preambles, which frequently imply everything for everyone, irrespective of conflicting prior promises.

Using the nation's body of law as the official pronouncement of the national interests provides great flexibility. If it is judged, for example, that a certain provision is important enough to prescribe in precise detail, the law can be so written. Thus the designated coastal lead agency should codify the provisions of existing federal laws in a manner that will improve their usefulness in coastal zone management. For example, an assembly of selected portions of the U.S. code within a coastal framework would be one way to accomplish this task. The lead agency should be an advocate of coastal interests. It should seek funds and knowledge and be a prime proposer of coastally sensitive legislation through which more effective services, inducements and constraints can be applied to the coastal zone.

The federal government should delegate most of the authority for coastal zone management to the coastal states. The Tenth Amendment to the U.S. Constitution provides that the states have all authority not specifically granted the federal government, but through increasingly aggressive legislative and executive interpretation, abetted by confirming judicial interpretation, the strength of this amendment has withered.

The state is in the desirable middle position between the objective remoteness of the federal government and the parochial familiarity of the locality. Although the effects of good use and misuse are transmitted elsewhere along the coast, the range and transmissibility of major effects is often within the state's geographic span of control. The state also has an established administrative machinery. With the delegation of authority comes the restrictions within which that delegation must operate and the goals that are sought by the delegator. The federal agency has a responsibility to see that these restrictions and goals are known and fully incorporated by the states, and that they are adjusted from time to time to reflect what is considered in the national interest. The federal lead agency should also assist the states in carrying out coastal zone management. Grants, research, technical assistance and services, uniformity of standards in mapping, inventorying and information access, and many other forms of help can be developed. The most important assistance, however, is financial. Many who have had intimate association with cooperative federal-state ventures have ob-

The Coastal Zone Workshop recommends
The state coastal zone authority should be established as an independent agency, with its expertise and primary responsibility exercised in cooperation with other state agencies involved in the coastal zone. Management programs should view the coastal zone as a complete natural system and not be restricted by political boundaries. Incentive policies, as well as regulatory powers, should be used to improve the management of the coastal zone. Local governments should be strongly encouraged to evolve their own local plans and programs within the guidelines of the state coastal zone program, while citizen's advisory boards should be used to gain public participation in the policy-making process.

served that the state participation frequently degenerates into one of unequal partnership, primarily because the federal agencies in their desire to contribute can almost always mobilize more funds and talent than the less affluent states. If the federal lead agency is unable to capture substantial funds and funnel them to the states, then most states will become political sham as "focal points" for coastal zone management.

The federal lead agency can delegate the authority to the states, but the states have the responsibility for seeing that coastal zone management succeeds. The federal agency must set standards of performance to include timetables for reaching milestones, scope and time span of the state's coastal plan, degree of comprehensiveness, degree of integration with inland and marine perspectives, uses to be considered, degree of public exposure, and other features. The progress of the national coastal zone management programs should be subject to continuing review by an independent public advisory board appointed by the President.

10.3.2 State Responsibilities

The state as the focal point of coastal zone management must bear the responsibility of seeing that the four functions of management are discharged, either by itself or by enlisting the cooperation of others. It must define its coastal zone. It must understand its coast—socially, economically, and environmentally. It must develop a plan (or a family of plans) that articulates objectives and selects time-phased ways of achieving these objectives, and it must ensure that that plan is implemented.

Operating within the national interest to the extent that this interest is defined at the federal level, the state has the overall responsibility for reaching conclusions on what is wanted or not wanted, on what is necessary or undesirable. The formation of regional units based on natural geographical areas, either interstate or intrastate, should be encouraged whenever possible.

To serve as the focal point in coastal zone management, the state or group of states must explicitly designate a coastal zone agency. The coastal zone agency may be entirely new or assembled from existing units. It must be able to enlist the participation of federal, regional, local, and nongovernmental representatives. Most of the tasks described previously at the federal level also apply at the state level with some modifications. Delegation of decision-making authority at the state level should be done in accordance with a state coastal zone plan. The state role differs from the Federal role significantly in that, for the most part, the state is responsible for the following two major tasks.

1. *Understanding its coast.* The state must acquire considerable knowledge about the social, economic and environmental (physical, chemical, biological) aspects of its coastal zone and the interrelationships between and within each of these characteristics. Coastal problems and usage conflicts must be perceived. These conflicts are by no means unique to the coast, but they tend to be more acute there because uses of the coastal zone must be compatible with both the landward and the marine environments, each with its special set of requirements. The socioeconomic and natural order tend to concentrate demands along the coast. Man focuses his population and energies there. To marine life, the coastal zone is a special nursery, feeding ground, and habitat.

Use conflicts in the oceans and on the land can often be minimized by keeping them spatially separated and controlling their external effects. Near urban centers, land available for competing uses tends to expand as the square of the distance from the urban center, but shoreland can expand only linearly. For example, a city can increase its total available land by a factor of sixteen by pushing out its limits by a factor of four, say from five to twenty miles from its core. But to get the same sixteenfold increase in its shoreline, the coastal city must expand sixteen times as far, or from five to eighty miles from the core.

Conflicts abound, not only among coastal uses, but also between coastal and inland uses. Estuarine fishermen and recreationers will have their own ideas of what the quality and quantity of inflowing riverine waters should be. People living upstream may have a very different opinion if they use these waters primarily for agriculture, water supply, waste dilution, flood control, or transportation.

The feasibility of substituting inland for coastal resources should be studied when considering these conflicting relationships. Many demands for coastal resources can be satisfied inland in some other way. For example, protein production, petroleum, sand and gravel, solid waste disposal, disposal of liquid wastes, water-based recreation, pipelines, and a host of land uses, such as airport sites and certain types of residential, commercial, and defense developments, can all be satisfied to some extent inland. Some programs, such as those regulating river flow, have a mixed relationship. When extremes of high and low flows are dampened out, they affect the coastal uses in a variety of ways as described in Chapter 6. The effect may be advantageous or harmful depending upon the influence of those environmental extremes upon coastal marine life, sediment control, coastal erosion, turbidity, and shipping. Nonmarket factors such as external effects, indivisibilities, and irreversibilities are not unique to the coastal zone, but they do have a special relevance at this interface between the continent and the sea.

Knowledge gaps are sometimes startling. Probably most significant is lack of knowledge of the requirements for the living resources of the sea. Can they be made more plentiful by potential changes within man's control, or does the present regime or a past one happen to be ideal? Current knowledge is still inadequate to evaluate the incremental effects of change and the total market and nonmarket costs and benefits associated with these incremental changes. Without such knowledge, society will have difficulty arriving at a rational trade-off between landward and seaward values. To acquire this knowledge, the state should sponsor relevant research and provide for the continuing education and professional development of its coastal zone leaders. The state should develop and maintain a classification and inventory system that describes its coastal resources.

The inventory should be maintained in sufficient detail to permit periodic evaluation of the changes being caused by nature and man and the impacts of these changes. Practically, such detail and the computer programs required to manipulate it are extremely expensive. An attainable compromise is a broad classification system (framework) with attention given to filling only those parts of the framework that turn out to be significant in resolving major problems.

2. *Developing a coastal zone plan.* Comprehensive planning is essential for adequate coastal zone management. Without a comprehensive plan and the insights developed in preparing one, the coastal zone manager is reduced to *ad hoc* responses at the random initiative of others. Armed solely with intuition, good intentions, and a staggering administrative process, he can prevent or delay, but he cannot cause or lead. The state must first define the kind of a coastal zone that is wanted and then try to find acceptable ways of achieving this objective. The process involves organizing the planning effort, development of objectives, definition of alternative solutions, and formulation of a program.

In organizing the planning effort, the scope of the plan must be defined; for example, the time frame might emphasize the actions required in the next five years or the next twenty years. The geographic area has been defined earlier in our definition of the coastal zone and for state management purposes the linear scale is defined by state boundaries. Strong consideration should be given to exterior factors and the degree to which they produce impacts on the coastal zone. The functional framework should identify a set of coastal uses, such as extraction of living and nonliving resources, waste disposal, recreation and aesthetics, transportation, land use, national defense, and ecological balance. The activities performed to enhance these uses might also be identified by research, mapping, engineering, construction, and shore stabilization. It would be desirable if many of these parameters were defined at the national level by the lead agency to establish a reasonable degree of uniformity. Precedents and examples abound, including the programs of the U.S. Water Resources Council and the U.S. Bureau of Outdoor Recreation.

The planning leadership should focus at the level of government commensurate with the individual state's needs. Some states may decide that a state-wide comprehensive coastal plan may not be the most desirable, and that interstate regional mechanisms may be more appropriate. For example, the New England states or the states bordering the Chesapeake Bay might want to develop a regional coastal zone plan employing an existing interstate mechanism, such as the New England River Basin's Commission or a new one for Chesapeake Bay. New York

The Coastal Zone Workshop recommends
The federal and state governments, acting together, create regional councils to assist in carrying out the national coastal zone policy. Such councils would work in concert with federal and state agencies in advising on regional problems of national interest and implement appropriate policies where consensus exists between federal and state governments.

might want two plans—one for its Atlantic coast and one for its Great Lakes coast. California appears to prefer a set of coastal plans developed by groupings of counties and strung together under the harmonizing effect of state guidelines and state-level approval. Rhode Island, Delaware, and Hawaii might combine their existing state-wide water-and-related-land-use plan with their coastal zone plan into one overall plan. There is controversy regarding the role that regional government should play in the planning and management process. Perhaps the approach should be that discrete geographical areas sharing a common ecology, natural landscape features, hydrological attributes, or other factors should be established as regions, even though these regions may not correspond to present jurisdictional boundaries. However, creation of regions that do not correspond to previously established political boundaries will add a new layer of government and may compete with or duplicate existing functions. Perhaps the interstate region should serve as a coordinator of state and local interests, although this approach should be tempered as one becomes concerned with relatively small and easily accessible areas. Probably their initial role should be chiefly informational rather than directive to any degree—except as the states concerned may voluntarily wish to establish such a body between their coastal zone authority and the federal agency.

States with long coastlines and correspondingly large coastal zones may find the only practical way to collect, evaluate, and transmit the public viewpoint adequately is to proceed from an intrastate basis. States where coastal zone problems are reasonably uniform and of general public interest may find regions superfluous. This chapter emphasizes the general case where the coastal plan is drawn up at state level rather than discuss all conceivable variations. Most of the concepts developed for the general case will apply, with some adaptation, to special situations.

The state should obtain participation from all levels of government, in order to produce a more realistic, politically implementable plan with less expense, by encouraging the participation of experienced federal agencies, regional groups, and, especially, local governments. To obtain the latter participation, the states should consider incentive grants to local government expressly for this purpose. Both inland and marine-oriented expertise must be fully utilized. Most existing comprehensive studies emphasize inland considerations and tend to fade out as the rivers meet the sea. Marine planners and oceanographers, on the other hand, tend to view the coast as a special edge of their domains. It would be natural for them to advocate a set of restrictions on coastal usage that would enhance or protect marine values with less emphasis on the significance that these restrictions might have on man's socioeconomic welfare. Having both perspectives brought together increases the probability of harmonizing these essentials.

Strong public interaction is important at every stage of the plan, especially during the early "listening" stage, when the agency needs to learn what the people want. If the right tone can be established early and maintained, progress should be somewhat more rewarding in the later phases wherein public reaction to alternative solutions is sought. Participation should be sought from those whose interests are directly affected and also from those who appear to represent less direct public interests. Participation by organized nongovernmental groups representing conservation and industry should be sought. Equally important but more difficult to obtain is input from less organized groups such as land owners and seasonal recreationers. Public participation is so important that someone on the planning team should be specifically assigned to sustain it. Techniques such as public information and educational programs, public hearings, and opinion sampling are useful. Public participation can become a "nuisance" to a planner; but if the plan is not substantially influenced by informed public opinion, it will probably not become politically viable.

The first basic step in coastal planning is to define a set of objectives that express the people's needs. Our overall objective is to maintain and improve the coastal zone's usefulness for man, including the natural system upon which he depends, for now and the future, in an economically, socially, and environmentally acceptable way. Following this lead, then, we begin coastal planning with the people. The plan should integrate a flexible research capability since every comprehensive planning effort will encounter major knowledge gaps. What are the uses of the coast? What are the demands imposed on the coastal zone by each of these uses? How should these demands be expressed in terms of requirements for coastal resources? How adequately can the coastal zone meet these current and future requirements? What is the potential for increasing the carrying capacity for a given activity? What are the additional needs, the requirements that cannot be met by the present supply?

The process will require considerable professional skill and imagination. For example, the demands can be developed from listening to people and by observing their usage patterns in the past. Planning technology has developed many techniques for projecting future demand for most uses; but for some of the less publicly perceived uses, such as ecological balance, new techniques such as quantitative expert judgment will also be needed. Considerable thought will be required to translate demand into requirements. For example, to translate the demand for beach recreation expressed in annual beach days within twenty miles of a city into the amount of public beach required, the distribution of the attendance and a usage standard, such as the square feet of beach per person when the beach is most crowded, must be developed. A perception of local desires and the environmental tolerance of the resource is required. A planner may "solve" a beach recreation need by default. People would overcrowd the site or become exasperated at gaining access to it and this high "price" throttles the demand to a stabilized level. New York City can be cited as an example here. The city has done little in the past quarter century to recognize and plan for the intense demand for beach recreation. On the other hand, San Diego has done much to accommodate a rapidly rising demand under unchanging use standards by an imaginative and well-timed responsive beach improvement program (California, 1972).

Some conflicts can be recognized, and trade-off judgments made when overlaps in demand arise. Informed public opinion is particularly useful in this recognition. When it is very difficult to make some trade-offs, the conflict is not resolved, but one of the following techniques may be able to satisfy a surprising number of apparently irreconcilable conflicts. Once the planning objectives are defined it is necessary to seek ways to achieve them.

The second basic step in coastal planning is to develop alternative ways of achieving established objectives. There are so many possible ways to accomplish this that we can only provide a few illustrations. These techniques are grouped into two broad classes, structural and nonstructural.

Examples of structural techniques are waste water treatment and disposal, including ocean outfalls and "beneficial" use of the nutrients in waste water; shore stabilization through periodic sand renourishment, dune stabilization, vegetative plantings, breakwaters, jetties, groins, bulkheads, levelments, seawalls, and dikes; shore enhancement through beach widening with sand fill, selective ditching and diking, for example, to increase wildlife habitats, wetland augmentation and fertilization, artificial reefs, control of river regimes, hurricane barriers; and secondary controls such as the provision, or limitation, of roads, parking areas and public utilities. Secondary controls can significantly expand or depress opportunities for coastal recreation. Coastal roads can be designed with primary emphasis on preserving the shoreline aesthetics while also exposing the aesthetic appeal of the shoreline to the driver; those who wish to decrease their travel time can use the highways located farther inland.

The nonstructural techniques seek to encourage or discourage certain uses to meet the desired objectives. As with the structural techniques, we will illustrate their great variety by listing some of them but will not develop their individual nuances, which are generally well-explained in the literature. The techniques are listed in order of increasing imposition of public power over individuals. (1) Agreements involve varying forms of volunteer of the fee simple, to provide beach access or preserve wetlands, for example, lessor rights such as scenic easements and contract zoning whereby neighborhood groups can build a set of provisions developed to protect themselves and promote the best use of their property. (2) Public policy inducements involve the adjustment of tax rates to influence usage, cost sharing to include incentive grants to promote goals, use charges to control usage, and planning maps to guide developers and minimize speculation. (3) Regulative controls include zoning which influences desired usage patterns favoring coastally dependent uses such as ports, beach recreation, or shoreline aesthetics, over other coast-desiring uses such as residential development or location of industry. Zoning can also limit use in hazardous or environmentally sensitive areas. Zoning ordinances, permits with their associated impact statements, orders, pesticide controls, and other regulatory techniques are, in general, founded on the police power of government to protect the health, safety, morals and general welfare. A planner should be cautious in depending upon regulative controls too heavily, as they introduce the probability of expensive, protracted legal debate and confusion. More direct methods, such as acquisition of the fee simple, are often less costly in the long run. (4) Compulsory taking represents the strongest form of public powers over individuals, examples of which are condemnation and inverse condemnation.

The power of the techniques listed above can be greatly reinforced if used in combination. For example, to meet recreational beach objectives, a planner can choose various combinations of techniques such as protecting existing beaches from erosion, enhancing existing narrow beaches with sand fill, creating new beaches along rocky or muddy shorelines, improving parking or other limiting facilities, providing improved roads to new beaches or failing to improve roads to existing saturated beaches, acquiring more beach or beach access, imposing user charges, such as parking meters, or providing low-cost public transportation, improving water quality, or providing grants to beach localities to improve public usage.

The third basic step in coastal planning is to formulate a program consisting of a time-phased set of actions selected to achieve stated objectives. The program can be formulated in an orderly manner in several phases. First, the previously developed objectives are mated with the solution techniques, preferably showing a combination of ways to achieve each objective. Then a preferred solution is selected, and its characteristics are listed such as cost, time, institutional and legal implications, and external consequences. Second, the objectives and preferred solutions are formed into a coherent set. Particular attention is placed on multi-use concepts, exclusive-use requirements, and the external effects of the preferred solutions. All solutions, whether structural or nonstructural, have external consequences, and the planner must recognize them, since external effects can often defeat other objectives. The conflict may be resolved by falling back upon a secondary solution for one or both of the conflicting objectives, or abandoning one or both of the objectives in whole or in part. The conflicting need for exclusive-use beach space by bathers and surfers might be resolved by allocating a more remote beach to the less-popular use all the time, and one of the more accessible beaches at off-peak hours. This is an example of a trade-off honoring a spectrum of users in a way most of them would consider equitable. Depending upon the state's concepts of land-use controls, the entire coast might be "allocated" to various uses or combinations of uses. Third, the objectives and associated solutions are ranked in priority on the basis of funds available. Cost-benefit considerations, distribution efforts, equity, execution time including design and sequential interrelationships between projects must be considered. Responsibilities for accomplishing each objective are designated and schedules are developed. Legislative and reorganizational needs, if any, are identified, and the program is explained to the public. Methods of financing the program are reexamined and decisions made on alternative approaches such as cost sharing, to include grants, user charges, and bonds, which is a way of transferring an equitable portion of the costs to future beneficiaries.

Implementation of the program includes the provision of the required public services, the inducement and control of the nongovernmental sectors in meeting the program objectives, and the feedback of new information and lessons learned into a continuing adjustment of the program and general plan. Further discussion of program implementation is found in Section 10.4.

10.3.3 Local Responsibilities

It is important to stress that it is at the local level where the impacts, good or bad, are primarily felt and where most of the everyday decisions must be made and carried out. Presumably, there will emerge from the state level, after vigorous participation by local governments, a blueprint strategy or comprehensive plan with supporting legislation (see Chapter 11), authority, funding, staffing, and guidelines. If the principle of delegation of authority is emphasized by the state, the local governments will make all management decisions as long as they are in harmony with the state plan. They will anticipate review by the state to see that they stay within these bounds. To the extent that local communities feel the plan inadequate or prejudicial to their values, they must initiate the presentation of their case before the state.

While the state develops the comprehensive coastal zone plan, local government is a prime participant, especially during the formulation of those parts of the program that apply to the local level. The state should seek local participation and provide grants to make it feasible. During the planning and the implementation, local government should advocate its own interests. It should offer resistance to the imposition of state-directed restrictions on local usage unless those restrictions can be demonstrated to reflect major external considerations extending beyond local borders. However, it may choose to accept voluntarily state-suggested restrictions on their own merits. The basic reason for this posture is to preserve the principle that the people living in a locality ought to be the ones to make the local decisions, unless it is clearly demonstrated that these decisions significantly affect the well-being of others outside the locality. If localities pursue this principle, the state's attention will be fixed where it ought to be—on the major issues, those that have significant interlocality or regional impact. Local planning may be considered as "special," "limited," or "single" interest planning with respect to comprehensive area planning.

Local government should in its planning process, identify those areas that are primarily of local concern, and using state and federal guidelines participate with higher levels of government in the determination of those areas which are of state, regional, or national significance. With cooperation and assistance from the state or federal government, local government should conduct an inventory in a detailed manner to include such facts as soil types, vegetation, slope, specific kind and density of structures, and use patterns. A determination of the capability of the environment or the resource to support proposed uses should result, and using the public hearing process as a tool, preparation of the local plan should begin. This plan will include such factors as aesthetics, convenience, transportation, or density, all of which go to make up the lifestyle people choose. With the will of the people clearly in mind, the planning process would then examine the existing and proposed land and water uses for their relevance to these expressed desires. Conflicts will be identified, if present, and brought to public attention by appropriate means. Evaluation of these conflicts by the same local unit and their attempted resolution would follow—with development of alternative solutions to the problems. These alternatives will be recognized by the state coastal zone management authority, and they would be expressed in the form of programs which, when implemented, would provide for the accomplishment of objectives, goals or missions. Local programs, having been developed within federal and state guidelines, and having been subjected to the public hearing process before adoption, should now be retested against the state-federal criteria and presumably should be acceptable with only minor modifications. The concept of detailed planning at the local level will ensure the consideration of local interests, even though in some cases regional, state, or national interests may necessarily take precedence when the comprehensive area plan is developed in its final form. If the public review process is done properly, the will of the people should be evident and incorporated in the guidelines. By the time the planning operation at the local level occurs, the path should already have been broken, and the chances for early compatibility of local and state guidelines should be enhanced.

Among the concepts which should be considered at any level of coastal planning is that of dependency—that is, the favoring of, or the assigning of priority to, that use that depends for its success upon some physical, sociocultural, or economic attribute of the coastal zone. Following this screening for dependency, the proposed use should be evaluated with respect to its impact upon the economic, physical, sociocultural environment and positive benefits established as a requisite for approval.

Evaluations, in order to be significant, must include a systematic monitoring process, which must provide for feedback into the planning management system over a period of time. There must be continuous evaluation of the results or consequences of prior decisions. This will provide a firm basis for system updating, of both inventory and guidelines, and policies and criteria. Outside influences or new parameters should also be inserted into the system, in part through the action of a technical advisory committee. The elements of the planning process then would reduce to establishment of objectives at the active planning levels; inventorying present and projected land and water uses; determination of the will of the people concerning the planning area; determination of projected demand; identification of conflicts and problems; alternative solutions to the problems; and an implementation procedure, including appropriate agencies to take the action. This process should work at all levels of government and should constantly make provision for public review and comment through a constant feedback into the system from the monitoring activity.

Delegation of considerable authority to the local level is appealing in some states because of geographic features or a tradition of decentralization. Examples are New York and California. New York's coastal zone consists of both a highly urbanized, saltwater coast and a long, comparatively unpopulated, freshwater lakefront some 300 miles away. California has recently demonstrated a strong tendency toward decentralized control of its 1350-mile coastline by forming "regional" coastal zone commissions (California, 1972).

The states themselves are in the best middle position to make the judg-

The Coastal Zone Workshop recommends
The establishment of regional *Coastal Zone Centers* to develop and coordinate natural science, social science, and legal research and to provide relevant information about the coastal zone to government agencies and the public. These centers should cooperate with existing research organizations to resolve basic questions of the environment in that region, help appraise management techniques, and provide inventories of coastal resources. Coastal Zone Centers should be established in regions corresponding to the major types of coastal environments and may be international in character.

ments as to the extent to which they will regionalize or localize their basic authorities. For example, the federal level may offer inducements, but unless the state itself perceives the need for and seeks interstate regional integration, the regionalization would probably not be viable.

10.3.4 The Regional Coastal Zone Centers

The regional coastal zone centers proposed by the workshop would have as a primary function the accumulation and distribution of information. No planning process can be comprehensive unless sources of viable information are available and up to date. Basic information must be an integrated input into the management system if the system is to be dynamic and address itself intelligently to the problem.

Formation of the centers will be useful in all sectors including the academic community, the general public, and all levels of government. In order to be manageable in size and scope, information should be primarily directed into three levels—federal, regional, and state. Many federal agencies and some states have data accumulation capabilities that are used in the implementation of their responsibilities. A basic problem is one of economics of scale, since data synthesis and collation could clog any information system unless the center has the hardware and software capability to assimilate all data generated for it.

A regional center then would be in a position to help in the coordination of data accumulation, while helping any user in the search for specific information. A list of examples of possible use would be—

—A congressional committee in search of data that would substantiate federal information in the development of national policy.

—A coastal zone manager in search of backup information to help in the decision-making process.

—A scientist in search of state-of-the-art information to help coordinate a research proposal.

—A coastal zone user in search of information or sources thereof that would help in the development of an impact statement.

Five well-defined needs exist for regional information at the national policy level of responsibility that would be of value in the formulation of federal policy: (1) environmental base-line information from federal and state sources should be interfaced and interacted; (2) base-line information requires updating periodically to help modify national policy where necessary; (3) an inventory of areas that have a significance of greater impact than at the state level is needed to help to identify areas of national significance; (4) the screening and collation of relevant social and legal information will also help in the development and the modification of national policy; (5) collation of research output and coordination of new research requirements addressed to coastal zone problems.

There are basically four well-defined needs for regional information at the federal level of responsibility to aid in the development of guidelines: (1) environmental base-line information would help in federal regulatory evaluations; (2) collation of relevant environmental social and legal information will help in the development of guidelines for total impact statements; (3) monitoring of the regional environment will be necessary to evaluate and update guidelines or redefine areas of environmental concern; (4) the collation and interactions of state-of-the-art information could be used as backup to specific base-line information at the federal level.

There are four well-defined needs for regional information at the state level of responsibility that could be provided by such a system: (1) regional base-line environmental information is needed to help develop an integrated set of statewide guidelines; (2) monitoring information should be collected and collated to help update base-line information and modify statewide guidelines when necessary; (3) relevant state-of-the-art information is needed as backup to specific base-line information both at the regional level and state level; and (4) relevant social and legal information should be collected and collated to help in the development and evaluation of state level total impact statements.

The monitoring capability of a system of regional centers would help in the development of a warning system. It has been clearly identified throughout the Coastal Zone Workshop that regional and national

The Coastal Zone Workshop recommends
The creation of regional and national monitoring systems to collect continually chemical, physical, and biological data with a capacity to give advanced warning on conditions that may be hazardous to the ecosystem of the coastal zone.

monitoring systems are needed to collect chemical, physical, and biological data continually with a capacity to signal any unusual conditions, which are critical or subcritical, though additive in nature.

Technological developments will require the creation or increased sophistication of many chemical or mechanical products with untested though possibly deleterious potential for environmental degradation. A regional information system could help in the development of guidelines and regulatory standards that would not be a direct cost to a state-level authority. Such a system must interact at three levels of government to assure its viability. This would include interaction with each state within the region, all other regional systems, and all federal systems.

There are several other specific areas where regional centers would be useful. These centers should (1) define information needs of all sectors; (2) coordinate in the development and refinement of other information systems at all levels; (3) offer incentives for federal systems to interact; (4) evaluate guidelines for creation of format for mapping of coastal resources for states in a region, to be comparable throughout the nation; (5) provide source or sources of information to aid in evaluating appeals at both the state and federal levels; and (6) provide backup information in evaluation of trade-offs. These needs demonstrate the scope necessary to develop a coordinated management process.

10.4 The Management Process

The coastal zone authority should serve as an innovator, a proponent of new actions, and active advocate of compatible programs, whatever their source. The role of proponent should be based upon a constant evaluation process with respect to how efficiently long-range planning goals are being met. Direct action could result from relaxing or strengthening regulations designed to control land and water uses. Encouragement of timely projects should be designed to be compatible with the programs devised to achieve the coastal zone objectives. Specific ways in which the proponent role can be carried out include the following.

1. *Allocation by permit.* The management process indicated by the diagram in Figure 10.1 is based on the workshop recommendation that impact statements be required of all projects, defined in the broad sense of the term, which desire to locate in the coastal zone and which may produce a significant alteration of existing social, economic, and environmental conditions (see Section 10.4.1).

2. *Exercise of zoning powers to favor desirable uses.* This would be a direct application of authority to both exclude specific uses from designated areas while encouraging other specific uses to locate in those areas. Such action could include the designation of unique areas. Positive designation of select valuable areas has the advantage that not only will the resource per se be maintained in a desirable state, but that adjacent proposed uses will be made conditional upon their compatibility with the environmental quality of the designated areas.

3. *Fee simple acquisition.* The coastal zone authority would act by direct purchase to acquire land in full title and thus gain control of uses. Money would be available from general funds, land issues or special tax levies.

4. *Purchase and lease back under restriction.* This has the advantage of continuing present compatible uses with at least a partial return of investment to the coastal zone authority. Concessions generally should be negotiated and managed under strict public interest clauses.

5. *Acquisition of selected, specified rights (easements).* In this case, outright purchase and thus high initial outlay is avoided; however, specific desirable attributes are obtained such as the right of public access, preservation of unique scenic or ecological areas, and creation or maintenance of open space.

6. *Compensable regulation.* This approach will in effect guarantee that a land owner will obtain at least a minimum price agreed upon beforehand, when he decides to sell. In return he must comply with agreed-upon stipulations of desirable uses of his property.

7. *Taxation.* By this means, the property owner may be compensated for holding the land in desirable use through provision of tax relief. This may take such forms as tax deferral, tax exemption, or reduced assessments.

8. *Restoration or renovation requirements inserted in permits.* Such

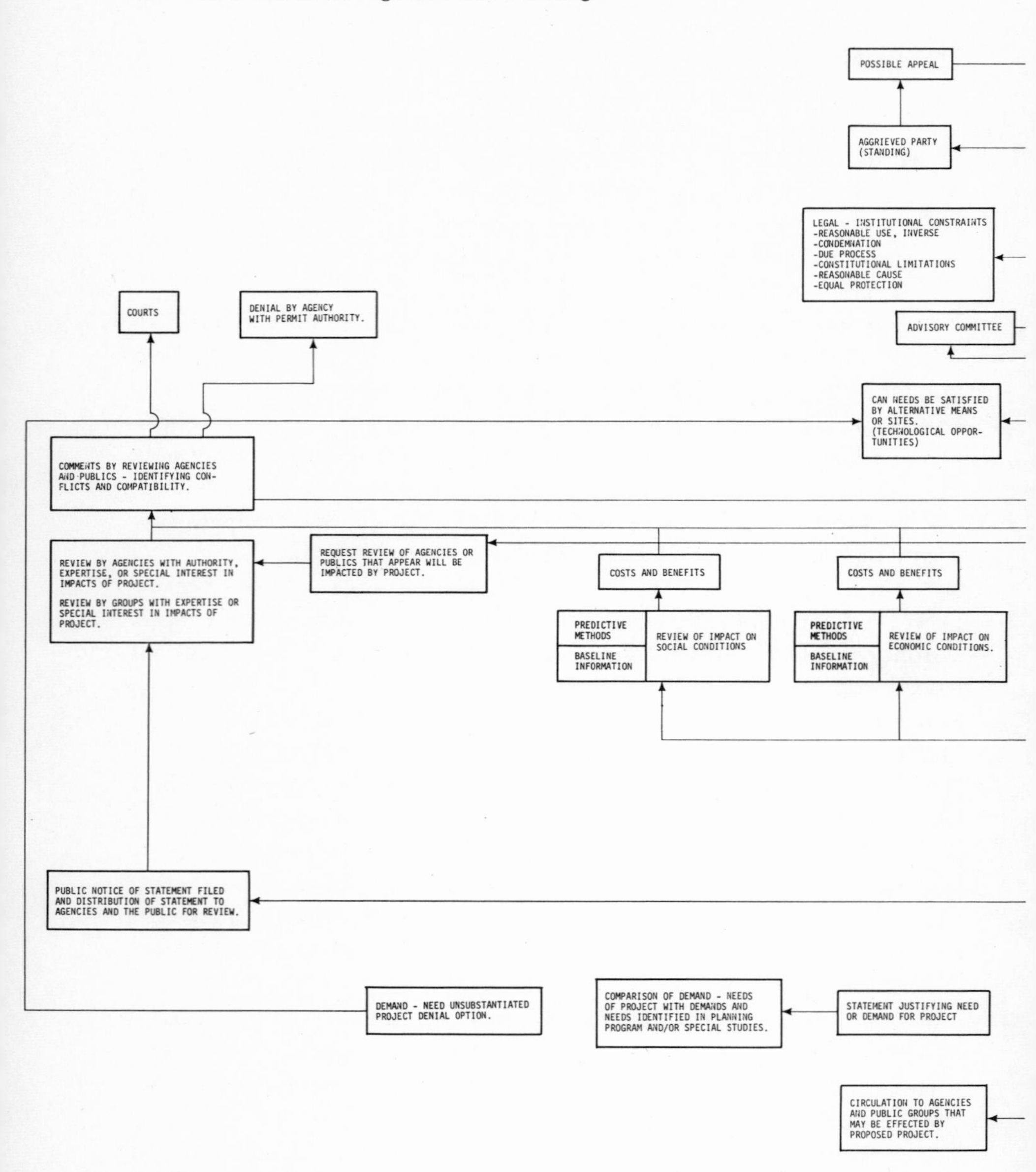

Figure 10.1 Elements of a coastal zone planning program.

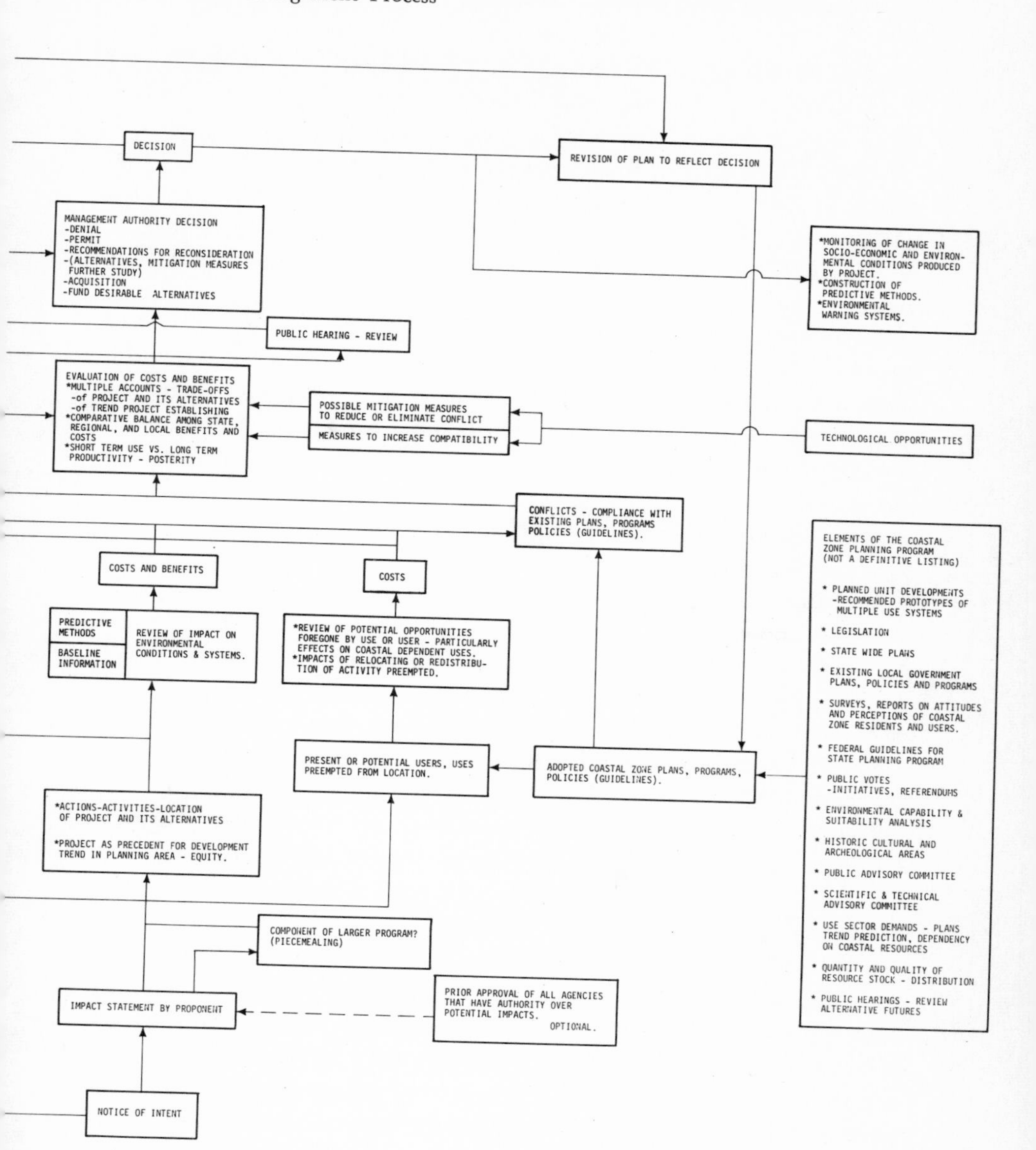
DECISION
REVISION OF PLAN TO REFLECT DECISION
MANAGEMENT AUTHORITY DECISION
-DENIAL
-PERMIT
-RECOMMENDATIONS FOR RECONSIDERATION
-(ALTERNATIVES, MITIGATION MEASURES FURTHER STUDY)
-ACQUISITION
-FUND DESIRABLE ALTERNATIVES
*MONITORING OF CHANGE IN SOCIO-ECONOMIC AND ENVIRONMENTAL CONDITIONS PRODUCED BY PROJECT.
*CONSTRUCTION OF PREDICTIVE METHODS.
*ENVIRONMENTAL WARNING SYSTEMS.
PUBLIC HEARING - REVIEW
EVALUATION OF COSTS AND BENEFITS
*MULTIPLE ACCOUNTS - TRADE-OFFS
-of PROJECT AND ITS ALTERNATIVES
-of TREND PROJECT ESTABLISHING
*COMPARATIVE BALANCE AMONG STATE, REGIONAL, AND LOCAL BENEFITS AND COSTS
*SHORT TERM USE VS. LONG TERM PRODUCTIVITY - POSTERITY
POSSIBLE MITIGATION MEASURES TO REDUCE OR ELIMINATE CONFLICT
MEASURES TO INCREASE COMPATIBILITY
TECHNOLOGICAL OPPORTUNITIES
CONFLICTS - COMPLIANCE WITH EXISTING PLANS, PROGRAMS POLICIES (GUIDELINES).
COSTS AND BENEFITS
COSTS
ELEMENTS OF THE COASTAL ZONE PLANNING PROGRAM (NOT A DEFINITIVE LISTING)
* PLANNED UNIT DEVELOPMENTS -RECOMMENDED PROTOTYPES OF MULTIPLE USE SYSTEMS
* LEGISLATION
* STATE WIDE PLANS
* EXISTING LOCAL GOVERNMENT PLANS, POLICIES AND PROGRAMS
* SURVEYS, REPORTS ON ATTITUDES AND PERCEPTIONS OF COASTAL ZONE RESIDENTS AND USERS.
* FEDERAL GUIDELINES FOR STATE PLANNING PROGRAM
* PUBLIC VOTES -INITIATIVES, REFERENDUMS
* ENVIRONMENTAL CAPABILITY & SUITABILITY ANALYSIS
* HISTORIC CULTURAL AND ARCHEOLOGICAL AREAS
* PUBLIC ADVISORY COMMITTEE
* SCIENTIFIC & TECHNICAL ADVISORY COMMITTEE
* USE SECTOR DEMANDS - PLANS TREND PREDICTION, DEPENDENCY ON COASTAL RESOURCES
* QUANTITY AND QUALITY OF RESOURCE STOCK - DISTRIBUTION
* PUBLIC HEARINGS - REVIEW ALTERNATIVE FUTURES
PREDICTIVE METHODS
BASELINE INFORMATION
REVIEW OF IMPACT ON ENVIRONMENTAL CONDITIONS & SYSTEMS.
*REVIEW OF POTENTIAL OPPORTUNITIES FOREGONE BY USE OR USER - PARTICULARLY EFFECTS ON COASTAL DEPENDENT USES.
*IMPACTS OF RELOCATING OR REDISTRIBUTION OF ACTIVITY PREEMPTED.
PRESENT OR POTENTIAL USERS, USES PREEMPTED FROM LOCATION.
ADOPTED COASTAL ZONE PLANS, PROGRAMS, POLICIES (GUIDELINES).
*ACTIONS-ACTIVITIES-LOCATION OF PROJECT AND ITS ALTERNATIVES
*PROJECT AS PRECEDENT FOR DEVELOPMENT TREND IN PLANNING AREA - EQUITY.
COMPONENT OF LARGER PROGRAM? (PIECEMEALING)
IMPACT STATEMENT BY PROPONENT
PRIOR APPROVAL OF ALL AGENCIES THAT HAVE AUTHORITY OVER POTENTIAL IMPACTS.
OPTIONAL.
NOTICE OF INTENT

conditional permits will ensure that otherwise acceptable operations, such as mining and logging, would not have undesirable residual effects. Clean up, burying, and reforestation are examples.

9. *Special user taxes.* This may involve the levying of a use tax upon certain activities such as motorcycling, horseback riding, bicycling, golfing, and swimming in order to pay the costs of needed special facilities or mitigation measures or acquisition of additional special-use areas.

10. *Coordination of public and private interests into a cooperative project.* Since private and public ends need not be in conflict, coordination should be encouraged in order to achieve long-range planning goals faster, at reduced costs, and with the advantages of a single coordinated approach with respect to permits, impact statements, and special studies.

11. *Solicitation of private groups and individuals for funds for desirable uses.* Funds from private foundations and indivduals who have a special interest in some feature of social or natural resources can be made available. Not the least of the advantages is that of flexibility and the opportunity to act rapidly. Outright donation, expert advice, special studies, or loans may be sought. Private agencies which can act in time of crisis to purchase an area in question and retain it until public funds are available, can make a great contribution to the powers of a coastal zone authority.

12. *Active support, through the public media, of projects and programs for desirable uses.* The coastal zone authority can significantly influence public opinion through use of public hearings and public statement to advocate legislative action and ballot propositions.

13. *Support of studies and efforts to design desirable prototype structures and uses.* The coastal zone authority should sponsor or prepare plans or studies using models, sketches, descriptions, and graphic presentations as fully as possible in describing the effects of such activities as housing, rent, and marinas. Construction of actual prototype structures or projects should be encouraged in order to demonstrate desirable uses and marshal support for accepted planning programs. These prototypes may also serve as essential elements in the environmental monitoring program, since they may be used as test devices and reference points against which to measure changes and impacts. This may be especially useful if they are used as check points in the implementation of long-term planning to represent different stages of development.

The Coastal Zone Workshop recommends
Improved environmental impact statements should be prepared for each new or additional activity and structure in the coastal zone to determine the extent to which they would have social as well as environmental effects. More stringent requirements for the preparation, detail, and use of such statements in making specific decisions should be developed.

14. *Support of research efforts by other agencies.* This can be a very effective process since research projects commonly require a practical orientation or backup. While this backup may be regarded as inadequate for a single applicant, if research promises to assist in the solution of management problems, then authority support should be forthcoming.

10.4.1 Impact Statements

The proponent of an activity should submit an impact statement substantiating the need for the project and a full explanation of its consequences and an evaluation of alternate actions. The following questions should be answered. Is the basis for predicting the need for the project accurate? What is the validity and accuracy of factors such as population trends, consumption patterns, community values, changes in technology, and market analysis? In this context, industry should be encouraged to disclose proprietary information to facilitate evaluation. Is the project premature? Is it a speculative venture and not a response to a demonstrated public need?

The project should also be examined to determine if it is a component of a larger long-range program. Proponents will often break a major program into many individual projects. An impact statement is then prepared for each of these smaller, often insignificant in scale, projects. The classic example is two separately constructed sections of a highway which upon completion force a connecting section to be routed between them. A coastal zone management authority should examine how a project will relate to a longer term program and what the impact implications of the total program will be. Presumably the program in question, such as the highway, will be compared with the related element in the coastal plan, such as transportation, to determine the nature and degree of conflict or compliance.

According to Figure 10.1, two distinct considerations emerge from the impact statement. On one level the project is described in terms of the activities, actions, and location it will require and the associated impacts on environmental, social, and economic conditions. On the second level, the project should be examined in terms of the development trend it may promote for the surrounding region, or governmental unit. Generally, this may occur in two ways. A proposed project, if approved, may set a precedent for the future development trends. For example, what would be the cumulative impacts if the density of a proposed housing development on a particular site were projected to all privately held coastal lands in the region with similar site characteristics? A housing development of 25 units on 10 acres may not cause significant impacts. However, if all developers were allowed to construct projects to the same density on land with similar site characteristics, as might be their right under equal protection provisions of the law, the cumulative impact could significantly stress existing environmental, social, and economic conditions. Development trends may also be stimulated by a project's requirements for public and private services or by a project's provision for development opportunities. A housing project may require the construction of a new sewage treatment plant which in turn may stimulate further residential growth. A new or enlarged highway will increase the desirability of adjoining lands for development. In order for a coastal management authority to predict accurately development trends based on a project proposal, a computer-mapping capability will be required. The impact statement would address both the impacts caused by a project's direct preemption of existing or potential use and users from its intended site and future direct and indirect impacts on existing social, economic and environmental conditions that characterize the project's location.

Determination of the impacts resulting from preemption of use or users from a project site depends on the ability of the coastal zone management authority to identify the value of the particular location to other uses or users and the effects of relocation or redistribution of the use or users preempted. Identification and evaluation of both these preemption impacts will require reference to the use and user demand analysis, resource stock, and environmental capability elements developed by the planning process.

Assessment of impact, whether by the coastal zone management authority or the proponent agency, can be viewed as a three-step process. The initial step is identification of condition changes that are likely to occur in response to a project's actions or activities. The next step is prediction of the degree and spatial dimensions of the condition changes identified. The predicted condition changes are then described in value terms, usually distinguished as costs or benefits, not only in monetary terms but also taking account of population affected, amenity values, and environmental impact.

The coastal zone authority will require a checklist procedure to determine whether the proponent has identified all the potential impacts of the project and evaluate the adequacy of the description of the identified impacts. Several impact identification procedures have already been developed as a means of systematically relating a project type (highways and housing development, for example) of actions or activities (for example, impervious surfacing and grading vehicles) to the impacts they normally generate, such as increased runoff and stream peak flow, vehicle noise, turbidity, and sedimentation of water body (Leopold et al., 1971; Sorensen, 1971; Battelle, 1971).

Identification of potential impacts is a comparatively simple procedure when compared to the task of predicting the probable degree and dimension of condition change. The coastal management authority could accept the impact predictions, but the management authority should also conduct its own impact prediction for those changes that the statement failed to consider or that were inadequately analyzed by the proponent. At present, agencies reviewing impact statements rarely conduct their own predictive assessment of potential changes because of lack of time, money, and staff capabilities.

Prediction of impacts is restricted by lack of sufficient time and funding to acquire necessary knowledge of base-line conditions that define a project's setting. Data and information are rarely gathered over a long enough period of time to permit accurate prediction of how a condition may change in response to a proposed action. If a project is permitted by the management authority, monitoring should be required to detect changes and to permit development of better predictive methods. In order to conduct more accurate impact assessments in the future, prediction should be made by the coastal zone management authority of the probable location of specific project types, so that necessary base-line data and information on possible condition changes can be collected well in advance of its need.

The amount of research necessary to do a thorough analysis of base lines or changes is often greater than the proponent agency or organization can afford to spend on the particular project, or the cost of acquisition of necessary data and information is thought not to be worth the potential cost of the particular impact identified. Methods and devices capable of providing satisfactory accuracy to the predictive assessment of a condition change need development. There may be a complete lack of predictive models, or the models available may not be sufficiently accurate. In many cases, only a rough estimate that describes the upper and lower bounds of the potential changes is possible. In a relatively few situations, there are fairly good modeling techniques for predicting condition changes. Simulation and statistical models have been developed for those processors readily subject to quantification, such as hydrology, climatology, demography, and economics (see Chapter 12).

The coastal zone management authority should have several means available for determining the value of the impact. The most direct means of valuation is a comparison between the potential impact and the coastal zone management authority's own adopted plans, programs, and policies to determine the degree of conflict or compatibility. The significance of the conflict or compatibility identified by such a comparison depends on how explicitly the values of the existing conditions have been assessed by the coastal zone management authority's planning program. The management authority should conduct or encourage studies that seek to put values on environmental and social conditions that are highly likely to be impacted by future projects. One obvious example of such a study would be to place open-space valuations on privately owned, low-slope, underdeveloped areas with a view of the ocean. Another means of determining costs and benefits would be offered by recording the spatial occurrence of a resource on a coastwide basis which would indicate its rarity or uniqueness.

A basic feature of the impact statement process is the review of the document by other agencies and public groups that have designated authority, special expertise, or interest in the conditions modified. The impact statement is submitted directly to the relevant agencies and affected publics and the coastal zone management authority for review. The management authority should also be encouraged to communicate with the agencies and groups to determine if they are aware of the impacts that their own review has identified and may seek further measurement of cost and benefit trade-offs through the process of public hearings and/or public review by an ad hoc or regularly functioning advisory committee (Figure 10.1). Once the trade-offs of costs and benefits have been made explicit, an informed and rational decision should then be possible. The coastal zone authority would be expected to have several decision-making options in addition to the direct action of permit or denial. The most common alternative is to recommend that the proponent modify the project in a certain manner and submit the revision for reconsideration. The coastal zone authority may give a conditional permit that requires the project to comply with certain stipulations during its construction and operation.

A variety of legal and institutional constraints will limit the coastal zone authority's decision-making powers. Most of the constraints will result directly from the enabling statutes that created the coastal zone authority and specify with more or less exactitude its operating procedure. These types of constraints are readily subject to modification or elimination through ordinary legislative process. Constitutional constraints on decision-making powers, both in the U.S. Constitution and various state constitutions, are less numerous, but more important since they are less subject to change.

The fifth and fourteenth amendments to the U.S. Constitution provide that governmental agencies shall not deprive persons of property without due process of law. Notice and a reasonable opportunity to be heard are generally recognized as requisites of procedural due process. But courts have not found such procedural protection to be invariably required. The compromise reached is to mandate constitutional notice and hearing when an administrative agency promulgates rules affecting the property or rights of a limited number of persons, but not to require notice and hearing when an administrative agency promulgates rules of general application. The notion of due process also has substantive overtones. Denial or restriction on the use of private property raises the constitutional question of whether the action amounts to a "taking," or inverse condemnation, of private property for which just compensation must be paid. For example, if the decision is made that a privately owned wetland should be preserved intact, the question arises as to whether government must compensate the owner for the reduction in private value of his wetland. The owner of the wetland should be allowed "reasonable use" of his property by zoning or land use regulations. The fourteenth amendment to the United States Constitution also provides that state or local governments shall not deny any person within its jurisdiction "equal protection" of the law. The basic notion is that persons similarly situated should be treated equally. If a project is denied that is similar to projects that have been consistently permitted in the past by the coastal zone authority and due cause is not demonstrated to support the decision, court challenge may be sought to declare the denial as an unreasonable, arbitrary, or capricious exercise of administrative powers.

The impact statement allocation process may appear to be a time-consuming and expensive operation. At least three means are evident to simplify and expedite the process.

1. Standardized formats or workbooks should be developed to assist in the preparation and review of impact statements for those project types that are commonly proposed for location within the coastal zone, such as tourist, commercial development, marinas, and channelization. The Bureau of Reclamation is developing an "environmental evaluation system" for water resources projects, primarily impoundments (Battelle, 1971). The Atomic Energy Commission is currently exploring the feasibility of developing a systematic methodology for assessing the impacts of power plants through a contract with Teknetron Corporation. Representatives of the Environmental Protection Agency have suggested that future National Environmental Policy Act guidelines include conceptual frameworks for analysis of major types of projects supported by the agency (Orloff, 1972). The project type formats will allow both the proponent of a project and the coastal zone authority to focus quickly on the principal impact issues and avoid needless exploration of irrelevant or second-level considerations.

2. The planning program should establish impact budgets for specific areas within the coastal zone. Areas that have been determined to have a direct influence on the quality of a desirable coastal zone resource would be identified, and investigation would then be made of how surrounding land-use activities in the area of influence accumulatively produce an impact on the desirable resource qualities. Budgets would then be established within the area of influence to limit the degree and location of activities, based on the resource's capability to tolerate the cumulative impact. An example of an effective application of the impact budget concept is the Lake Tahoe Regional Plan (Pepper, 1972). A project proponent would have the burden of proof to demonstrate to the coastal zone authority that condition changes generated by the project would not exceed the impact budget for the area.

The Coastal Zone Workshop recommends
The protection from environmental degradation of those coastal wetlands and estuaries that are highly productive habitats, spawning areas, or nurseries for aquatic life or contain rare and endangered species. Only coastal activities that will not markedly degrade the diversity and productivity of the existing ecological system in these areas should be permitted.

3. Many of the specific impacts of any project will be intensively reviewed by agencies that have expertise or authority for a particular condition change such as water or air quality. The coastal zone authority may considerably lessen the task of reviewing impacts by depending on the review comments from other agencies. Instead of attempting to review all the impacts of a proposed project, the coastal zone authority could concentrate on those impacts that no other agency or public would be reviewing thoroughly.

10.4.2 Examples of Possible Managerial Guidelines

The following examples of guidelines that a state may issue to local government as criteria for their formulation of regulations and performance have been extracted from documents issued by the states of California (California, 1972), Washington, and Florida (Florida CCC, 1971), and under contract for Nassau and Suffolk counties of Long Island (Ellis, 1969). The examples have been grouped broadly by degrees of specificity. Table 10.2 presents examples of state programs and relevant legislation or court cases.

Coastal wetlands, bays, lagoons, and estuaries that have been determined by the state coastal zone authority to be highly productive habitats or spawning areas or contain rare and endangered species should be protected from irreversible environmental degradation and altered only on the finding that no significant impact will occur. Watersheds that affect the environmental quality of these areas should be planned and managed so as to ensure that activities that would produce or contribute to irreversible environmental degradation will not occur.

Consideration should be given to the location of recreational facilities, such as playing fields and golf courses, and other open areas, such as cemeteries, which use large quantities of fertilizers and pesticides in their turf maintenance programs, to prevent these chemicals from en-

Table 10.2. Examples of State Programs

State	Type of program	Relevant legislation and court cases	Date
Alaska	Constitution provisions	Section VIII (Natural Resources) Alaska Constitution	1958
	Resource management	Alaska Land Act	1959
California	San Francisco Bay Conservation and Development Commission	1. Act 1165 (McAteer-Petris Act) as amended	1965
		2. Act 713 (McAteer-Petris Act—major amendments)	1969
	Thermal power-plant siting	Departmental Regulation as amended	1965
	Comprehensive ocean area plan	Act 1642 (Marine Resources Conservation and Development Act of 1967)	1967
	Beach access	1A. *Gion* v. *Santa Cruz*	1970
		2A. *Dietz* v. *King*	1970
Connecticut	Wetlands	1. Act 695 (preservation of tidal wetlands)	1969
		2. Act 536 (wetlands acquisition powers)	1967
Delaware	Heavy industry-port facility ban	1. Act 175 (Coastal zone act)	1971
	Coastal zone management plan		(Begun 1970)
Florida	Aquatic preserves	1. Resolution of Nov. 24, 1969 of the Board of Trustees of the Internal Improvement Fund	1969
	Beach area protection	1. Act 280 (Regulation of Coastal Construction and Excavation)	1971
	Coastal zone plan	1. Act 259 (Coastal Coordinating Council)	1970
Georgia	Wetlands	1. Act 1332 (Reid-Harris Bill, Coastal Marshlands Protection Act of 1970)	1970
Louisiana	Legislative committee	1. Senate Concurrent Resolution #3 (Joint Legislative Committee on the Environment)	1970
	Marine resource management comm.	1. Act 35 (Louisiana Advisory Commission on Coastal and Marine Resources)	1971
Maine	Wetlands	1. Act 348 (coastal wetlands regulation) as amended	1967
		2. Act 572 (coastal wetlands regulations)	1971
		1A. *State* v. *Johnson*	1970

Table 10.2 (*Continued*)

State	Type of program	Relevant legislation and court cases	Date
	Industrial, subdivision, power-plant siting	1. Act 571 (site location regulation)	1970
	Oil transport	1. Act 572 (coastal conveyance of petroleum)	1970
	Coastal development plan	(Period of development 1969–1973)	
Maryland	Wetlands	1. Act 241 (wetlands)	1970
		2. Act 242 (wetlands)	1970
	Power-plant siting	1. Act 223 (power-plant siting)	
	Comprehensive sewage treatment	1. Act 240 (Environmental Service Act)	1970
	Comprehensive resources plan		(Begun 1969)
Massachusetts	Wetlands	1. Act 426 (Jones Act—Coastal Wetlands Dredge and Fill Law of 1963) as amended	1963
		2. Act 768 (Coastal Wetlands Protection Act)	1965
		3. Act 220 (Hatch Act—Inland Wetlands Dredge and Fill Law of 1968)	1968
		4. Act 444 (Inland Wetland Protection Act of 1968)	1968
		1A. *Commissioner of Natural Resources* v. *S. Volpe and Co.*)	1965
	Estuarine area management plan	A1. Executive Order No. 59 (Commission on Ocean Management)	1968
	Conservation commission	1. Act 223 (Conservation Commission Act)	1957
		2. Act 517 (Massachusetts Self-help Act)	1960
Michigan	Shoreland management	1. Act 245 (Shorelands Management and Protection Act of 1970)	1970
Minnesota	Shoreland management	1. Act 777 (regulation of shoreland use and development)	1969
Mississippi	Marine Resource Management Council	1. Act 293 (Marine Resources Council)	1970
	Regional organization	1. Act 517 (Gulf Regional District Act)	1971
New Hampshire	Wetlands—dredging and filling controls	1. Act 215 (regulation of dredging and filling in and adjacent to tidal waters)	1967

Table 10.2 (*Continued*)

State	Type of program	Relevant legislation and court cases	Date
		2. Act 284 (regulation of dredging and filling in and adjacent to public waters)	1967
		3. Act 274 (regulation of dredging, filling, construction, etc., in surface waters)	1967
New Jersey	Wetlands	1. Act 45 (Green Acres Land Acquisition Act of 1961)	1961
		2. Act 46 (Green Acres Bond Act of 1961)	1961
		3. Act 27 (Wetlands Act of 1970)	1970
New York	Wetlands	1. Act 545 (Long Island Wetlands Act)	1960
	Coastal management unit	1. Act 715 (Division of Marine and Coastal Resources)	1969
	Power-plant siting	1. Act 294 (power-plant site acquisition) amends the Atomic Space and Development Act 210–1962	1968
North Carolina	Wetlands	1. Act 791 (dredging and filling regulation)	1969
		2. Act 159, Sections 6 and 7 (dredge and fill law) amends Act 791–1969, Coastal Wetlands Act	1971
	Beach area protection	1. Act 237 (sand dune protection)	1965
	Coastal zone plan	1. Act 1164 (estuarine zone study)	1969
Oregon	Beach access	1. Act 601 (beach bill) as amended	1967
		1A. *State ex rel Thorton* v. *Hay*	1969
	Coastal Zone Management Act	1. Act 608 (Coastal Zone Management Plan)	1971
	Power-plant siting	A1. Executive Order 01-069-25	Dec. 11, 1969
	Coastal construction moratorium	A2. Executive Order 01-070-07	March 3, 1970
Rhode Island	Wetlands	1. Act 140 (Coastal Wetlands)	1965
		2. Act 26 (intertidal salt marsh) as amended	1965
	Coastal management	1. Act 279 (Coast Resources Management Council)	1971

Table 10.2 (*Continued*)

State	Type of program	Relevant legislation and court cases	Date
Texas	Beach access	1. Act 19 (open beaches bill)	1958
		1A. *Seaway Co.* v. *Attorney Gen.*	1964
	Natural Resources Council	1. Act 417 (Natural Resources Interagency Council)	1967
		2. Senate Concurrent Resolution No. 38 (coastal zone study)	1969
	Sale and lease moratorium	1. Act 21 (submerged lands moratorium)	1969
	Marine Resources Council	1. Act 279 (Council on Marine-Related Affairs)	1971
Virginia	Wetlands	1. House Joint Resolution 60 (Wetlands Commission Study)	1971
Washington	Coastal management	1. Act 286 (Shoreline Management Act of 1971)	1971
		2. Initiative 43 (Shoreline Protection Act)	(to be voted on Nov., 1972)
		1A. *Wilbour* v. *Gallaher*	1970
	Thermal power-plant siting	1. Act 45 (1st extraordinary session—Thermal Power Plant Siting Act of 1970)	1970
Wisconsin	Shoreland zoning	1. Act 614 (Water Resources Act of 1965)	1965

Source: Bradley and Armstrong, 1972.

tering public water. If this type of facility is approved on a shoreline location, provision should be made for protection of public water areas from drainage and surface runoff.

Groundwater is a valuable natural resource which should be held in trust. Therefore, all activities or practices that will further degrade the quality, or seriously reduce the quantity, of that resource should be regulated. For example, ocean outfalls and extensively paved areas both can contribute to the reduction of the groundwater resource.

The state's policy is to protect and preserve historic sites and properties including buildings and objects of scientific and historical value relating to the history, government and culture of the state (Chapter 267, Florida Statutes).

Approved shoreline dredge and fill projects should be designed so that no significant damage to existing ecological values or natural resources, or alteration of local currents will occur, creating a hazard to adjacent life or property, or destroying ecological values or natural resources.

The Corps of Engineers should develop a preliminary classification system for dredging applications to be implemented on a trial basis on Long Island, using (1) size (volume) of the project, (2) environmental quality at the removal and disposal location, (3) timing, (4) reversibility, and (5) compatibility with the comprehensive development plan as classification characteristics. The Corps of Engineers should distribute the preliminary classification system along with the public notices. The widespread shoreline regression should be accepted as a long-term natural phenomenon beyond practicable capability to control; consequently, land-use control techniques and management practices that will influence occupancy and development of the threatened areas should be emphasized so as to minimize losses associated with this regression.

No coastal development, whether public or private, should unreasonably deny the public access from state or local highways to state tidelands. Accessways should be so located and controlled as to prevent concentrations of use pressures that would degrade the environmental qualities of the coastal area.

Residential development of coastal lands should be allowed as a desirable use in those circumstances where certain conditions are met. These conditions would include, but are not limited to, proven existence of a demand for residential housing on the location proposed; dedication of extensive open space areas; provision of additional public accessways to shoreline; design, siting, and density of units compatible with coastal landscape and carrying capacity. The development should encourage economic stability, produce long-term net fiscal benefit to the local government or district, and should not preempt or adversely affect coastally dependent uses.

Coastal areas subject to environmental hazards, such as floodplains, tsunami impact, storm impact, landslide and seismic stress zones, and actively eroding shorelines should be identified and retained in open-space uses, or permit uses that will not be endangered by the environmental hazards or that will not create demands for major alterations to the natural environment as protection against such hazards. Local authorities and developers should be alerted to the environmental dangers associated with additional future development in "conflict" areas, and the redevelopment of such areas after storm damage should be kept to a minimum.

All debris, overburden, and other waste materials from construction should be disposed of in such a way as to prevent their entry by erosion from drainage, high water, or other means into any water body.

10.5 Comparison of Proposed Federal Legislation with Workshop Recommendations

The Coastal Zone Workshop has taken the recommendations developed through an analysis of the coastal zone situation and compared them with two examples of pending federal coastal zone management legislation. Although the recommendations and approaches were developed independently, there is a good degree of correlation (see Table 10.3).

There are additional specific points that were not dealt with in the legislation and, as a result of the workshop's evaluation, were deemed necessary. These are as follows.

1. There will be a need to differentiate categorically between federal and state management programs so that areas of national and/or interstate concern can be dealt with where they should be—at the national level.

2. A new process control board should be set up on at least a regional basis whose responsibility it would be to evaluate any new mechanical or chemical process or action that might have a potentially detrimental effect on the resource base. Approval would be necessary before commercial clearance is authorized.

3. A warning system should be developed as part of a regional information network that would be capable of evaluating levels of concern and predicting critical and subcritical, though additive, situations beforehand.

4. The management program should look at all alternatives in developing a management plan, including incentive grants, and should only use regulatory means for resource allocation when it is necessary and not as a path of least resistance.

5. Specific funds should be set aside by the lead federal agency to allow for the creation of an interdisciplinary task force whose responsibility it will be to recommend national policy and supply a high level of expertise to assist the states in the development of state management programs.

These five points, though specific in nature, are in general concurrence

Table 10.3. Comparison of Proposed Federal Legislation with Workshop Recommendations

H14146 (3-28-72)	S3507 (3-28-72)	S638 (2-8-71)	Considered by Coastal Zone Workshop
Why			
National interest	x	x	yes
Variety of valuable resources	x	x	yes
Increasing demands	x	x	yes
Competing demands	x	x	yes
Ecologically fragile	x	x	yes
Irretrievable damage or loss of values essential to well being	x	x	yes
Damaged special characteristics	x		
Present institutional arrangements inadequate	x	x	yes
Key is exercise of full authority by states in cooperation with federal and local governments of other vitally affected interests over decisions of more than local significance	x		yes
unified policies	x		yes
criteria	x		yes
standards	x		yes
methods	x		yes
processes	x		yes

Key to more productive use; introduction of a management-system permitting conscious and informed choices among alternates		x	yes
Social concerns not otherwise expressed			yes—explicitly
Coordination of research (i.e., via regional research centers)			yes—explicitly
Better economic development, i.e., aim at higher economic efficiency			yes—explicitly
What			
A National Policy to—			
preserve, protect, develop, restore for future generations	x	x	yes—restoration not explicit
encourage and assist states through grants	x	x	yes—explicitly
encourage federal cooperation and participation with state, local and regional agencies	x	x	yes—implicitly
encourage participation of all (including public) in planning and cooperation among states and regions for implementing programs	x	x	yes—explicitly
How			
Management Program Development Grants			
Secretary of Commerce	x	x	not stated
Components of program—	x		
identification of boundaries	x		yes—implicitly
definition of permissible land and water uses	x		yes—implicitly
inventory and designation of areas of particular concern	x		yes—explicitly

Table 10.3 (*Continued*)

H14146 (3-28-72)	S3507 (3-28-72)	S638 (2-8-71)	Considered by Coastal Zone Workshop
identification of means of use control	x		yes—explicitly
guidelines on priority of use	x		yes—explicitly
specific identification of lowest priority uses	x		no
description of organizational structure proposed	x		yes
statement of responsibilities and interrelationships of area under state and regional agencies in the management process	x		yes
reasonably demonstrate use to develop a management plan and program		x	
Administrative Control and Authority			
Subcontract authority to local, area-wide, or interstate agency (LAI)	x	only 1	not specified
3 yearly grants	3	3	not specified
2nd and 3rd year contingent on performance progress	x	x	not specified
Federal evaluation of performance against "how" guidelines	x	x	yes—explicitly
State adoption of prepared program with opportunity for full participation by—			
relevant federal agencies	x	x	yes—explicitly
state agencies	x	x	yes—explicitly
local governments	x	x	yes—explicitly

regional organizations	x	x	yes—explicitly
port authorities	x	x	yes
other interested parties, public and private	x	x	yes—explicitly
State coordination with existing LAI plans	x	x	yes—explicitly
State development mechanism for continued coordination	x	x	yes—explicitly
State held public hearings	x	x	yes—explicitly
State governor review and approval	x	x	no
Single state agency review and administer planning grant	x	x	yes—explicitly
Existence of state authorities to—			
manage in accordance with program including:	x	x[a]	yes—explicitly
administer use regulations	x	x	yes—explicitly
control development to insure compliance	x	x	yes—explicitly
resolve competing use conflicts	x	x	yes—explicitly
acquire fee simple less than fee simple interests in property through condemnation or other means	x	x	yes—explicitly
State consideration of national interest in siting other than local in nature	x	x	dichotomy of concerns
State procedures for preservation or restoration of specific areas	x	x	dichotomy of concerns
State provide techniques for control of use, and	x	x	yes—explicitly
establish criteria and standards for local implementation			yes—also federal for federal lands

Table 10.3 (*Continued*)

H14146 (3-28-72)	S3507 (3-28-72)	S638 (2-8-71)	Considered by Coastal Zone Workshop
subject to administer review and enforcement of compliance	x	x	yes—explicitly
direct use planning and regulation	x	x	yes—explicitly
administer review for consistency of program with approval power after public notice and opportunity for hearings	x	x	yes—explicitly
Use not unreasonably restrictive for regional benefit	x	x	yes—coastal zone center explicit
State may amend program	x	x	yes—explicitly
State may partially implement within a total plan	x	x	yes
Interagency Coordination			
Views of federal agencies affected adequately considered	x	x	yes
Federal agency activities consistent except for overriding national-international goals	x	x w/o exception	yes
Federal agency new projects consistent (except national defense)	x	x	yes
Federal award of permits or licenses require certification of state for approval for every use derived outside the management program	x	x	yes
Federal establishment of coastal zone management advisory committee to recommend and consult on coastal zone policy	x	x	yes—but user and technical advisory boards

Federal government acquire, develop, and operate estuarine sanctuaries	x	x	yes
Federal program for management of area between 3 and 12 mi.	x	x	not considered
develop for benefit of wide range of uses	x	x	not considered
application to high seas and underlying seabed coordinated with coastal state	x	x	not considered
Federal mediation of differences with states in cooperation with executive office of President	x	x	not considered
Marine Sanctuaries			
Federal designation of Marine Sanctuaries for purpose of—			
preservation of values	x	x	yes
restoration of values	x	x	yes
Federal program needs state concurrence to extend to state waters	x	x	not considered
Federal government negotiate international ramification	x	x	not considered
National Coastal Resources Board			
coordinate between Federal agency programs within coastal zone	x	x	user and technical advisory boards
mediate federal and state program differences	x	x	user and technical advisory boards
forum for appeals by aggrieved parties	x	x	environmental court

[a] Also to borrow money and issue bonds for land acquisition or restoration projects with some federal guarantee.

with the intent and scope of both pieces of legislation and as such could be developed as part of the national policy.

10.6 Functional Flow of Coastal Zone Management: A Summation

The principles of coastal zone management within the various roles of government can be explained by an overlapping of well-defined sets of responsibilities. Conceptually, management activities should extend from federal policy to local implementation levels with principal planning responsibility resting with the state (Figure 10.2). In this context, one can think of the overlapping between these activities as transfer functions; federal guidelines would be between policy and plan; between plan and implementation and enforcement would be state established standards and regulations.

A functional flow description of the overall management process should be expected to have a number of feedback and control capabilities. Long-term effects can be provided for, and process for appeal of decisions made possible at every level of decision making. The following explanation is organized by major activity and formulated in a way that associates the written word with its functional path. In this way, the role of feedback throughout the entire process should become readily apparent.

The flow starts with the formulation of policy (see Figure 10.3). Since the coastal zone is a national resource, a national coastal zone management policy should be established by the federal government responsive to the needs and values of the citizenry as reflected by—

1. The Congress;
2. The executive branch of government;

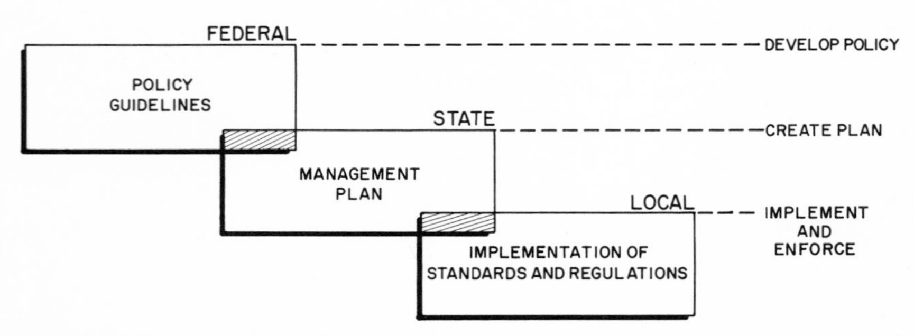

Figure 10.2 Division of responsibility in coastal zone management.

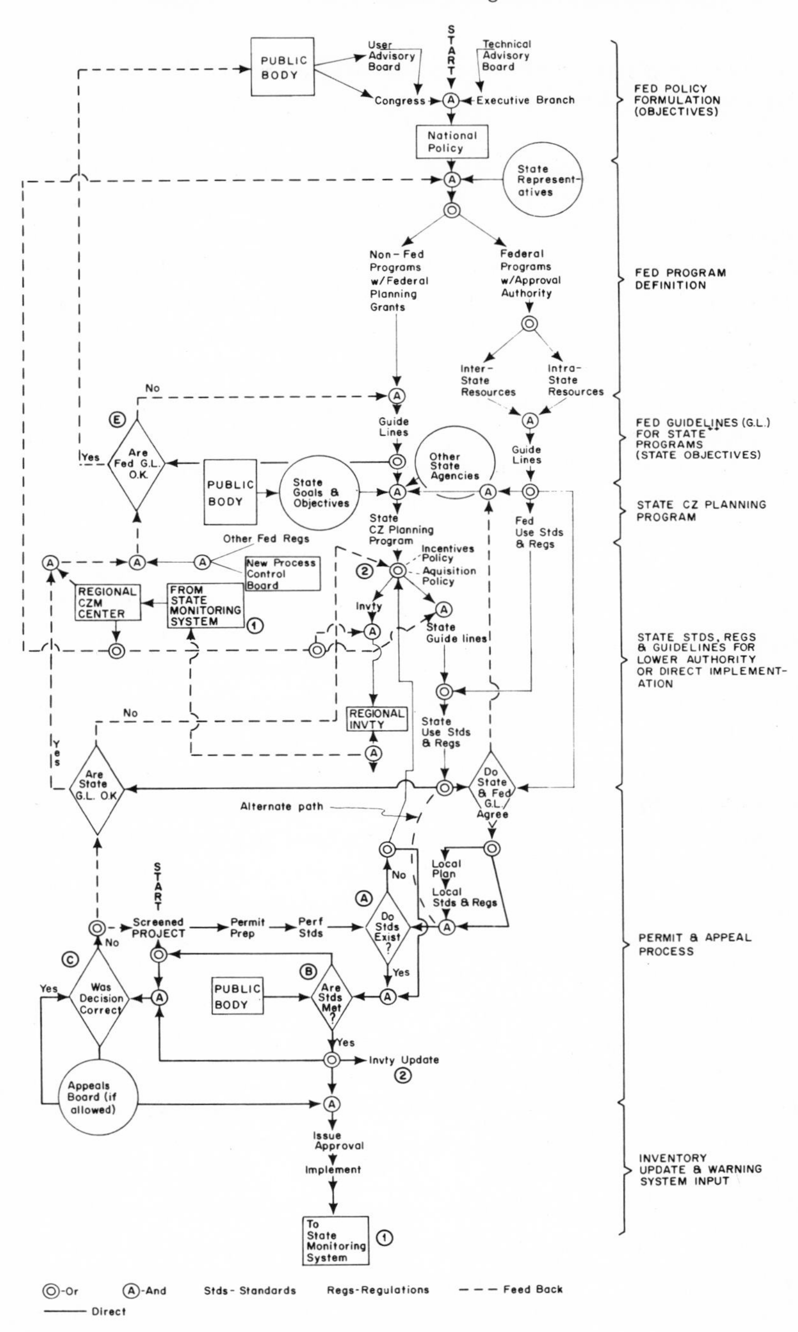

Figure 10.3 Coastal zone management flow diagram.

3. An independent advisory board of representative users;
4. A technical advisory board.

The latter two groups would exercise their function through either Congress or the executive office. As previously stated, such policy should probably be administered by a single federal agency with coordinating responsibility over other federal agencies with coastal zone responsibilities. A clear definition is mandatory to distinguish between issues of federal concern wherein the federal government will retain approval authority and those which are of principal concern at other levels of government. Coastal resource systems that are of statewide concern within the definition of the law will be eligible for consideration as part of a program supported by federal planning grants and coastal resource systems that span interstate jurisdiction and coastal zone resources that are of natural significance, even though only contained within a single state's boundaries, shall be defined by Congress and be under federal jurisdiction.

The responsibility will rest with the federal government to issue program guidelines for the benefit of those concerned with the development of coastal zone management programs. In the case of grant-supported programs, the guidelines will reflect the funding opportunity which should not be interpreted as federal control of state responsibilities, but as an opportunity for states to undertake comprehensive coastal zone planning programs with federal assistance.

The states, either individually or in regional groups, will develop comprehensive coastal zone management programs under the direction of a single coastal zone authority or coastal zone management agency with sufficient authority and financing to accomplish its task. Although functionally independent of other agencies of the states, the coastal zone management agencies will utilize information and expertise of other agencies as appropriate and be responsive to both federal guidelines and declared state goals and objectives. State use standards and regulations will be promulgated and will contain those items necessary to implement programs over which the federal government retains approval authority.

In the case of programs for which the federal government retains approval authority, the federal agency will move directly to the issuance of use standards and regulations. These will then be provided to the appropriate state agency for incorporation into state regulations. At this point in the flow, a test of adequacy should be performed to verify agreement between state and federal guidelines. The absence of verification would require recycling through the state process. If agreement exists, the next stage can be initiated.

States may use incentive programs to promote and encourage the development of local coastal zone programs along with the transfer of certain decision processes to the local authority. State standards and regulations will be used to establish guidelines for local plans and, subsequently, local standards and regulations. A permit process that can be initiated at any of a number of sequential levels with varying depth and breadth of detail should be designed to answer the following questions:

1. Is there a standard or regulation appropriate to a proposed use?
2. Is the standard or regulation met, or compliance with objectives otherwise determinable?
3. Has the decision regarding compliance been fairly rendered?

A prospective user requesting approval for a coastal zone activity must prepare a permit request complete with performance standards and an analysis of non-coastal zone options for achieving the same objective. If the prospective use does not already have standards established, the state management authority will undertake their development, and these standards will be evaluated for implementation approval. Decisions from this process will be subject to appeal for review by the state coastal zone authority with modifications to appropriate guidelines as necessary to reflect precedents and technological change at the highest level for which standards are inadequate. If during the course of appeal, it becomes apparent that federal guidelines derived from national policy do not appropriately reflect the public interest, then recourse through public pressure upon Congress is indicated as an outcome of E in Figure 10.3. The flow involved is illustrated by the series of decision points lettered A–E in the figure.

Provision will be made for updating the inventory of coastal zone usage for each of the participating states offering implementation of acceptable programs. A standard format prepared at the federal level for mapping, resource inventory, and other data collection will be used for this purpose. Provision should also be made for a reaction warning system in the form of a regional coastal zone center which, together with appropriate process information and information from other federal agencies and the states, will facilitate the modification of program guidelines in response to unanticipated deleterious effects arising from coastal zone activities and to unanticipated new demands.

References

Battelle, 1971. Battelle Columbus Laboratories. Design for an environmental evaluation system (Columbus: U. S. Bureau of Reclamation).

BCDC, 1965. San Francisco Bay Conservation and Development Commission, Established by Act 1162-1965.

Bradley, E. H., and Armstrong, J. M., 1972. A description and analysis of coastal zone and shoreland management programs in the United States (Ann Arbor: Univ. of Michigan), Sea Grant technical report no. 20, MICHU-SG-72-204.

California Dept. of Fish and Game, 1968. Fish and wildlife resource planning guide, 2nd edition (Sacramento: Calif. Dept. of Fish and Game).

California Dept. of Navigation and Ocean Development (California), 1972. Comprehensive ocean area plan (Sacramento: Calif. Dept. of Navigation and Ocean Development).

Council on Environmental Quality, 1972. Quiet revolution in land use (Washington, D.C.: U.S. Govt. Printing Office).

Delaware, 1971. Act 175-1971, Chapter 70, Title of, Delaware Code.

Ellis, R. H., 1969. The development of a procedure and knowledge requirement for marine resource planning (Hartford: Marine Resources Council, Nassau-Suffolk Planning Board), The Travellers Research Corp. TRC report no. 7722-3471.

Florida Coastal Coordinating Council (Florida CCC), 1971. Coastal zone management in Florida—1971 (Tallahassee: Florida Coastal Coordinating Council).

Hawaii, 1961. Act 187-1961, The State Land Use Law of 1961 (Chapter 98H, revised laws of Hawaii).

Leopold, L. B., Clark, F. E., Henshaw, and Balsley, J. R., 1971. A procedure for evaluating environmental impact (Washington, D.C.: U.S. Geological Survey), Geological survey circular 645, 13 pp.

Michigan, 1970. Act 245-1970 (The Shorelands Management and Protection Act).

Minnesota, 1969. Act 777-1969 (Regulation of Shoreland Development).

Orloff, N., 1972. Suggestions for improvement of the environmental impact statement program, *Environmental Impact: Analysis, Philosophy and Methods,* edited by R. B. Ditton and T. I. Goodale (Madison: Univ. of Wisconsin Sea Grant Publications), pp. 29–41.

Pepper, J. F., 1972. An approach to environmental impact evaluation of land plans and policies: the Tahoe Basin planning information system (Berkeley: Dept. of Landscape Architecture, Univ. of California).

Ralezel, J., and Warren, B., 1971. Saving San Francisco Bay: A case study in environmental legislation, *Stanford Law Review 23*(2): 349–366.

Sorensen, J. C., 1971. A framework for identification and control of resource degradation and conflict in the multiple use of the coastal zone (Berkeley: Dept. of Landscape Architecture, Univ. of California).

U.S. Department of Interior (USDI), 1970. *National Estuary Study* (Washington, D.C.: U.S. Govt. Printing Office), 7 volumes.

Washington, 1971. Act 286-1971 (The Shorelands Management Act of 1971).

Wisconsin, 1965. Act 614-1965, Section 22, Water Resources Act of 1965 (Section 59.971 of the Wisc. Statutes).

11 Legal Aspects of the Coastal Zone

11.1 Introduction

"Everybody talks about the coastal zone, but nobody does anything about it." With apologies to Mark Twain, we suggest that some things have been done and others can be done. With the premise that the environment is here for man, the focus of this chapter is on the legal decisions which affect the coastal zone.

We look to law to secure rights of access to the coast and its waters and to provide for orderly development. These and other functions of law in the coastal zone deal with a wide variety of events involving coastal resources with considerable variation in detail in the laws applicable in different states and localities. Issues of access alone are quite varied, ranging from claims of riparian property owners to the use of water areas for navigation, fishing, and other purposes to claims of the general public to use the tidal zone and beaches for recreation.

The concentration of population along the coast, intensification of uses of coastal resources in some areas, shoreline development, and waste disposal all have environmental consequences. The need for assessment of the environmental impact of man's activities in the coastal zone is greatest where there is the most pressure for extension of metropolitan-type activities of use and development into areas of the coast that have received comparatively little use or development. Local policies which permit too intensive use or development often cause environmental impacts on the coast which affect the ecology and resources to the detriment of the wider communities of the state, region, and nation.

The interrelation of law and government also reflects the fact that in most of our coastal states the local governments are dependent on properties and activities which generate tax revenues, the so-called "tax base." Industrial development may broaden this base, but it also places diverse types of waste disposal pressures on treatment facilities and on estuarine areas into which the waste is discharged directly or after some degree of treatment. Different local government, and even state government, situations can substantially influence the applicable regulations and their enforcement.

In this chapter we will attempt to (1) assess present law, (2) describe and consider major problems in the distribution of decision-making authority, and in the present law dealing with man's uses of the coastal zone and protection of the coastal environment, (3) examine the legal strategies or means by which coastal zone management may be made effective, (4) assess some current trends and proposals, and (5) elaborate several recommendations for action.

11.2 Goals and Objectives

The most general goal for the purposes of appraising present decision authority and legal regulation is to achieve a comprehensive, integral, and effective management of the coastal zone employing strategies of social influence and legal regulation in the alternative or in combination for optimum use, conservation, and preservation of the coastal zone.

11.2.1 Use and Development

In order to make coastal resources widely available in a democratic society it is necessary to—

1. Maximize access by the general public to the coastal zone, avoiding total exclusion by a line of private ownership.
2. Promote public use by establishing beach and other public recreation areas, and by encouraging boating, fishing, and other common uses of coastal resources.
3. Balance residential, industrial, and other shoreline uses to secure maximum benefits from the coastal zone with minimum adverse impacts on the coastal environment and minimum conflicts among the various users.
4. Establish priorities for marine/estuarine oriented uses, to encourage the location in the coastal zone of those which depend upon, or would produce the most benefit from, coastal location.
5. Maximize individual freedom in the use of the coastal zone and accomplish management with a minimum of regulation and administrative constraints.

11.2.2 Conservation of Coastal Resources

For those situations in which some use should be admitted or encouraged, but the environmental impact regulated or controlled, it will be necessary to—

1. Manage man's impact so as to maintain, or if possible enhance, the coastal environment.
2. Manage the specific exploitation of renewable resources to achieve maximum sustainable yield or maximum economic yield.
3. Manage the specific exploitation of nonrenewable resources for the

---

The Coastal Zone Workshop recommends
The immediate intact preservation of selected natural land and water areas in shoreline and estuary regions of the United States valued for their unique ecological character. Such areas should be severely restricted from any private or public coastal zone activity.

---

long-term interest of society as well as for achievement of an acceptable level of environmental impact of the exploitation process.

4. Manage the exploitation of resources by equitable allocation among those with a proprietary interest, and by avoiding waste.
5. Protect endangered species and impacted areas of critical environmental concern pending restoration of natural processes.

11.2.3 Preservation of the Resources of the Coastal Environment

Locations for the protection of indicated species, processes, and areas would require prohibition of use and total limitation of man's impact. Such areas could be directed toward—

1. Preservation of unique or irreplaceable species.
2. Preservation of critical or endangered species, processes, and areas.
3. Preservation of selected areas for aesthetic and future enjoyment purposes.
4. Preservation for educational and future research purposes.

11.3 The Distribution of Authority to Make Decisions

The determination of who can make authoritative decisions for what events in which areas is basic and important. These decisions about who has decision-making authority for what subjects is called the "effective constitutive process." This allocation or distribution of authority is a product not only of our constitutions, federal and state, but also of legislative, executive, and judicial decisions. Sovereignty, territorial boundaries, property rights, police powers, impact-territoriality jurisdiction, and numerous other important legal doctrines and concepts find their limits, or are subject to some special applications, in the coastal zone. A complete description of the processes by which the constitutive law is made must take into account state and local government organization and powers, which are subject to considerable variation among our coastal states; our national constitutive process; and even the international process which establishes the limits of national jurisdiction. Natural environmental processes regularly cross the coast, as do man's uses, but decision-making authority is defined by both functional and geographical limits. These grants and limits of authority are here examined separately.

11.3.1 Who Has Authority to Make Decisions?

International law recognizes nation-states as having sovereign competence over their territory, their nationals and national corporations, ships flying their flag, and also over some events which, though occurring outside, have effects within their territory. National jurisdiction, subject to innocent passage for navigation purposes over a marginal belt of water areas off the coast, is well recognized. The types of offshore zonal jurisdiction and their geographic limits will be discussed below. The nation-states, and their constituent elements according to their internal distribution of decision-making authority, today have the major role in making decisions in the coastal zone. This role is, howeever, limited somewhat by international customary law and agreements, which must be respected, such as innocent passage of foreign flag ships, refuge rights permitting ships in distress the use of ports, and agreements concerning fisheries and conservation (1).*

Only a very small percentage of coastal zone decisions concerns the international-national distribution of authority. If effective international law were established for comprehensive regulation of pollution of the ocean from land-based sources, this situation would probably be reversed, as many coastal activities have some impact on estuaries and on the ocean. Some of the proposals made at the 1972 United Nations Conference on the Human Environment call for international control of marine pollution. But it is precisely because such a regime would be expected to have very substantial effects on the decision-making authority of nation-states that the establishment of such measures is unlikely in the near future. Some present and proposed activities in the coastal zone do involve international authority, however, such as the enclosure of water areas for aquaculture, which must be reconciled with innocent passage rights. For such matters, decisions made at national and lower governmental levels must be compatible with international law.

The federal system of government is often described as a three-layer cake—federal, state, and local levels of government. At each level, the instrumentalities of government have geographically defined boundaries. The fundamental allocation is between the states, which possess sovereign powers, and the United States, which has major powers enu-

* Numbers refer to references at end of chapter.

merated and implied in our federal constitution. The reserved sovereignty of the coastal states includes the broad jurisdictional concept of police powers—powers to regulate for purposes of public health, safety, morals, or general welfare. Conservation jurisdiction and zoning powers are often delegated to local governments by the states and are founded upon these police powers. The federal government possesses exclusive constitutional powers for foreign relations and defense, and has an exclusive or preemptive admiralty and interstate commerce jurisdiction. The distribution of authority for land, water, and submerged land areas is summarized in Figure 11.1.

Though man's uses and their environmental impacts regularly cross the coast, it is precisely at the coastline that the federal-state cleavage of authority is greatest. The sovereignty basis of state jurisdiction, and the navigation jurisdiction of the federal government, meet at the coastline. The fact that most coastal states delegate zoning powers to county and municipal governments accentuates the problem of the division of authority in the coastal zone. The major share of decisions for events on the land side of the coastal zone is local, while most decisions for events on the water side are federal and state (if the state chooses to act, and to the extent it is not preempted by federal law). The problems stemming from the divisions of authority are compounded by the fact that, in many instances, states have delegated some of their police powers in "home rule" charters of local governments, and have established port authorities, flood control districts, and similar bodies. Some of these delegations of authority are difficult to modify after local institutions have been established around them.

Federal jusidiction over navigable waters, based on the commerce clause, has been broadened considerably beyond "navigation" in recent years. Federal water pollution control legislation, in which federally approved state guidelines are presently enforced by the states, is premised upon the commerce power, because water pollution can have an effect upon interstate commerce. The 1899 Rivers and Harbors Act, which includes the Refuse Act prohibition against placing "waste" in navigable waters, has been revitalized by two types of decisions.

In the case of *Zabel* v. *Tabb* (2), the 5th Circuit Court of Appeals held, in 1970, that the Army Corps of Engineers, in deciding on permit applications for construction or dredging and filling in navigable waters, must take conservation and environmental factors into account as well as navigation considerations. This result was based on the Fish and Wild-

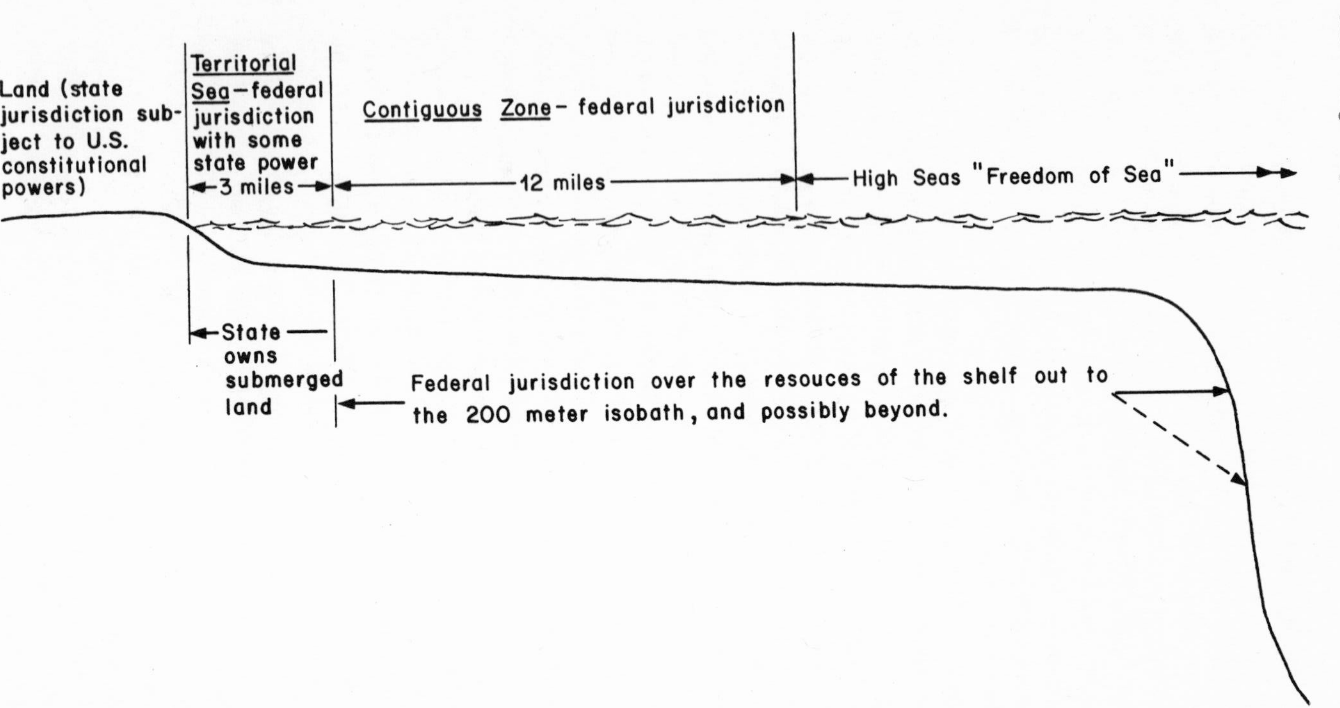

Figure 11.1 A schematic diagram of jurisdiction in the coastal zone.

life Coordination Act of 1958 (3) and the National Environmental Policy Act of 1970 (4), both of which require coordination of federal agencies which have different responsibilities. The Supreme Court declined to review the *Zabel* case, letting the Court of Appeals decision stand. The U.S. Army Corps of Engineers activated a permit application procedure in 1971 based on the Rivers and Harbors Act for the deposit of waste into navigable waters. The early decisions defining "waste" adopted a broad meaning for the term. This federal waste permit procedure is presently operating alongside state enforcement of water pollution laws.

A fundamental problem for coastal zone management is this discontinuity of jurisdiction at the coast. The predominant regulation at the local level of activities on land and the broad federal and state powers in water areas are indicative of numerous conflicts. Local governments are usually dependent upon the tax base, and therefore on use activities and development, for economic viability. Many land-based activities have effects in water areas, such as by construction or deposit of waste in navigable waters, and these activities are subject to direct federal control. Direct state control also usually may be exercised through enforcement of water quality guidelines. Within these limits, indicated by the discontinuity of jurisdiction, our environmental law protects the coastal zone. However, the distribution of authority to different government levels gives rise to significant difficulties in selecting the location of power plants and other major facilities along the coast.

Discontinuity of jurisdiction is also present at the interface of water and submerged land. The federal navigation power and the recent expansion of environmental protection of water areas are not accompanied by paramount jurisdiction over submerged lands along the coast. Within the three-mile limit (nine miles for Texas and the Gulf Coast of Florida), the 1953 Submerged Lands Act (5) has given states jurisdiction over the submerged lands and their resources. Except as sold or otherwise alienated by the coastal states (sometimes events early in the state's history may be controlling), the state has exclusive jurisdiction over these submerged lands. State approval is required before the Army Corps of Engineers will approve construction in the navigable waters of a state, both because the construction is in an area of state jurisdiction, and because federal water quality legislation requires state certification that there is reasonable assurance that all applicable water quality standards will be complied with.

In the past, the discontinuities of jurisdiction where land, water, and submerged land meet left pronounced gaps in legal regulation, as no government level had responsibility for the major areas of the coastal environment. With the advent of increased regulation of man's activities for purposes of environmental protection, these different bases of jurisdiction are leading to numerous instances of overlap and conflict in regulatory purpose. The federal-state-local distribution of decision-making authority is interrelated with political processes at each government level, and it affects different interest groups in a variety of ways. These discontinuities have presented vexing problems to proponents of coastal zone management. Proposed solutions are to coordinate the decisions at the various levels, or to federalize, or state-ize, all coastal legal problems. The coordination approach has met with only limited success, because final responsibility is usually not clarified.

Imposing control from federal or state levels also has limitations, both of a constitutional and political character. Local governments tend to resist displacement of their powers by state control. The legislative response to state coastal zone planning in California, which this year supplied funds for inventories of coastal resources, but not for recommendations on coastal problems, can be viewed as an expression of local concern over a shifting balance of power in the coastal zone (6). A viable approach to the problems of distribution of authority must involve all levels of government, with the present powers of state and local government continued to the extent possible within an overall management program. The technique of regulatory guidelines, prescribing use, development, and environmental protection criteria, and associated with funding availability, appears to be the most practical approach.

11.3.2 Decision Authority in Different Geographical Areas

International law authorizes national jurisdiction over land territory, the territorial sea, contiguous zones, and the continental shelf. As stated in the previous section, international law establishes limits of national jurisdiction but does not determine who, or which governmental agency, within the nation-state exercises that jurisdiction. The 1958 Geneva Convention on the Territorial Sea and Contiguous Zones (7) codified this national jurisdiction over territorial sea areas subject to the innocent passage rights of foreign flag ships. The breadth of the territorial sea was not fixed in the convention, but three miles is the limit on which there last was consensus. Though a majority of nation-states claim twelve miles as territorial sea today, and some as much as 200 miles, several major maritime powers, including the United States, have thus far refused to recognize a territorial sea broader than three miles.

The base lines from which the territorial sea and other geographically defined maritime zones are measured are the ordinary low-water line along the coast, and straight closing lines across the mouths of rivers, bays with a maximum closing line of 24 miles except for historic bays, and in certain other geographic situations where the coastline is deeply indented or has fringing islands. These international rules on base lines are applied in the United States, as are other provisions of international law operating as the law of the land.

The contiguous zone in international law is an additional zone of limited jurisdiction beyond the territorial sea to a maximum of twelve miles from the coastal base lines. Jurisdiction for customs, fiscal, immigration, and sanitation regulations are authorized by the Geneva Convention. More than 80 countries now claim exclusive fishing rights in such an area, and it is generally accepted that the nation-state may exercise exclusive control over fishing in the contiguous zone. For example, parties to the European Fisheries Commission recognize the right of member nations to establish an exclusive six-mile fishing zone plus an additional six-mile fishing zone restricted to the convention nations (8).

The continental-shelf jurisdiction of nation-states was established after the 1945 Truman Proclamation and extends, according to the Convention on the Continental Shelf, in the "adjacent" seabed and subsoil to the 200-meter isobath, and beyond that limit to the depth of "exploitability." As exploitability advances with improving technology, an awareness of expanding national jurisdiction has led to proposals for an international jurisdiction of the seabed.

An international conference on the law of the sea is scheduled for 1973. The United States has requested that the conference consider the width of the territorial sea, free transit through international straits, limited preferential fishing rights beyond national jurisdiction, the limits of the continental shelf, and the establishment of an international regime for deep seabed resources, with an international seabed authority to administer development of the deep seabed and with authority to make rules for a "trusteeship" zone extending from the 200-meter to the 2500-meter isobaths at the edge of the physical continental shelf.

While most proposals for coastal zone management in our nation adopt a seaward boundary at the limit of coastal state jurisdiction in order to avoid international complexities, we should consider some of the features of the seaward limits of jurisdiction. The working definition of the coastal zone for the Coastal Zone Workshop includes estuarine and ocean areas where the impact of natural coastal processes and man's coastal activities is felt. It seems clear that the functional coastal zone extends beyond the three-mile territorial sea, and even the twelve-mile contiguous zone, in many areas.

The discussion of federal, state, and local jurisdiction along the coast in the preceding section sums up the distribution of decision-making authority within our nation for the territorial sea areas. The deep seabed issues are beyond coastal zone management. What, then, of the contiguous zone and the continental shelf?

Three of the enumerated bases of jurisdiction in the contiguous zone—customs, fiscal, and immigration—are exclusively federal. The "sanitation" jurisdiction was not defined in the convention, and the first effort to exercise it was in the Water Quality Improvement Act of 1970 (9). In that act, the United States claimed jurisdiction to regulate oil spills from ships up to twelve miles from shore. This federal legislation preempts coastal states in matters of oil pollution from ships engaged in navigation. It is not established, however, whether the "sanitation" provision in the convention can be considered to cover the enforcement of water quality standards by coastal states (other than for oil spills from ships) out to twelve miles. Probably the states are now foreclosed by federal law from these matters.

Fishing jurisdiction in the contiguous zone is exercised by the federal government, although no clear preemption of potential coastal state jurisdiction on this subject has been established. The result is that there are three geographic zones for fisheries jurisdiction at present: coastal state jurisdiction within the territorial sea, federal jurisdiction from three to twelve miles from the base line, and internatinoal jurisdiction (freedom of fishing by foreign flag ships except as limited by international agreement) in the area beyond (10). Fish stocks move across these zones with little regard for the niceties of legal regulation. Another problem is created by movement of fish stocks along the coast from the territorial sea of one coastal state to another. Neither state regulation nor interstate compacts for fisheries regulation has been particularly successful (see Chapter 2). To overcome the problem of the confined geographic limits of fisheries jurisdiction, and to provide for effective management, there are proposals for federalizing fisheries jurisdiction. A paramount

---

The Coastal Zone Workshop recommends
The conduct of a comprehensive investigation by the federal government, in concert with state agencies, into the present management of coastal fisheries and an appraisal of the policies and costs of existing programs. The inquiry should include thorough study of the merits of limited entry to fisheries and lead to an effective national fisheries management policy under the aegis of the federal government. Fishery conservation on the high seas beyond national jurisdiction should be vigorously pursued by the federal government and the right of access to coastal resources by domestic fishermen must be preserved.

---

federal jurisdiction over coastal fisheries could be based upon the movement of fish stocks across the several jurisdictional boundaries and upon the commerce clause since some of the fish are sold in, and the process of fishing affects, interstate commerce.

Distribution of authority to make decisions for submerged land areas has been the subject of numerous decisions by the Supreme Court and Congress. In *U.S.* v. *California* (1947) (11) the court held that the coastal state had no jurisdiction beyond the coastline. Congress, in the 1953 Submerged Lands Act (12) provided that the state's boundaries extended three miles from the coast, except where more extensive historic boundaries could be shown. Also, in 1953, the Outer Continental Shelf Lands Act (13) established an exclusive federal leasing program for minerals, including oil and gas, for areas of the continental shelf in United States jurisdiction beyond the coastal state boundaries. A number of states have made claims to exercise some jurisdiction in this area, and litigation is now pending in the Supreme Court against most of the Atlantic coastal states and Alaska. The states' claims are argued generally on the basis of ancient grants or historic water jurisdiction, the essential issue being whether the Outer Continental Shelf Lands Act, or other federal actions, have prevented the coastal states from now exercising any jurisdiction beyond their boundaries or the territorial sea.

Another geographic aspect of jurisdiction is the problem of lateral extent of the areas of local and state governments. Determined usually by political considerations with little regard to natural processes and the patterns of use of the coast, the patchwork of municipal and county lateral boundaries makes uniform coastal management difficult. Treatment of these problems of the lateral limits of jurisdiction, like the problems of discontinuity at the land and water interfaces described previously, must involve recognition of the existing units of government and their integration into a coastal zone management program by means of a workable approach.

11.4 The Present Law Applicable in the Coastal Zone

The term "coastal zone" was coined by the President's Commission on Marine Science, Engineering and Resources (14). Although it is a useful concept, particularly in the natural and social sciences, it does not correspond to the rules of governmental authority and legal jurisdiction, as noted previously. Indeed, most of the law which relates to the coastal zone was developed long before the commission's report was filed in 1969. Legal problems relative to navigation, boundaries, commerce, fisheries, and land usage date back to the earliest times, and present principles of ownership are recognized by the commission to date back "at least to the Magna Carta." The advances of civilization and technology have, of course, given rise to new uses and their associated legal problems.

The legal rules applicable in the coastal zone are scattered throughout a broad range of traditional legal subjects, many of which are of general impact, without regard to the proximity to a coastline. For example, title to real estate, trespass, nuisance, zoning, easements, trusts, protection of public health and welfare, admiralty jurisdiction, revenue and taxation, equal protection of the law, civil rights, and many others are legal considerations that are not dependent on a locale on or near the shore.

Moreover, the sources of these doctrines are widely divergent, in both time and type. Many rules were inherited from the common law of England, but admiralty rules come from early civilizations around the Mediterranean, probably dating back to the ancient inhabitants of the Isle of Rhodes, and our present admiralty law resembles the civil law of Europe more closely than English common law. Federal and state constitutions and a bewildering variety of statutory enactments at all levels, down to local ordinances, supplemented by volumes of administrative regulations, books, and articles, and tens of thousands of court decisions, complete a montage worthy of depiction by a cadre of surrealistic painters and sculptors.

In order to come to a manageable grouping of legal considerations as they apply to the coastal zone, we have used a four-part outline based on current patterns of legal regimes. The classification of uses assembled in other chapters is cross-referenced in Table 11.1, which corresponds to the four major groupings of conflicting claims and decisions for the coastal zone: (1) shoreline use and development, (2) use of coastal waters, (3) use of the submerged land, and (4) environmental protection.

11.4.1 Shoreline Use

Normally, ownership of land is determined by state law. The statements and observations in this section are, therefore, general, and the reader is cautioned to check their applicability, or scope, for any particular state.

SCOPE OF PRIVATE AND PUBLIC USE. In common law, private ownership of land ends at the mean high-water mark, and title to the area between the mean high- and low-water marks (so-called tidelands) is held by the sovereign states. The location of the mean high-water line has been the subject of considerable litigation. In a leading federal case, the United States Supreme Court referred to: "the mean high tide line which . . . is neither the spring tide nor the neap tide, but a mean of all the high tides" (15). In other cases, a different line is used, for example, the vegetation line (16), the highest winter tide (17), and the mean higher high tide (18), and a few states use the low-water mark as the boundary (19).

Another dimension of the high-tide-line problem is that lengthy observation is necessary to fix it. Because of celestial influences, the mean tides are determined by a span of 18.6 years.

Historically, there were no special burdens placed on land or its owner merely because of location on the shore. Quite the contrary, as will be seen later, a riparian owner has special rights, even to the publicly owned area, which are unique to riverine and coastal riparian land and are superior to those of the public. In modern times, the main restrictions on private use and development are zoning and the exercise of police power, often in the form of pollution control or environmental protection regulations.

Sometimes, by usage over many years, the public may acquire rights in private property along the shore. This could happen, for example, when a well-defined pathway across private land is used by the public for many years as a means of access to the beach. The law of easements operates here. A broader principle of dedication to the public as a result of nonprohibited use over a period of years has also been applied. Legislation has also been used to declare public rights of access to beach areas (see Section 9.5.2).

Table 11.1. Cross References to Uses Defined in Other Chapters

1. Shoreline Use and Development
Habitation (Chapters 1, 5)
Land development (Chapters 1, 4, 5, 6)
Agriculture, industrial land use, power production (Chapters 1, 3, 5, 6, 8)
Government (Chapters 1, 5)
Preserves, conservation areas (Chapters 1, 2, 4)
Waste disposal (Chapters 1, 2, 5, 6, 7)
Ports, harbors, beaches, etc. (Chapters 1, 4, 6, 8)
2. Uses of Coastal Waters
Commerce, transportation, recreation (Chapters 1, 2, 4, 6, 8)
Food (Chapters 1, 2)
Water and chemical extraction (Chapters 1, 3)
Government (Chapters 1, 5)
Preserves (Chapters 1, 2, 4)
Waste disposal (Chapters 1, 2, 5, 6, 7)
3. Uses of the Submerged Land (water bottoms)
Mining and extraction, dredging, and filling (Chapters 1, 3, 6)
Food (Chapters 1, 2)
Preserves (Chapters 1, 2, 4)
Government (Chapters 1, 5)
Waste disposal (Chapters 1, 2, 5, 6, 7)
4. Environmental Protection
Water quality, waste disposal (Chapters 1, 2, 5, 6, 7)
Shore and beach erosion control; shore restoration (Chapters 1, 2, 5, 7)
Preservation (Chapters 1, 2, 4)

In common law, a "public trust" is imposed on tidelands even after they they are sold by the state to private owners (20). Here again there is no unanimity among the coastal states. In common law, the public trust was for the purposes of navigation, commerce, and fishing. Some states have expanded the concept to include such matters as conservation and bathing, while others have narrowed the limits of the trust (21).

Seaward of the boundary of private ownership and the tidelands area the Submerged Lands Act (22) recognizes state ownership of the seabed and subsoil out to the limit of three nautical miles from the coast (three leagues or nine nautical miles for Texas and the Gulf Coast of Florida). States may lease or sell such bottom areas, always subject to paramount federal law requiring a permit for construction in navigable waters and applicable federally approved water quality standards and other environmental protection criteria applied in the permit decisions.

RIPARIAN USES. Even though his property line may stop at the high-water mark, the littoral owner usually has riparian rights in and to water areas, the extent of those rights again being determined by state law. In general, and subject to permit requirements and other regulations, such rights include:

1. Access to and from the water by means of improvements such as piers and docks and dredging of channels to the land.
2. Navigation and fishing.
3. Aesthetic considerations, views and incidental uses.
4. Preference in development of submerged land.
5. Freedom from interference by neighboring owners.

These ownership rights are subject to limits in favor of neighboring owners, since one's actions may not interfere with neighbor's use, and public access and conservation.

PUBLIC USES. Some states have discrete legal rules that authorize public rights of access to the water area. However, many states do not have such rules, and the owners of the shoreline can cut off public access because it would involve a trespass across private property, unless the city or county acquires an access easement for the benefit of the public. Once on the beach or tidelands, and seaward of the property line, members of the public are free to travel laterally so long as they do not intrude on private property.

NATURAL CHANGES IN THE EXTENT OF THE LAND. The law divides natural changes into two categories: (1) those which add or detract from the dry land area, and hence riparian ownership, through the slow and continuing effects of waves, currents, wind, and rain, such as erosion and accretion, and (2) those which occur almost instantaneously, as through a cataclysm, for example, an earthquake or hurricane (avulsion). The general rule is that the landowner gains or loses (his boundary is changed) when the shoreline is moved through accretion, erosion, or reliction, but not when it results from avulsion.

11.4.2 Development; Man-Made Changes in the Coast

Bulkheading is the construction of a wall at the land-water interface. Its effect is to establish a definite line which is no longer dependent on the ebb and flow of the tide. Another effect very often is to expand the land area of the littoral owner by establishing the bulkhead line out in the submerged area and then filling it in. To forfend this latter procedure, some states have enacted laws concerning bulkheading. The only comprehensive legal regulation of this type of shoreline construction at the state level is in Florida, which provides procedures for establishment of bulkhead lines and regulates construction on state-owned submerged lands. State ownership of submerged lands and, in many states, the public trust doctrine, as well as zoning and other police powers, provide bases for state regulation of submerged lands development.

Waterways cut into land take a variety of forms, but the essential fact which may have legal consequences is that the water area is being expanded. If the extension is for a boat slip or marina, or the new water is otherwise navigable, the federal and state jurisdiction over navigable waters extends to these new waters. And, depending upon state law, the new submerged land may come into state ownership by the principle of dedication.

Dredging and filling typically involve pumping up material from the water bottom and depositing it in spoil areas in the water, along the shoreline to fill in to a bulkhead or to extend the beach without a bulkhead, on dry land above the high-water line, or to create new land—islands or peninsulas.

As the Submerged Lands Act confirms state ownership of the bottoms out to the three- or nine-mile limit, it is necessary to obtain permission from the state before starting to dredge. If the state has sold submerged land to a private person, he may still be unable to dredge material from it because of the "public trust" doctrine alluded to previously, environmental protection laws and regulations, or navigation considerations expressed in the permit authority of the United States Army Corps of Engineers.

With respect to the Army Corps of Engineers, the statute establishing the permit authority is the Rivers and Harbors Act of 1899. If, however, the operations cause turbidity or silting, they are covered by water quality and other environmental protection laws. Piers, jetties, docks, and groins all involve construction in navigable waters and are thus subject to the permit authority of the Army Corps. In most cases permits from local county or city commissions, zoning boards, and building departments are also required. Residential construction along or near the shore involves questions of setbacks, sea walls, the availability of sewage treatment and other public facilities, runoff, access to the beach, governmental services (schools, fire and police, roads, and so on), tax base, revenue needs, freshwater supply, population density, and environmental impact. Typically, these matters are handled on a local basis by zoning regulations and building permits, with some standards imposed on a county or state level, particularly with respect to pollution control.

Commercial and industrial installations are subject to the same regulations and, of course, additional rules because of possible discharges, uses of large amounts of water for cooling or washing, and undesirable aesthetic consequences. The regulatory pattern involves the same set of local, county, and state bodies, as well as the Army Corps of Engineers and the Environmental Protection Agency within their special spheres of jurisdiction.

The coastal states, as part of their reserved sovereignty under our federal system, have the power to provide for zoning under the broad police powers principle. It has been suggested that it may be wise to establish minimum (or maximum) standards at a state level, with a provision that the local body be allowed to impose stricter standards. The state can set aside certain areas for commercial and industrial development, for beaches or other public uses, and for environmental conservation and preservation, leaving the local authority no option to allow variances.

Public facilities and programs involve direct action by governmental bodies in the planning, approval, and construction of highways, bridges, canals, ports, harbors, marinas, levees, beaches, parks, and preservation areas.

While federal acts are not subject to state regulation, action by state or local bodies is subject to federal jurisdiction in navigable waters and interstate commerce, and to federal powers for defense and foreign relations, just to name a few of the sources of federal jurisdiction and powers.

11.4.3 Uses of the Water Areas

NAVIGATION. National and international public order for navigation provides that tidal waters, waters connecting to them, and other waters which may sustain interstate or foreign commerce are all "navigable waters" of the United States. Navigable waters extend from the mean high-water line to the seaward limits of jurisdiction. Beyond the territorial sea (three nautical miles at present), navigation is subject to the customary law of nations, in general, and to the 1958 Convention on the

High Seas (23) and other treaties in particular applications. Without regard to their location, every nation has jurisdiction over ships flying its flag.

States of the Union have a limited jurisdiction over foreign-flag ships while such ships are in internal or territorial waters. Most commonly this takes the form of a police power to take jurisdiction over crimes that are so serious as to "offend the peace of the shore." Boats which are registered or based in a state are subject to more substantial state regulations.

Admiralty law is mainly concerned with maritime contracts and torts —the shipping business and use of the waters for commerce. The Constitution of the United States and the Judiciary Act of 1789 place the exclusive jurisdiction over admiralty cases in the United States District Courts, saving to suitors certain remedies in the state courts.

Extraterritorial influence has been exercised outside of our territorial sea in at least two situations. Safety fairways are zones in the Gulf of Mexico in which the United States, through the Army Corps of Engineers, has designated certain areas as places where it will not grant permits for seabed construction (no oil rigs, and so on). While there is no direct legal regulation of navigation, the effect of the safety fairways is to influence admiralty decisions.

Sea lanes have been designated to help regulate traffic outside of some ports. For example, such lanes are designated in the English Channel and for areas extending more than 100 miles from New York Harbor. Again, these are intended as safety aids, not legal imperatives. They are an effective measure, however, because captains of ships can usually be depended upon to wish to avoid the risk of collision; and insurance companies and courts will assess negligence in determining who must pay damages for collision. Advances in man's capabilities and needs make necessary some accommodation even though navigation has always been the primary use of the sea, and the doctrine of "freedom of the sea" is repeated with an almost religious fervor. Oil and gas rigs, piers and jetties, offshore port facilities, even navigation buoys, all restrict navigation to some extent. But the restrictions are relatively limited and, in many cases, are directly beneficial to the navigational uses.

FISHING. Freedom of the sea in international law was traditionally equated also with freedom of fishing. With the overfishing of certain stocks and the dramatic total failure of the massive hatchery programs of the nineteenth and early twentieth centuries, some nations began a program of regulation within their own waters and, later, of regulation of the high seas fisheries by international agreements. The United States is a party to at least ten multilateral and numerous bilateral treaties regulating specific fisheries (24). The United States is also a party to the Convention on Fishing and Living Resources of the Sea adopted by the 1958 Geneva Convention. While these treaties apply primarily to areas beyond the twelve-mile limit, they can have relevance to the coastal zone because many fish migrate great distances with no regard for territorial boundaries, many hatcheries and nurseries are in the coastal zone, and zonal limits of three, and even twelve, miles from shore can be considered to be within the coastal zone.

The National Marine Fisheries Service conducts extensive research and advisory functions, but there is no comprehensive system of federal regulation of fishing. Up until now, the basic responsibility for fishery regulation within the coastal zone has remained with the states. Some observers have suggested that the federal government is the only really competent authority to regulate such matters, particularly in the case of anadromous or migratory fish and crustaceans. Fish-packing plants are subject to inspection and regulation by the United States Food and Drug Administration, as well as by state agencies.

State law, for the most part, consists of licensing and conservation efforts. Local regulation is mostly in the form of county or municipal health department rules regarding food for human consumption. Such rules, and their enforcement, are often spotty or nonexistent.

State governments have the power to regulate within their territorial limits and to control the activities of citizens of the state and enterprises operating there. Interstate compacts covering ecological regions may be one approach short of federal preemption, but to date the interstate compacts approach has not been effective for fisheries regulations.

Cables and pipelines occupy a favored status under both national and international law. In the coastal zone, they involve construction in navigable waters on the submerged land and on the continental shelf, and are subject to federal and state jurisdiction within the functional and geographic jurisdictions described above.

Freedom of research is generally considered to be included in the broad concept of freedom of the sea, although not specifically listed in the 1958

Convention on the High Seas. The Convention on the Continental Shelf (1958) specifically provides that nations shall not prohibit research in waters over their shelves which is undertaken by bona fide scientific teams and with a purpose of publication of the results. Since coastal states are given exclusive sovereign rights to explore and exploit their shelves, questions can arise about what is ocean research and what is shelf research. For example, which category governs a vessel which takes electronic soundings over a shelf to obtain readings of the bottom configuration? International problems of freedom of scientific research in the oceans are under consideration by the International Oceanographic Commission.

11.4.4 Exclusive Uses of the Seabed and Water Areas

EXPLOITATION OF LIVING RESOURCES OF THE SEABED. Under the 1958 Convention on the Continental Shelf, the coastal nation has the exclusive competence to exploit species which, at the harvestable stage, are either attached to the bottom or can move only in contact with the bottom. The Submerged Lands Act of 1953 (25) recognizes state ownership and control out to the three- or nine-mile limit. Beyond that limit, the federal government exercises jurisdiction out to the limit of the continental shelf. It is important to remember that this is different from the twelve-mile limit of national jurisdiction over free-swimming fish.

AQUACULTURE—THE WATER COLUMN. Some states, notably Florida, have enacted legislation permitting the sale or leasing of their submerged land, along with the rights to limit or prevent access to the water column and surface, such as by fencing; to add food for the species being cultivated; to poison or otherwise destroy predators of the cultivated species, including birds, which get into the restricted area; and to construct needful installations to operate and harvest the crop.

Florida has since amended its constitution to prevent sale of water bottoms, but leasing is apparently still permitted. Other states, Virginia and Maryland, for example, have constitutional provisions that appear to restrict such development.

EXPLOITATION OF NONLIVING RESOURCES. Oil, gas, and other minerals, as well as sand, gravel, shell, and other aggregates, occur on and under the seabed both within and without the three- or nine-mile limits of state jurisdiction. Both state and federal leasing programs have been undertaken along the Gulf, California, and Alaska coasts. However, most coastal states do not have mineral laws adequate to cover the exploration and exploitation of offshore minerals and related matters of environmental protection. Two types of controversies typically occur because of oil and gas rigs.

Federal admiralty law applies to maritime torts and contracts, for example, when a seaman is injured in the service of his ship or a charter party contract is violated, regardless of the location of the transgression. The civil and criminal laws of the adjacent state, as they existed in August of 1953, are applicable to nonmaritime civil and criminal acts, for example, when a roustabout is injured by tools falling from a derrick to the platform. The Outer Continental Shelf Lands Act, passed in August 1953 (26), contains this provision. Litigation is pending in the United States Supreme Court at this time to determine what rights the states may have, if any, beyond their boundaries and the territorial sea in the continental shelf (27).

CONSTRUCTION ON THE SEABED. Under the 1958 Convention on the Continental Shelf (28), coastal nations have exclusive sovereign rights to explore and exploit their continental shelves. As a domestic matter, the United States has retained federal jurisdiction over seabed construction, regardless of three- or nine-mile limits, through the extension of the permit authority of the Army Corps of Engineers under the Outer Continental Shelf Lands Act. Of course, if the construction is within state limits, permission must also be obtained from the state.

11.4.5 Environmental Protection and Enhancement

INTERNATIONAL AGREEMENTS APPLICABLE IN THE COASTAL ZONE. In the 1958 Convention on the High Seas (29) Articles 24 and 25 state:

Article 24

Every state* shall draw up regulations to prevent pollution of the seas by the discharge of oil from ships or pipelines or resulting from the exploitation or exploration of the seabed and its subsoil, taking account of existing treaty provisions on the subject.

Article 25

1. Every State shall take measures to prevent pollution of the seas from the dumping of radio-active waste, taking into account any standards and regulations which may be formulated by the competent international organizations.

2. All States shall cooperate with the competent international organiza-

* The word "state" in international agreements means "nation-state," as distinguished from states of the Union in the United States.

tions in taking measures for the prevention of pollution of the seas or air space above, resulting from any activities with radio-active materials or other harmful agents.

In the 1958 Convention on the Territorial Sea and the Contiguous Zone (30) article 24 states:

Article 24

1. In a zone of the high seas contiguous to its territorial sea, the coastal State may exercise the control necessary to:

a. Prevent infringement of its customs, fiscal, immigration or sanitary regulation within its territory or territorial sea;

b. Punish infringement of the above regulations committed within its territory or territorial sea.

The International Convention for Prevention of Pollution of the Sea by Oil (as amended, 1969) (31) applies to tankers of 150 gross tons or more and other merchant ships of 500 gross tons or more, with certain very limited exceptions. It forbids the discharge of oil or oily mixtures from a tanker within 50 miles of the nearest land. Discharges from ships other than tankers are also regulated, although they are not subject to the absolute 50-mile limit but to the less specific rule of "as far as practicable from land." Enforcement is to be handled by the state of the ship's flag.

UNITED STATES LAW. The Oil Pollution Act of 1961, as amended in 1965 (32), applies to ships of American registry and forbids discharges in United States waters. The 1899 Rivers and Harbors Act (33) forbids any kind of discharge or deposit of any kind of refuse material into the navigable waters of the United States. The Water Quality Improvement Act of 1970 (34) goes beyond merely forbidding discharges of oil or oily wastes and assessing the criminal penalties therefor. To be sure, there are prohibitions and penalties, but the really effective thrust is contained in the provisions requiring the polluter to pay cleanup costs—up to 14 million dollars in the case of a vessel, and up to 8 million dollars in the case of an oil rig. The Federal Water Pollution Control Act (first enacted in 1948 and amended many times up to and including 1970) (35) sets up the Federal Water Pollution Control Administration, which has undergone several metamorphoses and is now known as the Water Quality Office of the Environmental Protection Agency. In general, the system established is to impose upon states the obligation to adopt water quality standards that meet or exceed federal criteria. When approved, the state standards are adopted by the Water Quality Office (formerly F.W.P.C.A.) and become federal standards as well as state standards (36).

The Fish and Wildlife Coordination Act (as amended in 1964) (37) provides a scheme whereby all federal agencies, before they do, or issue permits for doing, anything that will have an impact on any body of water, must consult with the Fish and Wildlife Service of the Department of the Interior, and with the head of the state agency exercising administration of the wildlife in the area involved. As a result of this act, plus a "Memorandum of Understanding Between the Secretary of the Interior and the Secretary of the Army, July 13, 1967 (38)," a system has been established whereby the Army Corps of Engineers can and does refuse permits for work in navigable waters based on environmental considerations, even though there may be absolutely no effect on navigation.

The National Environmental Policy Act of 1970 (39) declares a national policy to encourage harmony between man and his environment; requires federal agencies to consider environmental factors and to file environmental impact statements with every recommendation or report for legislation and other major federal actions significantly affecting the quality of the human environment; directs that all federal laws and regulations be interpreted and administered in accordance with the policy set forth in the act; and creates in the Executive Office of the President a Council on Environmental Quality which is to assist in preparing the President's Annual Environmental Quality Report, gather environmental data, conduct research, review and appraise federal programs and activities, develop and recommend policies, and otherwise assist and report to the President on environmental matters. The Council has published "Guidelines for Federal Agencies" (40), annual reports for 1970 and 1971 (41, 42), and occasional separate reports on special problems (43–46).

STATE LAWS. As mentioned before, states adopt water quality standards which, upon approval, then become federal standards. Enforcement can then, of course, be undertaken on either or both levels. These standards vary widely from state to state and from place to place within states. Other phases of environmental protection that have attracted state attention range from detailed legislation with special regulatory mechanisms to little or no state regulation with only local zoning left to handle the problems (47).

The Coastal Zone Workshop recommends
The federal government lead in establishing regional *Coastal Zone Centers*. However, academic institutions, private foundations and enterprises, state governments, and granting agencies of the federal government should greatly increase their support of both fundamental and applied research in the natural sciences, law, and the social sciences in order to feed original information into the regional *Coastal Zone Center* and/or to public agencies for improving the management of the coastal zone.

11.5 Strategies

Due to the increased use of the resources of the coastal zone and the mounting impact resulting from increased use, there are pressures from both the government and private sectors for changes in the law of the coastal zone. The strategies employed for effectuating the desired changes are many and cover a spectrum of types of action, including broad-based social education programs, specific economic inducements, directed policy formulation, and legal procedures designed to bring a direct and immediate change in coastal resource allocation and environmental protection (see also Section 8.2).

Government employs many strategies to bring about change. The Sea Grant Program and the more recent Environmental Education Act are federal responses to the need for more understanding of problems of the coastal zone and development of practical solutions to these problems. Curricula in educational institutions reflect the desire for information on the problems of the coastal zone. Most governmental agencies dealing in coastal matters distribute educational information for general and technical readers.

The most critical function government agencies perform is the creation of a forum for the formulation of new policy. This strategy is fundamental to democratic processes, and is important to coastal zone matters because of the divergence of viewpoints about the best use of coastal resources and the necessity for wide support for new policies.

A typical governmental forum is the creation of a commission to address coastal zone problems. The Commission on Marine Science, Engineering and Resources has set the stage for coastal zone policies (48). Many state and local governments have created commissions designed to identify the interest and role of various governmental units in the development or conservation of their own coastal resources.

Legislative hearings are frequently the forum for discussing and proposing new coastal zone policies. In the federal government, hearings before congressional committees are frequently preceded by studies that suggest changes in law and public policy, and the hearings are called specifically to receive comments on the study. The Marine Science Commission's recommendations for federal legislation in the field of coastal zone management and the Department of the Interior's studies on land use policies were both aired before congressional committees to focus national attention on the issues and generate comment and criticism. A well-done study, properly disseminated, can exercise considerable influence on policy even without hearings. Such was the case with the National Shoreline Study and the Shore Protection Guidelines and Shore Management Guidelines (49) sponsored by the Army Corps of Engineers.

Conferences and workshops are techniques which can generate ideas and proposals for governmental and nongovernmental action. Once a governmental unit determines that some action should be undertaken, it may proceed in a number of ways to implement the policy chosen. With respect to the acquisition of property in the coastal zone, governmental bodies can act either by bargaining in the marketplace for lands or by the exercise of the power of eminent domain. An alternative to acquisition is acquiring a portion of the interest in the land, such as an open space easement or a surface rights easement for the management of wildlife. A more common governmental action is to induce the private sector to act in certain ways, or to regulate the private sector to ensure compliance with certain standards. A common inducement is federal spending, which acts to encourage states to create certain programs by offering federal funds.

A common governmental strategy for controlling the use of the coastal zone involves the judicial system for the determination of rights in coastal resources or for the enforcement of existing laws. All levels of government use this strategy. United States attorneys may proceed in court against those dumping waste into navigable waters of the coastal zone. Local district attorneys can bring actions into state courts for a determination of the proper use of the coastal zone by citizens. A common use of the courts is to fix functional and geographical limits of jurisdiction.

The preceding paragraphs have discussed governmental strategies frequently employed to bring about change in the use of coastal resources.

The private sector also exercises considerable influence over their use. Private strategies are as varied as the imagination of the individuals exercising them. By use of the media and educational forums, groups as disparate as oil refineries and conservation organizations spend many dollars to educate citizens and persuade them to accept their characterizations of the problems and potentials of the coastal zone. Private foundations spend money on the acquisition of property and the support of research and planning in the coastal zone. Private industry invests huge sums in promoting coastal development for homes and recreation, as well as creating the technology for profitable aquaculture operations in coastal areas, among other things. Research strategies are also quite important in elaborating and influencing coastal zone policy.

Political and judicial action initiated by the private sector has had significant influence in preparing the stage for coastal zone management programs. Suits brought by private environmental groups under the National Environmental Policy Act (50), testing the thoroughness of environmental impact statements have substantially influenced the pace and method by which coastal zone development and construction projects proceed, the types of projects being proposed, and the participation of citizens in the decision-making process.

Politically, these suits have often polarized communities. In one instance, industry groups countersued the environmental groups for damages resulting from the cancellation of an oil and gas lease sale. One constructive effect of these private actions has been the creation of environmental consciousness, which has permeated through all levels of decision making, especially decision making in the coastal zone. However, the problems of delay in projects that are finally approved and of differing and sometimes inconsistent environmental criteria applied by government agencies call for substantial improvements in elaboration of coastal and environmental policies and in the procedures of decision making.

11.6 Trends and Recent Proposals

The combined effect of the distribution of decision-making authority, the present law, and strategies of legal action discussed previously has culminated in trends and proposals pertinent to coastal zone affairs. Illustration of trends pertinent to coastal zone management is offered in the following list of current events. The enumeration does not purport to be complete and purposely deals with actions at other than the state level. More thorough attention is given to the land management proposals recently advanced at federal and state levels.

11.6.1 International Trends and Proposals

The 1972 United Nations Conference on the Human Environment dealt with all aspects of environmental problems, specifically addressing a proposed international wetlands convention and ocean dumping. There was agreement between the United States and Canada on regulation of pollution of the Great Lakes (51). There was agreement between the United States and the Soviet Union on the establishment of water quality programs. There were amendments to the IMCO convention on liability for oil spills (52).

Preparatory committee meetings for the 1973 Law of the Sea Conference have dealt with virtually all aspects of ocean law. There were further responses to unilateral nation-state claims to exclusive jurisdiction in high seas areas, with a United States agreement with Brazil allowing U.S. fishermen use of high seas areas claimed by Brazil as territorial sea. There was discussion of potential expansion of the use of the continental shelf for purposes other than resource extraction, such as for deepwater ports.

11.6.2 Federal Trends and Proposals

The most recent coastal zone management proposal (S3057 and HR 14146) provides grants to states to develop programs. These bills were passed in the Senate and House of Representatives, and a joint conference is pending. Federal water and quality control, presently exercised through the Refuse Act and the Federal Water Pollution Control Act, is currently under examination by Congress with consideration of proposals for further centralization of control in the federal government. National energy policy and the siting of power plants, legislation to establish positive ship control in congested areas, similar to control of aircraft around airports, and legislation to regulate waste disposal by all ships, United States and foreign, in territorial waters are all under consideration by Congress.

The development of deepwater ports is under study by the Federal Maritime Administration, the Army Corps of Engineers (53), and the Council on Environmental Quality. The Coast Guard has promulgated regulations providing for enforcement of oil-spill regulations in the contiguous zone.

In the area of research and training, the Sea Grant program continues to provide a major impetus for education, research, and advisory services in a wide range of disciplines pertinent to coastal zone management. The National Science Foundation and other agencies are providing increasing support for such activities.

11.6.3 Regional (Interstate) Trends and Proposals

River basin commissions, such as the New England and the Great Lakes River Basin commissions, provide technical services to the member states and have served as organizing agents for comprehensive surveys. They have become involved in particular river basin problems, such as power plant siting. Regional Development Commissions under the Economic Development Administration in the Department of Commerce are interstate agencies which have provided funding for marine science development, particularly in coastal plain areas. There has been continued operations of interstate compacts on fisheries and lateral ocean boundaries.

Regional planning and management commissions usually centered about a common resource and involving two or more counties have been established and expanded in the San Francisco Bay, Long Island Sound, and Lake Tahoe areas. Regional planning commissions are required within a state for it to receive planning grants from the Economic Development Administration. Water management districts operate in a number of multicounty regions to provide for water supply or floodplain management.

11.6.4 Local Trends and Proposals

Increased use is being made both of "new-town" concepts for fairly comprehensive regulation of major development projects, enforced through subdivision zoning or private contract rule making, and of county and municipal powers over shoreline and harbor development. Efforts of local government to opt out of statewide environmental controls are also increasing.

11.7 Critique of the Land Management Approach as Applied to Coastal Zone Problems

The problem of the interrelation of government levels in decision making on land use is dealt with by several proposals at federal and state levels. The recently enacted Florida Environmental Land and Water Management Act of 1972 is typical of this land management approach. Briefly stated, the development, conservation, and presentation purposes of this type of legislation are to permit the development of key facilities, such as jetports and highway interchanges, which are of state or regional interest; to provide for environmental protection and orderly land development under standards stricter than those that would otherwise be applied by local governments; and to set aside for preservation by purchase certain unique areas. The technique by which these purposes are to be accomplished is the selection and designation of some areas or decisions as being of "regional impact" or "critical concern." Decisions on these selected activities or areas would be made by higher governmental authority. In the Florida legislation, the areas which may be designated as areas of "critical state concern" are limited to 5 percent of the land of the state and comprising only areas (1) containing, or having a significant impact upon, environmental, historical, natural, or archeological resources of regional or statewide importance; and (2) significantly affected by, or having a significant effect upon, an existing or proposed major public facility or other area of major public investment; or (3) proposed major development potential, which may include a proposed site of a new community, designated in a state land development plan.

The new Florida Division of State Planning has been given the responsibility of administering the act and of making recommendations to the government and the cabinet. The cabinet will serve as the Land and Water Adjudicatory Commission for the designation of areas of critical state concern. After an area has been so designated, the local government entity having jurisdiction over it will be given an opportunity to write land development regulations for the area or to supplement existing regulations. After the existing or new regulations have been submitted to and approved by the Division of State Planning, the local government entity will administer them, unless it does so inadequately, in which case the state will enforce the regulations. In the event the local government does not respond with acceptable regulations, the Division of State Planning may promulgate the regulations it considers suitable.

A companion measure, the Florida Land Conservation Act of 1972, called a referendum on the issuance of up to $200 million in bonds for state capital projects for the conservation and protection of environmentally unique and irreplaceable lands, including coastal lands and marshes, wilderness areas, areas where development would require a remedial public works project to limit or correct environmental damage, and beach areas eroded or threatened by natural forces.

The land management approach thus operates by displacing local authority on matters designated to be of "critical concern" for a few selected actual or potential developments, and for a comparatively limited geographic area. If geographical limits, such as the 5 percent area, and functional limits of jurisdiction, such as "major public facilities," "major development potential," and "environmental impact," were not placed in the legislation, it would encompass virtually all land decisions now made by local governments. With local governments in most areas depending upon the tax base for economic viability, it is evident that there are major sources of political opposition to the land management approach.

In the coastal zone, federal and state jurisdiction is direct and substantial once we cross over into navigable waters. The interest of higher levels of government is generally limited to activities which are major public facilities or developments or which produce particular environmental impact. The construction of a pier or dock, or the filling of even a single acre of submerged land, can have substantial impact, either on navigation or on the local estuarine environment. Jurisdiction based upon the commerce clause, water quality legislation, and state police power in areas of common public use presently give state and federal governments power over the water side of these activities. Correlation of this jurisdiction with zoning and other regulations on the land side should be a two-way process exercised more or less generally throughout the coastal zone, and not confined to a limited area or to a few selected facilities or projects. Such correlation can be accomplished by planning and guidelines at federal and state levels. This is an expansion of the technique now employed in water quality regulation and on certain other environmental protection matters. The regulatory guidelines would, of course, be more specific at state levels than at the federal. They would ordinarily be less specific than zoning ordinances, but would provide regulation of such matters as setback lines, dune preservation, wetland preservation, the siting criteria for the major types of land-based coastal activities, development standards, and so on.

An example of the guidelines approach to coastal zone management is the establishment of minimum building floor elevations and construction standards for shore areas. The federal government might set as a guideline, and as a condition for receiving grants for coastal zone management, a requirement that the recipient state must establish a detailed regulation providing that buildings in its coastal areas must, for example, be adequate in elevation and construction to withstand a

---

The Coastal Zone Workshop recommends
The application of environmental quality standards and performance criteria based upon monitoring or surveys to be evaluated by all government agencies involved in the management of the coastal zone. They should take into account socioeconomic needs of the community, and resort to general regulations, zoning, and other codes only when necessary for compliance.

---

storm intensity expected once in a hundred years. A given coastal state would review the inundation and storm surge records for such a period and prescribe, say, an 8-foot minimum elevation for buildings in one area of its coast and a 10-foot standard in others, as appropriate. State and local governments could, of course, provide stricter standards than required if they choose.

A guidelines approach to coastal zone management could be accomplished along with a land management approach. However, limitations of the land management approach to the coastal zone must be noted.

1. Land management focuses on specific types of areas or projects of "critical concern," while coastal zone management deals generally with common use, public and riparian access, and numerous other problems of general concern.

2. Land management concentrates on an area or a project, while coastal zone management deals mainly with uses and functions, geographic zonation being only one factor, and often not the major one, in decisions on uses, developments, and environmental impact.

3. Land management approaches involve lifting selected important decisions to higher governmental levels, while coastal zone management approaches deal with a variety of natural processes and human use patterns which require a set of continuing management and regulatory policies and decisions rather than a single decision on preservation or development.

11.8 Development of Coastal Zone Policy

The necessity and urgency of the development of a clear national policy for the coastal zone are documented in the discussions of the Coastal Zone Workshop and in the preceding chapters. Man's uses frequently seriously affect natural environmental processes, and in doing so, they are foreclosing important future options of society. Local, state, regional, national, and international activities, commerce, and other interests converge in the coastal zone. Local ordinances and international law meet there. Without an affirmative national policy, many critical problems are either unresolved or are dealt with by relatively uncoordinated actions of many agencies at different government levels.

In order to develop a national policy on the coastal zone, it will be necessary to utilize the best scientific information available. This should be applied with due regard to the practical realities of management and legal problems, and take into account the economic, social, and psychological factors which are of great importance to people. This will be difficult to accomplish, and the Coastal Zone Workshop has recommended that a task force be established for this purpose. The charge of the task force would be twofold: (1) to assist in developing national policy and in "designing the national program" for coastal zone management, and (2) to develop regulations and "model guidelines for state coastal zone management authorities." The presently pending federal legislative proposals on coastal zone management and on land use management do not provide adequately for the assignment of these policy responsibilities or for development of model guidelines.

The task force would be an autonomous multidisciplinary body in the Academies of Sciences and Engineering and could thus be independent from all government agencies. It would have a probable life of two years. Upon its expiration, implementation and further development of the policy and guidelines should be assigned to the appropriate designated federal agency. The roles assigned to the national academies are intended to involve the support of the scientific and engineering communities. However, the academies would provide only housekeeping services once the task force is constituted. The task force would be basically funded by the federal government, but could attract additional funding from other public and private sources. Creation of the task force by the President is intended to establish its significance for federal agencies as well as to indicate its independence from any particular agency. Specific problems which should be considered by the task force have also been identified by the Coastal Zone Workshop. A leadership role of the federal government is essential in order to deal effectively with the present problems of the distribution of coastal authority between government levels and agencies. As noted in Section 11.3, new approaches to the accommodation of federal and coastal state powers, as well as new balances of regional, state, and local authority, must be elaborated. The workshop recommendation provides that the task force should work with state, regional, and federal agencies and prepare specific plans for coastal regions. These regions for planning purposes presumably would correspond to the major types of coastal environments in the nation.

11.8.1 Adoption of Federal and State Guidelines

Another major recommendation of the Coastal Zone Workshop concludes that, "The integral element of the National Coastal Zone Policy should be the focus of management responsibility at the state level, with the active participation of local governments under federal policies that provide grants and set guidelines for creative and effective programs." A key problem for a coastal zone management program is ensuring its effectiveness through enforcement of its policies. Any system of guidelines is imposed from higher levels of government, but the salient feature is a degree of flexibility in regulations which leaves further specification to lower levels of government. Whole sets of legal problems revolve around the questions of who has authority to set guidelines, at what level they should be set, what binding force they have, if any, who is intended to be bound thereby, what kinds of inducements will enhance voluntary compliance, and what sanctions are applied in the event of noncompliance.

The idea of guidelines, rather than direct action or comprehensive regulation, flows from the conclusion that it is not wise simply to elevate every decision to a higher level of government. This, in turn, is based on the observation that it is necessary, in order to involve available knowledge and competent manpower, to have the active participation of all levels of government. Furthermore, there are distinct advantages of governmental action at the lowest level compatible with the attainment of the objective. Just the fact of being close to the scene in question may provide local bodies with awareness and insight into problems. These local bodies already exist, and it is not economical to superimpose a giant nationwide regulatory bureaucracy to deal with the environmental aspect of myriads of decisions about the coastal zone including licenses and permits of various kinds.

With these things in mind, it is proposed that the legal technique of guidelines serve as a major management tool for the coastal zone. Guidelines are national in effect, but provide for regional differences where appropriate; and they should be established at the federal level. States would set more detailed, and possibly more strict, guidelines in compliance with the federal standards. The federal program should be structured to require states to adopt guidelines which require further specification as appropriate to local conditions by their political subdivisons and other administrative units. These bodies, in turn, would be required to publish and enforce detailed guidelines for users and the public. Decisions on building permits, licensing requirements, ordinances, and other regulations presently made at local levels would be reviewed if not in conformity with the guidelines.

Under this approach, federal leadership is used to stimulate the whole program, but actual regulation and enforcement become more detailed and appropriate to regional and local conditions as one proceeds down the ladder of governmental hierarchies.

11.8.2 Research Needs

Under the common law system, it is necessary to research prior decisions on an issue to form a hypothesis, test the hypothesis against the collected patterns of decision, and finally apply the conclusions of law to the facts of the particular case. Because we operate within a federal republic based upon written constitutions at both federal and state levels, research into constitutional rules and modes and patterns of Congressional and state legislative approaches (and their success or lack of it) is necessary. Integration of modern scientific knowledge into the law is essential for proper coastal zone management.

Using the law and legal institutions as a creative, policy-applying force, instead of merely as a set of constraints, involves the highest level of legal knowledge and ability. These efforts should be supported by substantial grants from federal agencies (such as the National Science Foundation and the Sea Grant Program Office of NOAA), for their results can be extremely important to the resolution of coastal zone problems.

11.8.3 State Environmental Review Boards

The theory is to establish a decision-making body at the state level where the give-and-take of administrative procedures can be reviewed promptly, with a minimum of cost, and without the necessity of confrontation in judicial litigation. States have a variety of administrative structures, but they typically (with notable exceptions) involve delegation of the basic police power of the state to local (county, municipal, or special-purpose) units of government. Thus, many decisions which may have long-term effects on the coastal zone are made by such units as a flood control district, a municipal zoning board, or a town building commission.

The Coastal Zone Workshop recommends

Research in the legal, political, economic, and social aspects of the coastal zone should be directed toward the following types of problems:

a. Exercise of property rights in wetlands and shore areas;

b. Administrative and judicial enforcement of codes;

c. Statutory guidelines and their interpretations with respect to shoreline development;

d. The decision-making process for the coastal zone at local, state, and national levels;

e. Group interests and political pressures in coastal zone uses;

f. Value systems that affect management practices in coastal zone activities;

g. Cost-benefit analysis of ultimate uses of the coastal zone, including ecological effects;

h. Economic models for policy guidance in calculating inputs and outputs;

i. Economic factors and mixes in resource evaluation.

Some states have already provided for review of certain types of decisions either on the basis of a specific agency (for example, a zoning board of appeals) or by means of a generalized procedure (for example, a state administrative procedures act). In addition, such local decisions can, in some cases, be challenged in the courts. This brief discussion can indicate only the general picture—a picture in which the pieces are fragmented, differing environmental protection criteria are applied, and not all local decisions having environmental impact are entitled to the same type of review at a single (state) level. The proposal for environmental review boards could be implemented in many states by merely strengthening existing structures and procedures, in others by creation of a new state board or agency, depending on the present arrangement in the particular state.

11.8.4 Environmental Court

In broad perspective, the purpose of the proposed federal environmental court is to provide an alternative path for decision making in controversies involving actual or threatened environmental impact. As will be seen later, no one should be compelled to resort to this decision maker. Rather, the object is to provide an arena for those who wish to take advantage of its streamlined form and procedures, or who have

---

The Coastal Zone Workshop recommends
The development of *legal institutions and procedures* to make coastal zone management more effective. Substantial improvements in the existing types of decision procedures and laws are required and consideration should be given to—
Development of innovative approaches through new coastal land and water use accommodations;
Alternative means for the regulation of coastal development besides the taking of private property;
Improvement of statutes and administrative regulations for land, water, and submerged land activities;
Increased access of individuals, groups, and governmental units to administrative and judicial proceedings;
Establishment by state legislatures of Environmental Review Boards for appeals of local administrative decisions concerning activities that have coastal and environmental impact;
Establishment by Congress of an expert federal Environmental Court with broad jurisdiction over private persons, state, and local government agencies, and federal agencies in controversies involving coastal and environmental impact.

---

limited or no standing to pursue litigation in the traditional court system. The philosophy of the court should be oriented toward an inquiry into truth, not a reward for clever advocacy under an adversary system.

A cogent reason for establishing the environmental court would be to relieve the mounting pressure on the traditional court systems, both state and federal. Faced with already immense burdens and backlogs, such courts could be inundated by a tidal wave of environmental litigation. The answer to the simplistic argument that courts have been responding rapidly with temporary restraining orders and hearings on preliminary and permanent injunctions is that while the court is dealing with these things, its backlog of other cases is growing and growing. One solution would be to add additional judges to existing court systems instead of establishing a new environmental court. However, merely adding judges would not develop a uniform body of jurisprudence in these matters.

It is proposed that Congress clothe the environmental court with the broadest jurisdiction over persons and corporations, municipal and county agencies, officials, councils, boards, and so on, of every type, as well as state and federal governmental bodies and their officials.

The environmental court should have substantive jurisdiction and should be invested with power to handle all cases and controversies where someone, including a governmental body, feels aggrieved in a matter involving coastal or environmental impact. A person may feel aggrieved because he observes pollution or other action by a person or governmental body which is damaging to the environment; or he has been prevented from taking action on the basis of environmental considerations, for example, denied a zoning or building permit, or denied an Army Corps of Engineers permit; or he observes that a governmental body is permitting an improper use in violation of law or regulations, or in violation of an impact statement, or that the statement is false or faulty. A governmental body may observe that some other unit of the same or a different government is performing or permitting actions having an adverse effect on the environment; for example, a city might appeal a ruling of the Atomic Energy Commission.

Unlike the Federal District Courts, the environmental court should not rely on diversity of citizenship or a minimum money amount in controversy, as these are not appropriate. Better, the legislation establishing the court should include a provision to the effect that the right to a clean and viable environment is a federal right and that the coastal zone and other specified areas are national resources, and, further, that these should be protected and enhanced at the federal level. Congress should provide that the central fact essential to the court's jurisdiction in any case would be the substantial nature of the environmental consequences involved. A mechanical test would be impossible of definition, much less of enforcement.

Standing to sue has been a difficulty in a number of environmental cases (54). On the theory that the environment affects everyone, the environmental court should be open to everyone. The formalized niceties of some state and federal jurisdictional rules should have no place in this scheme. If a criminal offense is uncovered by the environmental court, the case, or the criminal aspect of it, should be referred to the appropriate United States attorney or state prosecutor for action.

The environmental court should be given the power to establish its own rules of procedure, because its philosophy, purpose, and operations differ in major respects from other courts. Congress should charge the court to operate under equitable principles, and not limit it by common law equity procedures or constraints. Three-judge panels should be assigned by a regional chief judge (or court admininstrator) from a list of available judges, with assignments made on the basis of the expert qualifications needed in the particular case. There should be no juries in the environmental court; the panel of expert judges should decide questions both of fact and of law. Bonds or other undertakings could be required to protect against frivolous cases.

Appeals would go directly to the United States Court of Appeals for the appropriate circuit. The environmental court should have access to all data of all agencies of government (state, federal, local), except that which is legitimately classified for national security or foreign relations reasons. Decisions of the environmental court must include a written statement of the scientific, engineering, or other factual data and conclusions on which they are based.

The environmental court should, at its discretion, be empowered to grant temporary restraining orders, preliminary and permanent injunctions, court costs, attorney's fees, all expenses for scientific and technical help which enabled the litigant to amass data in order to prove his case, actual money damages, and punitive damages, and permits, licenses, and orders, which should have been issued or to revoke or modify any which have been issued within a stated period of review.

Judges should be appointed by the President for life or for a long enough term to ensure that they have no motivation other than doing the appointed job. The legislation should specify that judges be selected on the basis of expertise in environmental matters as well as good legal credentials.

The environmental court would not have any appellate jurisdiction, so it could not review decisions of other courts, state or federal. Persons who are entitled to a jury trial under our Constitution could have the case removed to the appropriate United States District Court, but all remedies listed previously would still be available.

The right to use the traditional system would not be impaired; any plaintiff with standing could go to the appropriate state court or the United States District Court. Substantial research should be done on constitutional and other legal matters which are inevitably involved in a major restructuring of environmental litigation within our legal institutional framework. The origins and history of such special federal courts as the Tax Court, the Court of Claims, and the Court of Customs and Patent Appeals might well provide an insight into the considerations which would affect establishment of a specialized environmental court.

References

1. M. Whiteman, Vol. 4, *Digest of International Law,* U.S. State Department, 1965.
2. *Zabel* v. *Tabb,* 430 F. 2d. 199 (5 Cir. 1970).
3. 16 U.S.C.A. No. 661–666.
4. P. L. 91-190, 83 Stat. 853, 1 January 1970.
5. 67 Stat. 29; 43 U.S.C.A. No. 1301–1315.
6. California beach battle should make us thankful, *Capital Journal,* Salem, Oregon, May 30, 1972, Sec. 1, p. 4, Col. 2; and Is OCCDC dying?, *Newport News-Times,* May 25, 1972, p. 2, Col. 1.
7. 15 UST 1606, TIAS 5639, 516 UNTS 205.
8. P. M. Fye, A. E. Maxwell, K. O. Emery, and B. H. Ketchum, 1968. Ocean science and marine resources, *Uses of the Seas,* edited by E. A. Gullion (New York: The American Assembly), pp. 17–68.
9. P. L. 91-224.
10. See reference 8.
11. 332 U.S. 19.
12. See reference 5.
13. 43 U.S.C.A. No. 1331 *et seg.*
14. Commission on Marine Science, Engineering and Resources, 1969. *Our Nation and the Sea* (Washington, D.C.: U.S. Govt. Printing Office), including panel reports, vols. 1, 2, and 3; PL 89-454 (1966).
15. *Borax Consolidated, Ltd.* v. *Los Angeles,* 296 U.S. 10 (1935).
16. *Harkins* v. *Del Pozzi,* 50 Wash. 237 (1957).
17. La. Rev. Civ. Code, art. 451.
18. *Luttes* v. *State,* 159 Texas, 500 (1958).
19. See reference 14.
20. *Marks* v. *Whitney,* 491 P. 2d. 374 (1971).
21. See reference 14.
22. See reference 5.
23. 13 U.S.T. 2312.
24. See reference 1.

25. See reference 5.

26. 67 Stat. 462; 43 U.S.C. 1331–1343.

27. *U.S.* v. *Maine, et al.,* No. 35, Original, and *U.S.* v. *Florida,* No. 52, Original.

28. 15 U.S.T. 471.

29. See reference 23.

30. 15 U.S.T. 1606.

31. N. Healy, 1970. International Convention on the establishment of an international fund for compensation for oil pollution damages-1969 (9 ILM), *Journal of Marine Law and Commerce, 1:* 312; and L. Hunter, 1972, Proposed international compensation fund for oil pollution damage (unpublished manuscript).

32. 33 U.S.C.A. No. 1001–1015.

33. 33 U.S.C.A. No. 401–413.

34. See reference 9.

35. 33 U.S.C. 1151.

36. Approval standards are listed in 18 C.F.R. No. 620.10 and in *Env. Rep.* (Fed.) 31: 6101–6109.

37. 16 U.S.C.A. 661–666.

38. See reference 2.

39. 42 U.S.C. 4321–4347.

40. See, Executive Order 11514, March, 1970, *Env. Rep.* (Fed.) 71: 0121.

41. Council on Environmental Quality, 1970. *Environmental Quality, the First Annual Report* (Washington, D.C.: U.S. Govt. Printing Office).

42. Council on Environmental Quality, 1971. *Environmental Quality, the Second Annual Report* (Washington, D.C.: U.S. Govt. Printing Office).

43. Council on Environmental Quality, 1970. *National Oil and Hazardous Materials Contingency Plan* (Washington, D.C.: U.S. Govt. Printing Office).

44. Council on Environmental Quality, 1970. *Ocean Dumping* (Washington, D.C.: U.S. Govt. Printing Office).

45. Council on Environmental Quality, 1971. *Toxic Substances* (Washington, D.C.: U.S. Govt. Printing Office).

46. Council on Environmental Quality, 1972. *The Economic Impact of Pollution Control: A Summary of Recent Studies* (Washington, D.C.: U.S. Govt. Printing Office).

47. See generally, W. McNichols, Legal problems regarding the extraction of minerals (including oil and gas) from the continental shelf, University of Miami, Sea Grant Program, March 1971.

48. See reference 14.

49. U.S. Army Corps of Engineers, *National Shoreline Study: Shore Management Guidelines* (Washington, D.C.: U.S. Govt. Printing Office), 56 pp.

50. See reference 4.

51. R. B. Bidler, 1971. Controlling Great Lakes pollution: A study in U.S.–Canadian environmental cooperation (A background paper for the Arden House Conference on legal and institutional responses to problems of the global environment).

52. See reference 31.

53. U.S. Army Corps of Engineers, 1971. (a) Preliminary analysis of the ecological aspects of deep port creation and supership operation, IWR report 71–10, Natural Resources Institute, University of Maryland. (b) Foreign deep water port developments; a selected overview of economics, engineering and environmental factors, IWR report 71–11, Arthur D. Little, Inc.

54. See *McQueary* v. *Laird,* 449 F. 2d. 608 (10 Cir., 1971).

12 A Systems View of Coastal Zone Management

12.1 Introduction

An adequate systems model for coastal zone management does not now exist but the time appears ripe for its development. Such a model is necessary in order to handle the great variety of social and environmental data discussed in the previous chapters of this volume and to display such information in a form suitable for decision making. An essential requirement is to develop the capability of reaching rational decisions once the information is readily available. This will require various levels of analysis, various governmental levels of decision concerning the desirable objectives of coastal zone management, and various techniques for approaching and implementing processes that will reach the desired goals. The complexity of the system makes it necessary to develop a model that can be handled by rapid computer methods and that will be able to predict the end results of various alternate courses of action. Once a given course of action is decided upon and put into effect, a monitoring system will be required to indicate how well the model predicts conditions in the real world and to suggest alterations in the actions needed if the goals are not being achieved.

Useful models relevent to the coastal zone management problem are the urban planning simulation model of Ray and Duke (1968); the problem statement and solution proposed by Lee and Clayton (1972), despite the different object of their study; and the critical survey of estimating procedures for population parameters by Kleijnen, Naylor, and Seaks (1972). Forrester's (1971) model of world dynamics is an example of the kind of approach needed. His work represents an effort to take into account a multiplicity of inputs in order to assess the impacts of various alternative social actions. Refined management at the ecological and socioeconomic levels is, at best, a very long-term goal. Nevertheless, the progressive introduction of systematic approaches for better human decision making can begin now and would certainly result in substantial practical management improvements.

Throughout this volume the various factors that must constitute the information sources for the model of the system have been discussed. They range from the ecological to the social, economic and legal impacts and constraints on man's activities in the coastal zone. The flow charts for contaminants shown in Chapter 7 (Figure 7.1) and for information and management decisions discussed in Chapter 10 (Figure 10.3) would form part of the information sources necessary. A great deal of effort will be required to formalize the appropriate model, and it is suggested that this development would be an essential task for the recommended coastal zone centers (see Section 10.3.4).

The initial approach to the coastal zone management problem proposed below follows the recommendations of Quade (1968), and seeks to establish the conceptual basis for actual program development. The analysis develops a formal foundation outlining the necessity, nature, and variety of systems approaches, and combining scientific, legal, economic, and management information. The ingredients of the tentative management model discussed include general requirements; social, environmental, and technological factors; space-time considerations; and the integrated model.

12.2 Formal Analytical Foundation

Effective management of the coastal zone implies viable and economic integration of complex life systems and varied human activities within a confined area which is itself subject to the influence of major exterior forces, both from the land and from the sea. This level of complexity can be understood and evaluated for management purposes by harnessing contemporary mathematics in response to carefully defined social goals and within the limits of required constraints.

The greatest uncertainty in developing a systems model of the coastal zone is the prediction of human activities. Man responds to a wide variety of external forces which are variable in space and time and which are themselves largely unpredictable. For example, Chapter 9 has stressed that, even though we know that population pressures on the coastal zone will increase, we cannot predict where or how much with any certainty. Forrester's (1971) model of world dynamics is designed to account for these responses on a global scale, but uncertainties increase when smaller areas are considered.

As a general approach, the heterogeneous components of the coastal zone environment, including both the natural processes and the impacting patterns of human activity, require formal statement for each separately, as well as a progressive synthesis. The more definable and efficient the relationship, the better it can be handled discretely. The linkages between the results of discrete analyses and large aggregates are being explored and stated with increasing success. Samuelson's (1971) Nobel Prize lecture provides illustrations from economics and

Morris (1964) has made possible orderly ways of moving from the specificities of ordinary perception to rigorous mathematical hypotheses, and back to persuasion in the realm of social action.

The systems approach provides powerful means of summarizing vast bodies of information and of translating them into various levels of perception, analysis, and action. The use of these techniques risks losing concrete applicability with every stage of translation. Consequently, precise terms and procedures must be linked with continuous scrutiny by experts in each discipline—for example, specialists in biology, physics, engineering, or sociology—if the potentialities of systems analysis and integration are to be realized.

12.3 A Preliminary Management System Model

The general requirements for the development of an effective management system model for the coastal zone rests upon a proper conceptual base which must be both sufficiently general to provide a comprehensive framework and specific enough to be directed toward operational decisions. In the development of such a base, four steps are essential:

—The formulation of an appropriate problem statement;
—A procedure for problem attack;
—A substantive evaluation of the major groups of information input; and
—Continuing evaluation of the model and system integration.

The general problem is to determine how to proceed from the present state toward some desired future state with a high probability of attaining the goal, that is, to find the mechanism that maximizes the probability of achieving the desired state. This probability may be stated symbolically as follows (Charnes and Stedry, 1966, pp. 157–161; Lee and Clayton, 1972, pp. 397–398):

$$P\{S_i, t_i \rightarrow S_p, t_p \mid z, c\} \geq 1 - d,$$

where
P is the probability of goal attainment;
S_i is the initial or present state of the system being managed;
t_i is the intial time of the management period;
$\rightarrow$ is a transition mechanism; the action needed to achieve the desired goal;
S_p is the planned state of the system, or the goal defined;
t_p is the time when the plan is to be completed;

z is the aggregate of natural and social constraints which must not be violated in the transition;
c is the cost limit of incremental resources available to the planner; and
d is the uncertainty to be minimized.

The critical steps in planning for coastal zone management are the definition of the desired state (S_p), the time required to achieve it ($t_p - t_i$) and the determination of the transition mechanism ($\rightarrow$) which will maximize the probability of attaining the goal. As has been discussed previously (Section 8.2) there are three basic strategies for the use of the coastal zone resources: multiple use, exclusive use, or displaceable use. The first two of these cannot coexist in the same place and time in the coastal zone, and will thus require geographical separation. The use strategy chosen for any given segment of the coastal zone will determine the character of the desired state (S_p), though even in multiple use areas, application of modern or improved technology will permit attainment of a higher level of environmental protection.

The basis of management requires the specification of the current state, the desired state, and the difference between the two. No plan is possible without concrete objectives and assessments. For example, what kind of a coastal zone is wanted, and when? How does this differ from the coastal zone today? Two types of constraints exist: categorical limits (z) which are absolute, and conditional limits (c) which are related to some cost-effectiveness calculation. This distinction is of utmost policy importance since it defines the break between areas controlled by natural ecological constraints or legal regulations (until they are changed) and those open to bargaining. Increasing available funds (c) for management or control, for example, could be a mechanism to increase the probability of achieving the desired goal (S_p).

12.3.1 Definition of Desired States

It is important to recognize two conceptually idealized environmental states (Odum, 1969). The first is the natural system, developed through millions of years of evolution and adaptation and untouched by man. Such a system is generally characterized by diversity, some degree of self-regulation, and stability. It tends to be a conservative system in which nutrient recycling is highly efficient and with little leakage. Completely natural systems probably no longer exist, but close approximations are still present in fair abundance, including representatives of at least the major coastal ecosystems. The second idealized environmental state is that environment that is managed for optimum human use, whether it be harvest of living or nonrenewable resources, waste disposal, or recreation. Diversity and stability are of secondary concern, and regulation is a function of human management. Nutrient cycles tend to be more open, and nutrient loss through harvest and erosion must be made up through the application of fertilizer.

Clearly, the two idealized states cannot characterize the same ecological system, yet both states are needed for the long-range health of the overall managed system. Through large-scale planning, different sites must be allocated to and managed for one or the other idealized state so that both will be represented on a regional basis.

A greatly oversimplified diagram relating human uses of the coastal zone with environmental quality is presented in Figure 12.1. Human uses range from no use to total use in the left-hand column and these are related to the expected environmental quality ranging from pristine to exhausted in the second column. These are related to various environmental use–quality conditions on the right. As has been mentioned previously, many choices can be made concerning the desired state (S_p) to be chosen for management purposes depending upon the objectives or goals for the particular region. None of these states are absolute, but should fall between chosen limits. For example, when natural preserves are desired (see Section 4.3.3) human use should be severely limited and the chosen state might fall between the lines u_0 (no use) and u_1 (little use). On the other hand, complete utilization of the resources of the coastal zone would require a decision that the state should fall between the lines q_0 and q_1 representing exhaustion of the quality of the environment and its serious degradation. The one strategy illustrated provides a chosen state for optimum use which would leave the environmental quality in an acceptable but not a natural state. In this example, the early decisions made at time t_1 produced a further deterioration of the environment which would be identified by an appropriate monitoring system. Decisions at time t_2 are required to reverse this trend and return the environment to the chosen state. The arenas for decision making have been discussed in Chapters 9 and 10 and are indicated in the central portion of this diagram. The primary arena would be the federal-national policy for coastal zone management and would presumably cover the entire range for use and possible conditions in the coastal zone. The secondary arena would be at the state or regional level, which could cover the entire range of possibilities as shown for the federal policy, but the example chosen suggests the state

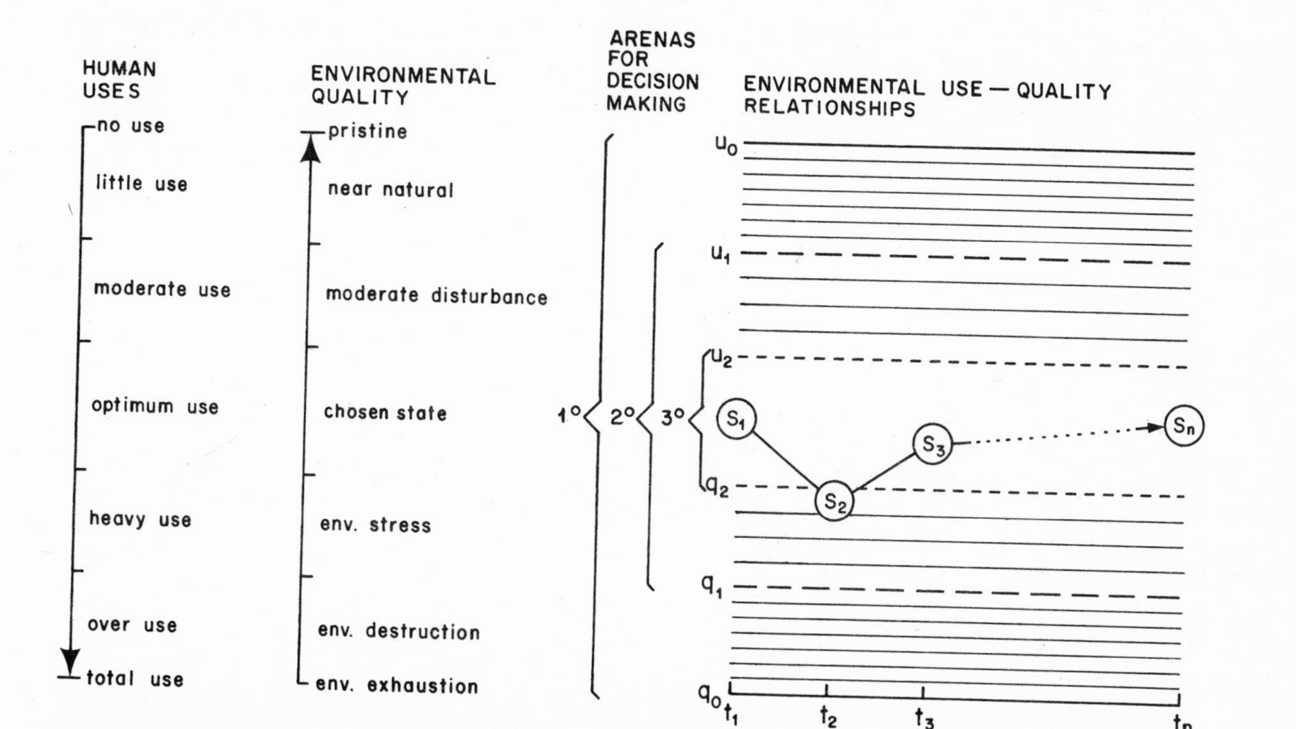

Figure 12.1 Environmental use—quality control track. As we pass into future time, the use-effect balance point ("state") will trace a line-path which, ideally, will be held within the conditional use and quality control boundaries (u_2 and q_2). The path could exceed these values at least for short time periods without significant permanent effect, but if it transgressed the categorical boundaries (u_1 and q_1) certain long-range options would be forfeited.

has made the decision that its coastal zone should be kept within the range of moderate to heavy usage. A third arena for decision making represents the local level where regulations and controls are enforced to ensure that the coastal zone remains within the bounds of the chosen state.

Management to maintain or restore a natural system is a difficult and uncertain task, and it is important here to examine the bases upon which such management must rest. Natural ecosystems include complex biological systems together with the physical environments with which they interact. Through evolutionary adaptation each population in the ecosystem has adopted a strategy for long-term survival in the face of the prevailing complex environmental factors. Species compete for survival and living space, and the balance between species is often delicate and based upon quite subtle factors (Mosser, Fisher, and Wurster, 1972).

In general, stress is imposed upon living systems by suboptimal environmental stimuli. It is usually measured in terms of systemic response but could also be related to the nature and intensity of the stress agents. Both general and specific stress responses are observable at the level of populations and ecosystems. Populations may explode, decline, or be eliminated. At the ecosystem level, generalized responses may be expected to show up as deviations in rate phenomena such as photosynthesis or respiration, and other gross ecosystem characteristics, such as diversity, stability, regulation, and efficiency of nutrient cycles. Specific responses may appear in relation to the success or loss of sensitive species, kinds of nutrients being lost, or changes in abundance or dominance of species.

12.3.2 Implementation of Change

The time needed to achieve the desired change is essentially a political, social, and economic problem. Obviously, if the requirements for the use of the coastal zone area are made more stringent, prior users must be offered a period of time in which to change their operations. The managing authority must have some measure of bargaining power, as well as access to one or more loci of decision making, such as markets, communications media, and the courts. The emphasis is upon actions, decisions, and processes, rather than upon regulations, which are considered to be constraints upon operational realities (Deutsch, 1963; March, 1966; and Burns, 1966).

The following human behavioral attributes may affect the rate of change.

—Values, decisions, implementing actions, and purposeful control are attributes of human actions. Clear distinction must be made between human decision processes and their supporting institutions and documents (Easton, 1966; Shimkin, 1968).

—Human behavior may be rational, but human judgment is often faulty. Human rationality is an expression of personal and group desires and needs. Some desires or uses reflect direct physiological necessities; others are symbolic means of gaining group approbation or other satisfactions (Foa, 1971; Helsen, 1959; Parsons, 1959; Pruzon, 1966).

—Human decisions are characterized by high rates of error and require complex self-correcting capability (Mengez, 1966; von Neumann, 1958).

—Uses and demands are ranked in value hierarchies (Strauss, 1963). Deeply held values and prejudices determine which life-styles are developed.

—Individuals and organizations characteristically maintain multiple use hierarchies which reflect identification with corresponding roles (Churchman and Emory, 1966).

—The pattern of individual or group behavior is generally consistent, but not immutable, over time. Experimental methods are available for the analysis of such patterns in both individuals and groups (Helsen, 1959; Osgood, 1963; Scheflen, 1968).

—Human beings seek four basic objectives: satisfaction of immediate wants, attainment of future goals, development of power among their fellows, and determination of the degree of support they can anticipate from others. It is not possible for any one decision at the same time and place to maximize all of these objectives (Parsons, 1959).

These various human attributes must be taken into account both in making the original decision of the desired state or goal to be achieved in any part of the coastal zone and also in the length of time that would be required to achieve the specified goal. For large areas and over long time horizons, the coastal zone agency must increasingly consider latent as well as active interests. The latter may usefully be categorized into a hierarchy of decreasing need for location on the coastal zone (see Chapter 9). First are those interests that are functionally bound to the coastal zone, such as fishing, recreation or ecosystem preservation. Second are projections of established interests or activities, for example, the changes in a coastal area that is developing from limited settlement to urbanization. Third are those potential interests empowered through law or financial equity but which are without immediate active expression, such as the potential navigability of a water body requiring dredging and harbor development. There are a host of direct and indirect active and latent interests in the coastal zone of lesser immediate importance but which may have to be considered in longer range or broader-scale planning.

The forecasting of trends in areas of social decision and technology is essential, yet exceedingly complex. The problems derive from three main sources: unreliabilities in historical records, difficulties in interpreting cause-effect relations in past trends, and discontinuities. The recent national trend toward the coastal location of population and resources is weak evidence for extrapolation into the future. During and following the period of World War II, the distribution of federal expenditures, primarily for space exploration and defense, has artificially intensified the coastalization process by providing job opportunities and has stimulated migration of employable people for non-coast-related industries (Shimkin, 1965; Lee, 1966). Very large capital investments have been made in coastal areas which represent long-term vested interests. High density coastal settlements have, in turn, led to local saturation and lateral coastwise spread of population, transportation networks, and associated activities with increasing environmental effects. This, in turn, has stimulated major financial interests to purchase large tracts of the scarce remaining coastal lands for future development. The coastal zone is, thus, passing from individual to corporate control.

Given these and other similar considerations, simple extrapolation from the historical record is an unreliable forecast tool. Econometric analyses from a well-studied base date to a defined time, not more than 20 years hence, are preferable. The 20-year limit reflects two constraints: migration peaks at ages 20–25, and the forecasting of mortality among those already born is far more reliable than the forecasting of future births. It is very unlikely that a significant association would be empirically determined between average population density in a particular coastal zone area and measures of impact intensity. Associated with, but distinct from population density, per se, is the set of population-maintaining activities, with transportation, recreation, and industrial operations being far more potent influences than single household units.

Having determined the initial state (S_i, t_i) and cognizant of the categorical constraints (z), the manager must evaluate necessary and acceptable goal possibilities (S_p, t_p) as the basis of a technological concept of implementation. The problem is one of an adequate decision-forming and implementation strategy which represents, in essence, balancing the preliminary version of (S_p, t_p) with socioeconomic conditional constraints (c) through the mechanisms of the marketplace, government administration, and the courts with judicious input through public information media. These will all influence the time required to reach a specified goal. From a systems standpoint, it is of utmost importance that data on these mechanisms be available to managers in forms suitable for rapid and reliable operational evaluation and use. What the manager seeks is the proper locus of decision making which can bring together, directly or through advocacy, the relevant interest groups so that tolerable and sufficiently binding decisions can be made with reasonable speed and at manageable cost.

The success of future efforts to manage the society-environment relationship will rest, in large measure, upon our understanding of the environmental systems themselves. Much has already been said about the nature of the environmental systems of North America (see Chapters 1 and 2), which need not be repeated here. What is necessary is a sharp focus upon those aspects of the environment which must be considered in relation to the perceived managerial needs.

12.3.3 Transition Mechanisms

The most important, and probably the most difficult step is to determine the transition mechanisms ($\rightarrow$) which will maximize the probability of achieving the stated goal (S_p). The modeling of implementing activities involves initial screening for general suitability and compatibility with local constraints. Basic distinctions must be made between the impact evaluation of displacement processes, such as construction and dredging, maintenance processes, such as traffic and sewage generation, and persistent effects of structures and of human modifications. Once the logical framework has been drawn for the entire process of coastal zone management, including goal setting, decision making, implementation and control, one should—

—Determine informational requirements in specific operational terms, including permissible errors;

—Establish a data bank, with input, coding, and retrieval components, of appropriate speed, capacity, and accessibility;
—Progressively evolve the necessary computational aids, such as modeling, simulation, and numerical specification of important subroutines. The problem definitions must interface with the most appropriate mathematical model (Formby, 1969; Shapiro, 1969; Pearson, 1969);
—Design the model in such a way so that various alternative methods of action can be evaluated and the probable impacts assessed;
—Provide the coastal zone agency with information necessary to assist in reaching useful decisions. A stage-by-stage development with much counseling with operating agencies and interest groups is essential.

A review of the analytical issues for the development of coastal zone models has revealed none that require more than modest modifications or limitations of existing techniques, especially those used in physics and operations research. Improvements in monitoring instrumentation will be necessary, particularly those essential for assessing the state of the ecological systems; present computer technology appears to be fully adequate to the required tasks.

In summary, the environmental manager must have access to detailed and reliable information concerning the state of the present environment (S_i, t_i), and he must be able to define a desirable future state (S_p, t_p), as well as an environmentally and socially tolerable pathway between the two states. This implies a level of managerial skill and environmental knowledge which is probably within reach, although certainly not available at the present time. The large uncertainties concerning the nature and indicators of environmental quality and especially concerning tolerance characteristics and symptoms of environmental stress, must eventually be reduced to acceptable levels if real environmental predictability is to be placed within the manager's grasp. In the interim, however, a reasonable level of expert judgment exists to permit the first stages of model development, especially when that judgment stems from specific information concerning the nature and environmental levels of potential stress agents.

12.4 The Integrated Model

A composite model is a set of component formulations which are uniquely associated with the variables of the universe under study. It must include directions so that the data may be ordered, have scales which permit the orderly ranking of data, permit the location or mapping of characteristics in terms of dimensions and scales, and it must include certain common denominators, such as energy, which permit transition between categories. In order to quantify, predict, and apply to real situations, the model components must be identified with quantitative values specific to the phenomena being considered.

The coastal zone management model must conform to the following quality criteria (Wagner, 1969).

Relevance. The model must be able to deal with both general management problems and those specific to the coastal zone. It must take these into account at all levels of decision making, local, state, and national (see Chapter 10); it must consider various types of interest, such as private, nonprofit, business, and general public, and various functions, such as physical assessment, administration, and impact analysis.

Validity. The model components, logic, computational methods, and outputs must be consistent, and the outputs must be judged by substantive specialists and relevant users to be usable in appropriate management operations.

Transition capability. The commonalities throughout the system, including the components, translations between stages, and mechanisms for generalizing inputs and specifically applying outputs, must be capable of providing direct response to user needs. In addition, the various stages of the system must be compatible and capable of being progressively linked.

Reliability. The system failures, malfunctions, and random errors must be maintained within specified narrow limits; also the relations between inputs and anticipated outputs, and between outputs and anticipated inputs must not exhibit systematic deviations in the judgment of qualified specialists, nor should such deviations be mechanically generated as revealed through computer program simulation tests.

Sensitivity. Access and response times, as well as output specificities, must correspond to defined user requirements.

Cost. Initial and operational costs must conform to budgetary limits, with some trade-offs between cost and performance constraints permitted.

Timeliness. User requirements must be met for each developmental stage, with close coordination in schedules for hardware, software, reference data, and training readiness. A continuity of services must also be assured.

In constructing a model, the following general strategy is suggested

(Wilson and Wilson, 1965). The entire process of model construction must be conceptualized in relation to components, processess, resources, and uses of the coastal zone. The model should be judged on the basis of feasibility, reliability, economic utility, and relative freedom from mechanical constraints. In its progressive development, the general design must constantly be monitored by competent judges, including social scientists, ecologists, and engineers, for its intuitive and practical validity. Throughout the construction, integration, and progressive development of the model, continual attention must be paid to repeated reviews. As appropriate, modifications must be made to ensure consistency.

A theoretical coastal zone management model is presented in Figure 12.2. It must be emphasized that this is one component of the integrated model which might consist of a score or more of similar components which would apply to different aspects of the management problem that would be interrelated in the integrated model. In phase A, the solid line illustrates the progression from the present state of the coastal zone (S_1) to the chosen state (S_2). The regulation points lead to and control human activities. The conditions of the coastal zone must be monitored to determine whether the objectives of increasing the probability of reaching the desired goal are being met. The monitoring process has a feedback as indicated by the long dashed vertical line to the translators and the systems analysis. The model must also provide for trade-offs as a result of differing activities which could lead to arrival at the desired state (S_2) by alternate routes as shown.

Since man's needs, both physical and psychological, involve broad areas that are imperfectly perceived, and since the human mind has qualitatively different capacities from those of a machine, it is of utmost importance that man control his machines rather than conform to them. This is imperative to any acceptable concept of coastal zone management. The theoretical basis for human control of complex systems is provided by adaptive control theory (Gaines, 1969). This involves a relationship between a controller, a controlled system, and a purpose, objective or goal. In operation, the controller implements one policy from a predetermined set of allowable policies. He also senses the effect of a given policy implementation, and, as appropriate, may subsequently select and implement an alternative policy to reduce deviation from the optimum pathway. Even nonlinear and noisy environments may be handled by such means. Memory of past effective policy performances

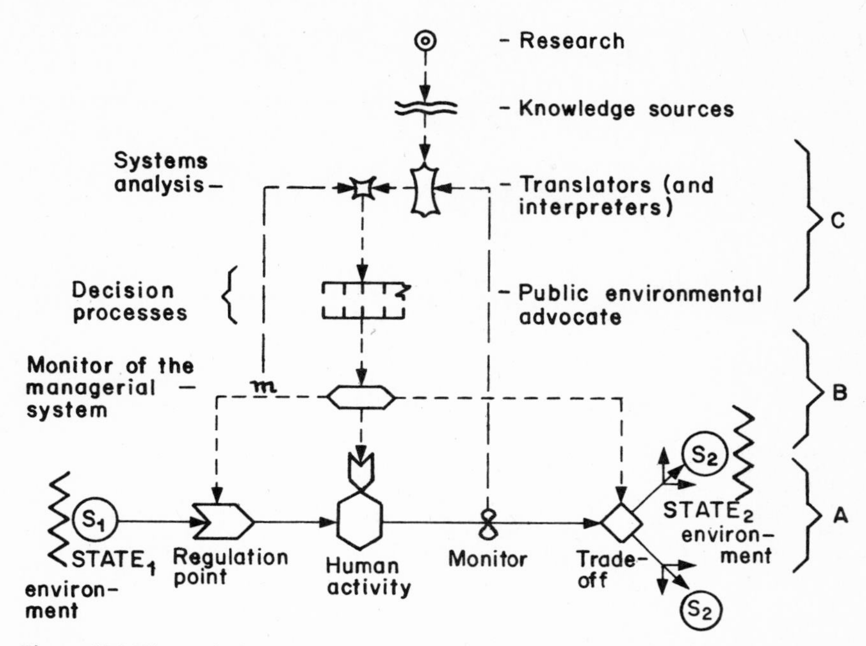

Figure 12.2 Theoretical *environmental management unit model* showing the three essential functional levels of such a system including possible components of each level.
A. (unbroken lines). This is the basic *use-impact level* which must be regulated through time so that in passing from one perceived environmental state (S_1) to another (S_2), the environmental use and environmental quality boundary conditions are not transgressed.
B. (short-dashed lines). This is the *environmental management level* which implements environmental decisions. Through technology, management can intervene at several points in the use-impact train to achieve desired society-environmental goals.
C. (long-dashed lines). This is the *referee level* wherein decisions are made concerning society-environmental goals. Ideally, the referee system has access to environmental and social information which is (a) sufficient for the decisions, (b) interpretable to all sectors of the referee system, and (c) displayed in such manner that alternative goals are clearly and perceptually linked with associated management strategy alternatives. At present the referee system is poorly developed, but it cannot long remain so if long-range environmental options are to be retained. There is considerable room for innovation at this level, and some innovative suggestions are included in the figure. However, the critical feature which must be developed is the integrated systems analysis program which can quickly organize, analyze, and display in simple meaningful terms the vast body of data upon which decisions must be made.

is incorporated into the system. Within the limits of knowledge and program capability, the model may be adapted to new circumstances on the basis of previous experience.

Phase C represents the analysis of the system which is modified by new research or the addition of readily available knowledge and must take account of the feedback from the monitoring of the regulations and human activities shown in stage A. The analysis of the system can, in turn, have additional input into various levels of the decision processes as illustrated. Ideally, the systems analysis level has access to environmental and social information which is sufficient to make or modify intelligent decisions about the desired state (S_2), and this information must be interpretable by all sectors of the referee system and displayed in a way so that alternative goals are clearly linked with associated management strategy alternatives.

Phase B, reflected in the short dashed lines, represents the management position of the system. Within the framework of information analysis and policy decision, it implements the procedures required to achieve the desired system state. As indicated, management can intervene at several points in the use-impact train to achieve desired society-environment goals.

The components of an integrated model would reflect all of the various uses and constraints, ranging from the ecological to the legal, which have been discussed in previous chapters in this volume. These would all be interrelated in a great variety of ways. For example, all of them would be taken into consideration in the definition of the desired state and would thus determine the characteristics of the regulations and of human activities.

An operational formulation of adaptive control theory has recently been made by Howland (1966). In this formulation, the specific research problems include partitioning the system into integrated levels, defining tasks and missions at each level, and identifying the fundamental system unit and components upon which overall system goal achievement ultimately rests. Important steps in the development of a functional model are to identify the key task of the fundamental unit selected, to select and measure dependent and independent variables that influence performance of the key task, to determine functional relationships between these sets of variables, and to validate the functions by testing their predictive capability.

At the present time, we are a long way from the desirable objective of effective technology as a subservient component in the larger system. Only now are we beginning to realize the necessity and capability for developing the framework of the total integrated system. Ahead lie development of appropriate decision-making bodies, effective managerial personnel, environmental hardware, including effective monitors, and environmental understanding. Time, expertise, and considerable financial support will be required. The first priority, however, must be the development of the capability of generating a coherent body of integrated information concerning the coastal zone and its management (Collier, 1970). All other actions must flow from this primary activity if coherence is to be achieved.

12.5 Future Prospects and Recommendations

The coastal zone problem may logically be reduced to an operation in quality control in which human utilization is balanced against environmental effect. It is a problem of optimization under constraints. These constraints themselves fall into two categories, conditional constraints, which although stressful are tolerable (at least in the short run), and categorical constraints which cannot be violated without permanent damage to the support system.

The total time horizon with which we are dealing is unlimited, yet it must be phased into foreseeable horizons for short-term planning. Therefore, our immediate objective in planning for a rational future is to establish the framework for a realistic pattern of decision making which is responsive to the many conflicts which are inevitable as demands exceed resources. Thus, as we proceed into the future by a successive series of approximately informed and rational decisions, we can periodically adjust coefficients of the use-environmental quality equations for achievement of successively perceived optimal states. These optimal states, in turn, will, it is hoped, reflect wiser and better informed public concensus.

The primary goal of the discussion of a systems approach has been to demonstrate both the necessity and feasibility of developing a model for evaluating the potential effects of conceived alternate courses of action as a component of the necessary decision-making processes. Another important goal has been to indicate the kinds of information input which will be necessary for successful development, evaluation, and use of such models. A third goal has been to indicate the type of managerial and technological framework within which the society-ecosystem relationship may be successfully controlled.

Even at a conservative appraisal of possibilities, systems approaches can certainly aid in the following ways:

—Bring orderly criteria evaluation into standard management practice;

—Develop common concepts, languages, and measures usable for progressive development of simulation and later to numerical approaches of monitoring evaluation and control;

—Bring to governmental authorities models of the natural environmental processes, appropriate institutions, and activities for complex management decisions, and

—Train professionals, officials, owners, and communicators, as well as the interested public to higher scientific awareness and greater competence in coastal zone management.

At best, systems approaches can establish the framework for handling the entire coastal zone management job.

Although it appears that the capability for development of the model is already within advanced technological reach, it is absolutely essential to attract to this job, especially in the development stage, the most competent national intellectual resources available, including representatives from industry and the defense agencies, as well as academic specialists in the areas of ecology, physical science, social science, and mathematics. It is finally anticipated that a large fraction of the model program, components, and design would exhibit major commonalities with a considerable range of other man-environment management problems.

On the above grounds, it is recommended that a major project be initiated in the near future for the development of an integrated systems analysis program concerning the coastal zone, which should include (a) development of a model for evaluating the potential effects of conceived alternate courses of managerial action; (b) determination and generation of necessary kinds of information essential to successful development, evaluation, and use of the model; and (c) development of the managerial and technological framework within which the society-ecosystem relationship may successfully be controlled. These three elements should be so programmed that all three converge at a future point to provide the many kinds of information required by decision makers attempting to achieve optimum use of the coastal zone with minimum environmental degradation.

References

Burns, T., 1966. On the plurality of social systems, *Operational Research and the Social Sciences,* edited by J. R. Lawrence (London: Tavistock), pp. 165–177.

Charnes, A. and Stedry, A. C., 1966. The attainment of organizational goals through appropriate selection of sub-unit goal, *Operational Research and the Social Sciences,* edited by J. R. Lawrence (London: Tavistock), pp. 147–164.

Churchman, C. W., and Emory, F. E., 1966. On various approaches to the study of organizations, *Operations Research and the Social Sciences,* edited by J. R. Lawrence (London: Tavistock), pp. 77–84.

Collier, A. W., 1970. Oceans and coastal waters as life-supporting environments, *Marine Ecology,* Vol. I, Pt. 1, edited by O. Kinne (London: Wiley-Interscience), pp. 1–94.

Darnell, R. M., 1970. Evolution and the ecosystem, *American Zoologist 10:* 9–15.

Deutsch, K. W., 1963. *The Nerves of Government* (London: Collier-Macmillan).

Easton, D., 1966. Categories for the systems analysis of politics, *Varieties of Political Theory,* edited by D. Easton (Englewood Cliffs, N.J.: Prentice-Hall), pp. 143–154.

Foa, U. G., 1971. Interpersonal and economic resources, *Science 171:* 345–351.

Formby, J. A., 1969. Control: Basic elements, *Encyclopedia of Linguistics, Information and Control,* edited by A. R. Meethom (Oxford: Pergamon Press), pp. 1–9.

Forrester, J., 1971. *World Dynamics* (Cambridge, Mass.: Wright-Allen Press).

Gaines, B. R., 1969. Adaptive control theory, *Encyclopedia of Linguistics, Information and Control,* edited by A. R. Meethom (Oxford: Pergamon Press), pp. 1–9.

Helson, H., 1959. Adaptation level theory, *Psychology, A Study of a Science,* Vol. 1, edited by S. Koch (London: McGraw-Hill), pp. 565–621.

Howland, D., 1966. A regulatory model for system design and operation, *Operational Research and the Social Sciences,* edited by J. R. Lawrence (London: Tavistock), pp. 505–514.

Johnson, P. L., 1969. *Remote Sensing in Ecology* (Athens, Ga.: University of Georgia Press).

Kinne, O., 1970. *Marine Ecology,* Vol. I, *Environmental Factors,* Part 1 (London: Wiley-Interscience).

Kleijnen, J. P. C., Naylor, T. H., and Seaks, T. Q., 1972. The use of multiple ranking procedures to analyze simulations of management systems: A tutorial, *Management Science 18:* B245–256.

Lee, E. S., 1966. A theory of migration, *Demography 3:* 47–57.

Lee, S. M., and Clayton, E. R., 1972. A goal programming model for academic resource allocation, *Management Science 18:* B395–408.

March, J. G., 1966. The power of power, *Varieties of Political Theory,* edited by D. Easton (Englewood Cliffs, N.J.: Prentice-Hall), pp. 39–70.

Margalef, R., 1963. On certain unifying principles in ecology, *American Naturalist 47:* 357–374.

Mathews, W. H., Smith, F. E., and Goldberg, E. D. (eds.), 1971. *Man's Impact on Terrestrial and Oceanic Ecosystems* (Cambridge, Mass.: MIT Press).

Mengez, G., 1966. The suitability of the general decision model for operational applications in the social sciences, *Operational Research and the Social Sciences,* edited by J. R. Lawrence (London: Tavistock), pp. 565–578.

Milgram, S., 1970. The experience of living in cities, *Science 167:* 1461–1468.

Morris, C., 1964. *Signification and Significance* (Cambridge, Mass.: MIT Press).

Mosser, J. L., Fisher, N. S., and Wurster, C. F., 1970. Polychorinated biphenyls and DDT alter species composition in mixed cultures of algae, *Science 176:* 533–535.

Odum, E. P., 1959. *Fundamentals of Ecology* (Philadelphia: Saunders).

Odum, E. P., 1969. The strategy of ecosystem development, *Science 164:* 262–270.

Odum, W. E., 1970. Insidious alteration of the estuarine environment, *Trans. Amer. Fish. Soc. 99*(4): 836–847.

Osgood, C., 1963. Psycho linguistics, *Psychology, A Study of a Science,* vol. 6, edited by S. Koch (London: McGraw-Hill), pp. 244–316.

Parsons, T., 1959. An approach to psychological theory in terms of the theory of action, *Psychology: A Study of a Science,* vol. 3, edited by S. Koch (London: McGraw-Hill), pp. 612–711.

Pearson, J. D., 1969. Control, hierarchical, *Encyclopedia of Linguistics, Information and Control,* edited by A. R. Meethom (Oxford: Pergamon Press), pp. 113–115.

Pruzon, P. M., 1966. Is cost-benefit analysis consistent with the maximization of expected utility?, *Operational Research and the Social Sciences,* edited by J. R. Lawrence (London: Tavistock), pp. 319–335.

Quade, E. S., 1968. By way of summary, *Systems Analysis and Policy Planning. Applications in Defense,* edited by E. S. Quade and W. I. Boucher (New York: Elsevier), pp. 418–429.

Ray, P. H., and Duke, R. D., 1968. The environment of decision-makers in urban gaming simulations, *Simulation in the Study of Politics,* edited by W. D. Coplin (Chicago: Markham), pp. 149–179.

Samuelson, P. A., 1971. Maximum principles in economics, *Science 173:* 991–997.

Scheflen, A. E., 1968. Human communication: behavioral programs and their integration in interactions, *Behavioral Science 13:* 44–55.

Shapiro, S., 1969. Control by the maximum principle, *Encyclopedia of Linguistics, Information and Control,* edited by A. R. Meethom (Oxford: Pergamon Press), pp. 101–106.

Shimkin, D. B., 1965. Human ecology and resource management: an application to the Great Lakes, *Proceedings, Eighth Conference on Great Lakes Research,* (Ann Arbor: University of Michigan Press), pp. 3–12.

Shimkin, D. B., 1968. The calculus of survival, *Medical Opinion and Review 4*(10): 47–57.

Strauss, G., 1963. Some notes on power-equalization, *The Social Science of Organizations,* edited by H. J. Leavitt (Englewood Cliffs, N. J.: Prentice-Hall), pp. 41–84.

Twiss, R. H., and Sorensen, J. C., 1972. *Methods for Environmental Planning of the California Coastline,* manuscript (Berkeley: University of California, College of Environmental Design).

von Neumann, J., 1958. *The Computer and the Brain* (New Haven: Yale University Press).

Wagner, H. M., 1969. *Principles of Operations Research* (Englewood Cliffs, N.J.: Prentice-Hall).

Walsh, J. J., 1972. Implications of a systems approach to oceanography, *Science 176:* 969–975.

Wilson, I. G., and Wilson, M. E., 1965. *Information, Computers and System Design* (London: Wiley).

General Bibliography

American Petroleum Institute and Federal Water Pollution Control Administration. *Proceedings of the Joint Conference on Prevention and Control of Oil Spills,* December 1969. Washington, D.C.: American Petroleum Institute, 1970.

American Petroleum Institute, Environmental Protection Agency, Coast Guard. *Proceedings of the Joint Conference on Prevention and Control of Oil Spills,* June 1971. Washington, D.C.: American Petroleum Institute, 1971.

Blumstein, A., Kamrass, M., and Weiss, A. *Systems Analysis for Social Problems.* Washington, D.C.: Washington Operations Research Council, 1970.

Bosselman, F., and Callies, D. *The Quiet Revolution in Land Use Control, a Summary Report,* prepared for the Council on Environmental Quality. Washington, D.C.: U.S. Govt. Printing Office, 1971.

Bradley, E., and Armstrong, J. *A Description and Analysis of Coastal Zone and Shoreland Management Programs in the United States.* University of Michigan Sea Grant Publ. no. MICHU-SG-72-204, Ann Arbor, Mich., 1972.

Brahtz, J. F., ed. *Coastal Zone Management: Multiple Use with Conservation.* New York: Wiley, 1972.

California Interagency Council for Ocean Resources. *Comprehensive Ocean Area Plan.* Sacramento, Calif., 1970.

Commission on Marine Science, Engineering and Resources. *Our Nation and the Sea,* report of the Commission on Marine Science, Engineering and Resources. Washington, D.C.: U.S. Govt. Printing Office, January 1969.

Commission on Marine Science, Engineering and Resources. *Panel Reports of the Commission on Marine Science, Engineering and Resources,* vols. 1, 2, and 3. Washington, D.C.: U.S. Govt. Printing Office, January 1969.

Council on Environmental Quality. *The Economic Impact of Pollution Control; a Summary of Recent Studies,* prepared for Council on Environmental Quality, Department of Commerce, and Environmental Protection Agency. Washington, D.C.: U.S. Govt. Printing Office, 1972.

Council on Environmental Quality. *Ocean Dumping,* a report to the President prepared by the Council on Environmental Quality. Washington, D.C.: U.S. Govt. Printing Office, 1970.

Ditton, R. and Goodale, T. *Environmental Impact Analysis: Philosophy and Methods,* proceedings of the Conference on Environmental Impact Analysis, Green Bay, Wisc., January 1972, Sea Grant Publ. no. WISC-SG-72-111, 1972.

Douglas, P. A., and Stroud, P. H., eds. *A Symposium on the Biological Significance of Estuaries.* Washington, D.C.: Sport Fishing Institute, 1971.

Ducsik, D., ed. *Power, Pollution, and Public Policy,* MIT Report no. 24. Cambridge, Mass.: MIT Press, 1971.

Florida Coastal Coordinating Council. *Coastal Zone Management in Florida—1971,* a status report to the Governor, the Cabinet and the 1972 Legislature. Tallahassee, Fla., December 1971.

Gardner, B. *The Crowded Coast: The Development and Management of the Coastal Zone in California.* Los Angeles, Calif.: University of Southern California Center for Urban Affairs, 1971.

Hood, D., ed. *Impingement of Man on the Oceans.* New York: Wiley-Interscience, 1971.

IDOE. *Baseline Studies of Pollutants in the Marine Environment and Research Recommendations,* the IDOE Baseline Conference, May 1972. Convener, E. Goldberg. New York.

Lauff, G. H., ed. *Estuaries,* publ. no. 83, American Association for the Advancement of Science, Washington, D.C.: 1967.

Livingstone, R. *A Preliminary Bibliography with KWIC Index on the Ecology of Estuaries and Coastal Areas of the Eastern United States,* U.S. Department of the Interior, Spec. Scientific Report—Fisheries no. 507. Washington, D.C.: U.S. Govt. Printing Office, 1965.

Marx, W. *The Frail Ocean.* New York: Ballantine Books, 1967.

National Academy of Sciences. *Marine Environmental Quality,* report of a special study held under the auspices of the Ocean Science Committee, NAS-NRC Ocean Affairs Board. August 1971, Washington, D.C.: National Academy of Sciences, August 1971.

National Academy of Sciences and National Academy of Engineering. *Background papers on coastal waste management, Vol. I,* National Academy of Sciences Committee on Oceanography and the National Academy of Engineering Committee on Ocean Engineering. Washington, D.C.: National Academy of Sciences, 1969.

National Academy of Sciences and National Academy of Engineering. *Waste Management Concepts of the Coastal Zone,* prepared jointly by the National Academy of Sciences Committee on Oceanography and the National Academy of Engineering Committee in Ocean Engineering. Washington, D.C.: 1970.

Radcliffe, D., and Murphy, T. *Biological effects of oil pollution—Bibliography.* A collection of references concerning the effects of oil in biological systems. USDI-FWPCA clean water series DAST 19. Washington, D.C.: U.S. Govt. Printing Office, 1969.

Study of Critical Environmental Problems (SCEP). *Man's Impact on the Global Environment.* Cambridge, Mass.: MIT Press, 1970.

U.S. Department of Interior. *The National Estuarine Pollution Study.* U.S. Department of the Interior–Federal Water Pollution Control Administration, report to the Congress. Washington, D.C.: U.S. Govt. Printing Office, 1969. 3 vols.

U.S. Department of Interior. *National Estuary Study.* U.S. Department of the Interior Fish and Wildlife Service, Bureau of Sports Fisheries and Wildlife and Bureau of Commercial Fisheries. Washington, D.C.: U.S. Govt. Printing Office, 1970. 7 vols.

U.S. Department of Interior. *Water Quality Criteria.* U.S. Department of Interior–Federal Water Pollution Control Administration. Washington, D.C.: U.S. Govt. Printing Office, 1968.

Ward, G., and Espey, W. *Estuarine Modelling: An Assessment.* Environmental Protection Agency, Water Quality Office. Washington, D.C.: U.S. Govt. Printing Office, 1971.

Author Index

Subject Index

References to tables are denoted by t; references to figures are denoted by f.